MW01096448

Metabolomics

Chapman & Hall/CRC Mathematical and Computational Biology

About the Series

This series aims to capture new developments and summarize what is known over the entire spectrum of mathematical and computational biology and medicine. It seeks to encourage the integration of mathematical, statistical, and computational methods into biology by publishing a broad range of textbooks, reference works, and handbooks. The titles included in the series are meant to appeal to students, researchers, and professionals in the mathematical, statistical, and computational sciences and fundamental biology and bioengineering, as well as interdisciplinary researchers involved in the field. The inclusion of concrete examples and applications and programming techniques and examples is highly encouraged.

Computational Exome and Genome Analysis
Peter N. Robinson, Rosario Michael Piro, Marten Jager

Gene Expression Studies Using Affymetrix Microarrays
Hinrich Gohlmann, Willem Talloen

Big Data in Omics and Imaging
Association Analysis
Momiao Xiong

Introduction to Proteins
Structure, Function, and Motion, Second Edition
Amit Kessel, Nir Ben-Tal

Big Data in Omics and Imaging
Integrated Analysis and Causal Inference
Momiao Xiong

Computational Blood Cell Mechanics
Road Towards Models and Biomedical Applications
Ivan Cimrak, Iveta Jancigova

Stochastic Modelling for Systems Biology, Third Edition
Darren J. Wilkinson

Metabolomics
Practical Guide to Design and Analysis
Edited by Ron Wehrens and Reza Salek

For more information about this series please visit: https://www.crcpress.com/Chapman--HallCRC-Mathematical-and-Computational-Biology/book-series/CHMTHCOMBIO

Metabolomics

Practical Guide to Design and Analysis

Edited by
Ron Wehrens
Reza Salek

CRC Press is an imprint of the
Taylor & Francis Group, an **informa** business
A CHAPMAN & HALL BOOK

CRC Press
Taylor & Francis Group
6000 Broken Sound Parkway NW, Suite 300
Boca Raton, FL 33487-2742

Printed on acid-free paper

International Standard Book Number-13: 978-1-4987-2526-2 (Hardback)

Library of Congress Cataloging-in-Publication Data

Names: Wehrens, Ron, editor. | Salek, Reza, editor.
Title: Metabolomics : practical guide to design and analysis / edited by Ron Wehrens, Reza Salek.
Description: Boca Raton, Florida : CRC Press, [2019] | Includes bibliographical references and index.
Identifiers: LCCN 2019006745 | ISBN 9781498725262 (hardback : alk. paper) | ISBN 9781315370583 (e-book)
Subjects: LCSH: Metabolism. | Metabolism--Research.
Classification: LCC QP171 .M38245 2019 | DDC 572/.4--dc23
LC record available at https://lccn.loc.gov/2019006745

Visit the Taylor & Francis Web site at
http://www.taylorandfrancis.com

and the CRC Press Web site at
http://www.crcpress.com

Contents

Preface vii
Editors xi
Contributors xiii

1. **Overview of Metabolomics** 1
Oscar Yanes, Reza Salek, Igor Marín de Mas and Marta Cascante

2. **Study Design and Preparation** 15
Pietro Franceschi, Oscar Yanes and Ron Wehrens

3. **Measurement Technologies** 35
Oscar Yanes, Katherine Hollywood, Roland Mumm, Maria Vinaixa, Naomi Rankin, Ron Wehrens and Reza Salek

4. **Mass Spectrometry Data Processing** 73
Steffen Neumann, Oscar Yanes, Roland Mumm and Pietro Franceschi

5. **Nuclear Magnetic Resonance Spectroscopy Data Processing** 101
Maria Vinaixa, Naomi Rankin, Jeremy Everett and Reza Salek

6. **Statistics: The Essentials** 129
Ron Wehrens and Pietro Franceschi

7. **Data Fusion in Metabolomics** 157
Johan A. Westerhuis, Frans van der Kloet and Age K. Smilde

8. **Genome-Scale Metabolic Networks** 177
Clément Frainay and Fabien Jourdan

9. **Metabolic Flux** 201
Igor Marín de Mas and Marta Cascante

10. Data Sharing and Standards 235
Reza Salek

11. Conclusion 253
All Authors

Appendix MTBLS1 259
Reza Salek and Jules Griffin

Appendix MTBLS18 263
Tilo Lübken, Michaela Kopischke, Kathrin Geissler, Lore Westphal, Dierk Scheel, Steffen Neumann and Sabine Rosahl

Index 269

Preface

The Goal of the Book

This book aims to provide a concise and yet complete overview of data analysis in metabolomics, concentrating on nuclear magnetic resonance and mass spectrometry, the two dominant detection techniques in the area. Like all -omics sciences, metabolomics is a multidisciplinary field, encompassing biology, chemistry, computer science, statistics and data analysis. DNA and proteins, related to genomics and proteomics respectively, are, in a sense, much simpler to measure than metabolites: They consist of sequences of letters from alphabets with limited numbers of letters. Metabolites cannot be enumerated in this way, and show a much bigger diversity in chemical and physical properties. This implies that different analytical techniques are necessary to obtain a complete picture of the metabolome of a biological system. The choice of analytical technique determines to a large extent what is measured, and some knowledge of the underlying (analytical) chemistry is more important than in transcriptomics, for example. This large variability in chemical structure and analytics also leads to a data processing pipeline that is much more diverse than what we see in other omics sciences and has to be adapted to the analytical techniques and experimental protocol used for each experiment. In that sense, metabolomics is not only a science, but also somewhat of an "art."

Textbooks in metabolomics typically dedicate a significant fraction of the content to the analytical techniques and treat data analysis as one of the many aspects of a metabolomics experiment. In this book, we have reversed the balance and focus mainly on the data analysis aspects. These are diverse and often complicated: Typically we have a lot of data from relatively few samples, something that makes the application of classical statistical procedures difficult. The experimental aspects are only covered to the extent that is necessary to understand the consequences of data handling, analysis and interpretation.

In this book, we are going further than just the identification of metabolic fingerprints, and try to interpret them in a biochemical context by constructing metabolic networks. The dynamic nature of the metabolism of the system under study is addressed through flux modeling. Finally, data standards and data sharing have become integral parts in publishing metabolomics results.

Who Should Read This Book

The intended audience consists of scientists searching to obtain a comprehensive overview of the field of metabolomics—typically PhD students or postdocs entering the field. Such people rarely have experience in metabolomics itself, but very often they do come with a (very useful!) background in related disciplines, such as chemistry, bioinformatics, data science or statistics. To cater to the different backgrounds of our readers, we have in several places included boxes containing very brief explanations of words or idioms that may not be familiar. Even though this may not be sufficient for the non-expert to obtain a full understanding, it should provide enough background to be able to grasp the basic principles, and some pointers to useful search terms. In addition, we have set up a supplementary material website containing many of the scripts, data sets and tutorials used in this book (see below). This allows the reader to get some hands-on experience, essential to grasp the finer points.

Software

Metabolomics is a field in rapid development. One reason is the continuous improvement in analytics: Each new generation of equipment brings higher sensitivities and better resolution. The same is happening in the software field. Not only do instrument manufacturers provide their own software packages, but software is also available from many different sources, commercial as well as open source. In this book, we focus on general principles and concepts, rather than on descriptions of particular software solutions. Where we do mention individual programs, we have given preference to software with a large user base, preferably open source. More details can be found on the supplementary material website at http://www.metabolomics-pgda.eu.

Example Data

In several chapters, we will refer to data from real-world experiments. One of the data sets (see Appendix MTBLS18) is combining metabolomics and transcriptomics measurements, and investigating the defense of the plant *Arabidopsis thaliana* against *Phytophthora infestans*, a pathogen causing late blight disease in potatoes. Raw data are available at the MetaboLights and ArrayExpress repositories and are described in more detail in Appendix MTBLS18 of the book. The MetaboLights identifier MTBLS18 will be used throughout the book when referring to this data set. A similar set for the field of NMR is formed by MTBLS1, described in Appendix MTBLS1.

Acknowledgments

The idea of the book was conceived during the annual EMBO Practical Course on Metabolomics Bioinformatics for Life Scientists, held from 2013–2018 at the European Bioinformatics Institute in Hinxton, UK, and continued at the International Agency for Research on Cancer, Lyon, France. Editions of the same course outside the UK have been organized in Brazil and South Africa. Each time, 30 young scientists, mostly at the PhD and postdoc level, were selected from many applicants (often we had 4–5 times more applicants than openings). Teaching these groups has proven to be a most satisfying experience, and feedback from the many excellent students in the course has been incorporated in this book. We thank them for the enthusiastic and constructive way in which they have participated. In this book, all authors contributed to several chapters and have read the entire book. We hope that, by doing so, the book is more coherent, and chapters are linked—hence, avoiding repetition. Finally, we would like to acknowledge EMBO for providing continuous support for the "EMBO Practical Course on Metabolomics Bioinformatics for Life Scientists." Without their generous support for the course, we would not have come together and collaborated on this project. Special thanks to the present and past speakers of the EMBO course (ordered alphabetically):

Raphael Aggio – School of Biological Sciences, University of Auckland, Australia

Stephan Beisken – EMBL-EBI, UK

Marta Cascante – University of Barcelona, Spain

Michael Eiden – Medical Research Council, Human Nutrition Research, UK

Jeremy Everett – University of Greenwich, UK

Pietro Franceschi – Fondazione Edmund Mach, Italy

Jules Griffin – University of Cambridge, UK

Jos Hageman – Wageningen University & Research, Netherlands

Kenneth Haug – EMBL-EBI, UK

Rick Higashi – University of Kentucky College of Medicine, USA

Fabien Jourdan – INRA, France

Gwenaelle Le Gall – Institute of Food Research, UK

Silvia Marin – University of Barcelona, Spain

Igor Marín de Mas – Technical University of Denmark, Denmark

Robert Mistrik – HighChem, Slovakia

Steffen Neumann – Leibniz Institute of Plant Biochemistry, Germany

George Poulogiannis – CRUK, UK

Naomi Rankin – University of Glasgow, UK
Age Smilde – University of Amsterdam, Netherlands
Christoph Steinbeck – University of Jena, Germany
Etienne Thévenot – CEA, France
Maria Vinaixa – SYNBIOCHEM-MIB, University of Manchester, UK
Ralf Weber – University of Birmingham, UK
Ron Wehrens – Wageningen University and Research, Netherlands
Johan Westerhuis – University of Amsterdam, Netherlands
David Wishart – University of Alberta, Edmonton, Canada
Oscar Yanes – University Rovira i Virgili, Spain

Editors

Ron Wehrens was an associate professor at the Radboud University Nijmegen (Netherlands) before founding the Biostatistics unit at the Fondazione Edmund Mach, San Michele all'Adige (Italy). Currently, he is Business Unit Manager at Biometris, Wageningen University & Research, Netherlands.

Reza Salek has worked in the past in the University of Cambridge and the European Molecular Biology Laboratory European Bioinformatics Institute (EMBL-EBI), Cambridge, UK. He is currently a scientist at the International Agency for Research on Cancer, an intergovernmental agency forming part of the World Health Organization of the United Nations, Lyon, France.

Contributors

Marta Cascante
Integrative Systems Biology, Metabolomics and Cancer Research Group
Department of Biochemistry and Molecular Biology
Faculty of Biology
University of Barcelona
Barcelona, Spain

Jeremy Everett
Medway Metabonomics Research Group
University of Greenwich
London, United Kingdom

Clément Frainay
Toxalim (INRA Laboratory)
Université de Toulouse
Toulouse, France

Pietro Franceschi
Computational Biology
Research and Innovation Centre
Fondazione Edmund Mach
San Michele all'Adige, Italy

Kathrin Geissler
Leibniz Institute of Plant Biochemistry
Department of Stress and Developmental Biology
Halle, Germany

Jules Griffin
Department of Biochemistry
Cambridge Systems Biology Centre
University of Cambridge
Cambridge, United Kingdom

Katherine Hollywood
Manchester Institute of Biotechnology
School of Chemistry
University of Manchester
Manchester, United Kingdom

Fabien Jourdan
Toxalim (INRA Laboratory)
Université de Toulouse
Toulouse, France

Frans van der Kloet
Swammerdam Institute for Life Sciences
University of Amsterdam
Amsterdam, Netherlands

Michaela Kopischke
Leibniz Institute of Plant Biochemistry
Department of Stress and Developmental Biology
Halle, Germany

Tilo Lübken
Leibniz Institute of Plant Biochemistry
Department of Stress and Developmental Biology
Halle, Germany

Igor Marín de Mas
The Novo Nordisk Foundation Center for Biosustainability
Technical University of Denmark
Kongens Lyngby, Denmark

Roland Mumm
Wageningen University & Research
Business Unit Bioscience
Wageningen, Netherlands

Steffen Neumann
Leibniz Institute of Plant Biochemistry
Department of Stress and Developmental Biology
Halle, Germany

Naomi Rankin
Glasgow Polyomics
University of Glasgow
Glasgow, Scotland

Sabine Rosahl
Leibniz Institute of Plant Biochemistry
Department of Stress and Developmental Biology
Halle, Germany

Reza Salek
International Agency for Research on Cancer
World Health Organization
Lyon, France

Dierk Scheel
Leibniz Institute of Plant Biochemistry
Department of Stress and Developmental Biology
Halle, Germany

Age K. Smilde
Swammerdam Institute for Life Sciences
University of Amsterdam
Amsterdam, Netherlands

Maria Vinaixa
Department of Electronic Engineering
Universitat Rovira i Virgili
Tarragona, Spain

Ron Wehrens
Business Unit Biometris
Wageningen University & Research
Wageningen, Netherlands

Johan A. Westerhuis
Swammerdam Institute for Life Sciences
University of Amsterdam
Amsterdam, Netherlands

and

Centre for Human Metabolomics
Faculty of Natural Sciences and Agriculture
North-West University (Potchefstroom Campus)
Potchefstroom, South Africa

Lore Westphal
Leibniz Institute of Plant Biochemistry
Department of Stress and Developmental Biology
Halle, Germany

Oscar Yanes
Department of Electronic Engineering
Universitat Rovira i Virgili
Tarragona, Spain

and

Spanish Biomedical Research Center in Diabetes and Associated Metabolic Disorders (CIBERDEM)
Madrid, Spain

1

Overview of Metabolomics

Oscar Yanes, Reza Salek, Igor Marín de Mas and Marta Cascante

CONTENTS

Introduction 1
Measuring the Metabolome 4
Targeted Versus Untargeted Metabolomics 7
Pathways 9
Conclusion 11
References 12

Introduction

Metabolomics, the science of small chemical molecules known as metabolites, is now finding applications that span almost the full width of biological and biotechnological research. These range from human, plant and microbial biochemistry to drug and biomarker discovery, toxicology, nutrition and food control, and include the discovery of preclinical biomarkers of human diseases such as Alzheimer's disease (Trushina et al. 2013; Mapstone et al. 2014), type-2 diabetes (Salek et al. 2007; Wang et al. 2011) and prostate cancer progression (Sreekumar et al. 2009), the identification of biomarkers of food intake (Bondia-Pons et al. 2013) and the understanding of how plants use small molecules in pathogen defense (Rajniak et al. 2015). Other examples of the potential of metabolomics are understanding how the modulation of diet may lead to changes in metabolism and histone methylation in cells (Mentch et al. 2015) and the elucidation of the interplay between gut microbiota and human physiology (Shoaie et al. 2015; Thorburn et al. 2015). Finally, the study of natural metabolic variation by integrating metabolomics with genome-wide association studies is important for progress in biological (Strauch et al. 2015) and medical research (Kettunen et al. 2012; Rueedi et al. 2014). This complements upstream biochemical information obtained from genes, transcripts and proteins, widening current genomic reconstructions of metabolism and improving our understanding of cell biology, physiology and medicine by linking cellular pathways to biological mechanisms.

BOX 1.1 GLOSSARY OF TERMS RELATED TO THE GREEK WORD 'METABOLĒ', MEANING CHANGE

- *Metabolism:* The biochemical processes that occur within a living organism or cell in order to maintain life.
- *Metabolite:* A biochemical compound of low/medium molecular weight (30–1500 Da approximately) involved in a biochemical change.
- *Metabolic pathway:* A series of chemical reactions occurring within a cell. These reactions are catalyzed by enzymes, where the chemical product (i.e., metabolite) of one enzyme acts as the substrate for the next reaction.
- *Metabolome:* The totality of molecular changes, formed by a large network of metabolites and metabolic reactions, where outputs from one biochemical reaction are inputs to other reactions.
- *Metabolomics:* The discipline of studying the metabolome: the comprehensive characterization and quantification of metabolites translating into the structure, function and dynamics of an organism or biological system.

BOX 1.2 GLOSSARY OF IMPORTANT BIOLOGICAL CONCEPTS

- *Genome:* An organism's complete set of DNA, including all of its genes.
- *Gene:* A locus (or region) of DNA that encodes a functional RNA or protein product; the molecular unit of heredity.
- *Epigenome:* The set of chemical modifications to the DNA (i.e., methylation) and DNA-associated proteins (i.e., histone modifications: acetylation, phosphorylation, methylation, ubiquitylation and sumoylation) in the cell, which alter gene expression.
- *Transcript (or mRNA):* The single-stranded copy of the gene, which next can be translated into a protein molecule.
- *Protein:* A large biomolecule consisting of long chains of amino acid residues. Proteins include such specialized forms as collagen for supportive tissue, hemoglobin for transport, antibodies for immune defense and enzymes for metabolism.

(*Continued*)

BOX 1.2 (Continued) GLOSSARY OF IMPORTANT BIOLOGICAL CONCEPTS

- *Transcription:* The first step of gene expression, in which a particular segment of DNA is copied into a complementary, antiparallel RNA strand called a primary transcript (mRNA).
- *Translation:* The process in which mRNA—produced by transcription from DNA—is decoded by a cellular ribosome to produce a specific amino acid chain, or protein.
- *Post-translational modification (PTM):* Refers to the covalent and generally enzymatic modification of proteins aiming at increasing their functional diversity after protein biosynthesis. Some of the most important PTMs include phosphorylation, glycosylation, ubiquitination, methylation, acetylation and lipidation.
- *Enzyme:* A protein that catalyzes chemical reactions, converting chemical substrates into different molecules, called products.
- *Biochemistry:* The study of chemical processes and the biological molecules that occur within living organisms.
- *Physiology:* The branch of biology dealing with the functions and activities of living organisms and their parts, including all physical and chemical processes.

Despite this great and widely shared interest, metabolomics has not evolved as fast as the other omic sciences. There are challenges along different points of the metabolomics workflow (from study design to data analysis) accounting for this situation. Unlike genes, transcripts or proteins, which are biopolymers encoding information as a sequence of well-defined monomers (namely nucleotides and amino acids), metabolites are chemical entities that do not result from a residue-by-residue transfer of information within the cell (some examples are shown in Figure 1.1). The great success in the characterization of genes, transcripts and proteins is a direct consequence of currently well-established technologies and bioinformatic tools allowing for the amplification and subsequent accurate characterization of the sequence of nucleotides and amino acids in these polymers. Metabolomics, in contrast, aims to quantify and identify metabolites. The extremely large physicochemical diversity of metabolite structures in living organisms results from a series of chemical transformations, often catalyzed by enzymes.

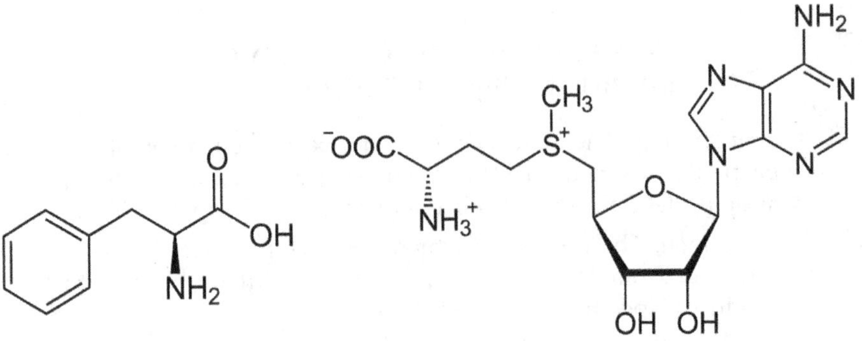

FIGURE 1.1
Examples of two metabolites: the amino acid phenylalanine (left) and the co-substrate S-adenosylmethionine (right) involved in methyl group transfers.

Currently, the analysis of metabolites can be carried out on a large variety of organisms and biological samples, including human and animal biofluids such as urine, blood serum and plasma; tissues and cell extracts; plant tissues and their extracts; model organisms such as C. elegans, nematodes, yeast or bacteria; and in vitro cell studies. Examples are endogenous metabolites that are naturally produced by an organism (such as amino acids, organic acids, nucleic acids, fatty acids, amines, sugars, vitamins, co-factors, pigments, antibiotics, etc.) as well as exogenous chemicals (such as drugs, environmental contaminants, food additives, toxins and other xenobiotics) that are taken up by an organism as result of interaction with its environment. The large physicochemical diversity of metabolites is reflected in a wide array of polarities, molecular weights, different functional groups, stereochemistry, chemical stabilities and reactivities among other important properties.

Measuring the Metabolome

No single analytical technology can measure all of these diverse compounds at once. This leads to metabolites being analyzed using several technologies and instrumental setups to increase metabolome coverage (see Figure 1.2), something remarkable that does not occur in genomics or proteomics. In this regard, two technological platforms are most commonly used to identify and quantify metabolites: nuclear magnetic resonance (NMR) spectroscopy (see Chapters 3 and 5) and mass spectrometry (MS; see Chapters 3 and 4), often coupled to chromatographic techniques such as liquid (LC-MS) and gas chromatography (GC-MS), and, to a lesser extent, capillary electrophoresis (CE-MS). The metabolomic dynamic range, the inherent chemical (e.g., mobile phase contaminants) and instrumental noise of the metabolite profile, the vast number of unidentified

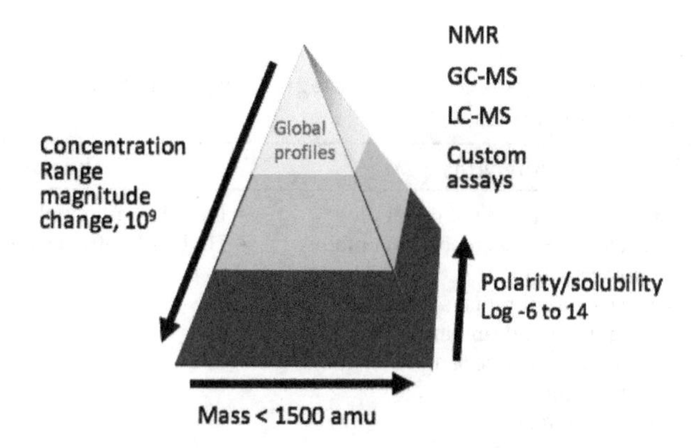

FIGURE 1.2
Several analytical techniques are often needed to cover the wide variety in metabolites and metabolite concentrations. The metabolite dynamic range is quite high in terms of size, concentration and physicochemical properties. Metabolite sizes vary by orders of magnitude (~100–1500 Da), concentrations range from the picomolar to millimolar ranges, and physicochemical properties such as stability, hydrophobicity, stereochemistry, polarity, volatility and pKa range can show huge differences. Typically, several different analytical assay types are needed for a wide coverage of the metabolome, even assuming extraction methods are optimal.

compounds present within single samples, and the rapidly changing temporal and spatial variability (flux) of the cell's metabolites complement present huge challenges. The dynamic range defines the concentration boundaries of an analytical determination over which the instrumental response is linear to the analyte concentration. It can be severely limited by the sample matrix or the presence of interfering and competing compounds. This is one of the most difficult issues to address in metabolomics. The presence of some excessive metabolites can cause significant or severe chemical interferences that limit the range in which other metabolites may be successfully profiled.

Another consequence of the physico-chemical diversity of compounds present in biological systems and of the corresponding richness in analytical platforms used to measure them is that identifying and characterizing the structure of metabolites has become one of the major bottlenecks in untargeted metabolomics. Chapters 3 through 5 provide an overview of the state of affairs for the two main analytical techniques, NMR and MS, respectively (Table 1.1). Even in cases where identification is not straightforward, one may deduce useful information: from statistical analysis of the results, one sometimes sees a particular signal, an unknown, to be correlated with a trait of interest. For instance, a metabolite may be found in (nearly) all samples from the treatment group, and not at all (or only rarely) in the control group. Clearly, to make these calls, a basic level of statistical knowledge is required. Statistics, and especially multivariate statistics plays an important part in visualizing the results of a metabolomics experiment. Chapter 6 provides an

TABLE 1.1

The Strengths and Weaknesses of Nuclear Magnetic Resonance and Mass Spectrometry for Metabolomics

	Strengths	Weaknesses
NMR	• The universality of metabolite detection (if metabolites contain hydrogens). • Easy and minimal sample handling and preparation. • Non-destructive sample technique. • Very high analytical reproducibility. • Robust quantitation.	• Low detection limits (micromolar). • High volumes of sample used.
LC-ESI MS	• Very sensitive. • Low volumes of sample used (nL to μL).	• Often poor chromatographic separation. • Ion suppression or lack of ionization for certain types of metabolites. • Metabolites must be extracted in a suitable solvent. • Destructive sample technique.
GC-EI MS	• Very sensitive. • Good chromatographic separation.	• Only suitable for volatile and thermally stable compounds. • Sample preparation typically requires chemical derivatization. • Destructive sample technique.

informal overview of some of the underlying concepts in statistics. Once the individual raw data files have been processed, it is time to put all information together. When several analytical platforms have been employed, one can interpret each one individually, but there are also some tools available to combine all data sets—a process that goes under the name of "Data Fusion," and that is described in Chapter 7. Altogether, this has possibly made metabolomics the most multidisciplinary among all other omic sciences, involving knowledge from engineering and signal processing, analytical and organic chemistry, biostatistics and statistical physics, to biochemistry and cell metabolism (Figure 1.3).

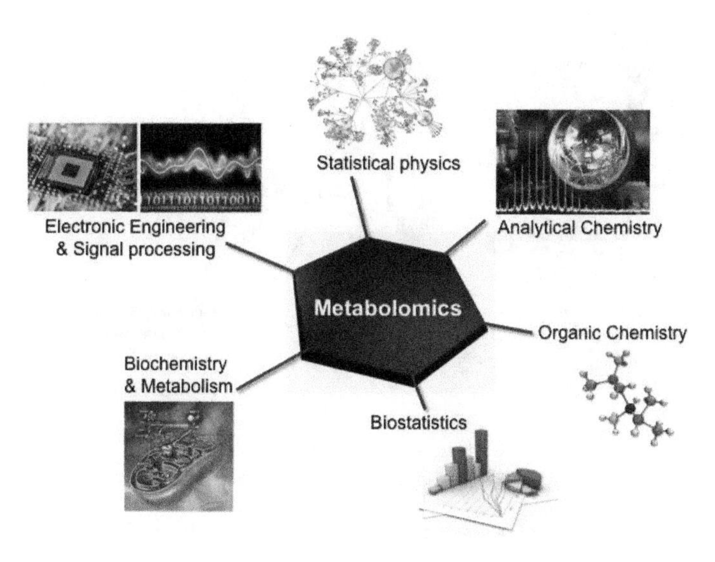

FIGURE 1.3
The multidisciplinary nature of metabolomics. Being able to measure a wide variety of metabolites in biological systems requires in-depth knowledge of analytical chemistry and physics to do the measurements, statistics to identify signals and find true differences, and biology to be able to interpret the findings. Bioinformatics is also used, particularly when dealing with large-scale data analysis, data mining, batch correction and quality control.

Targeted Versus Untargeted Metabolomics

The integration of these diverse disciplines is essential in order to successfully implement the two main accepted experimental approaches in the field: targeted and untargeted metabolomics. In general, focusing on a specific metabolite or a small group of distinct metabolites (targeted metabolomics) is associated with hypothesis-driven studies and involves the optimization of chromatographic and MS conditions (e.g., retention times, MS/MS transitions) using pure standards. The main advantages of targeted metabolomics are specificity, quantitative reproducibility and a relatively high throughput. Untargeted metabolomics studies, in contrast, are designed to profile simultaneously the largest possible number of compounds (i.e., from hundreds to thousands) and therefore they have the capacity to reveal previously unexplored biochemical

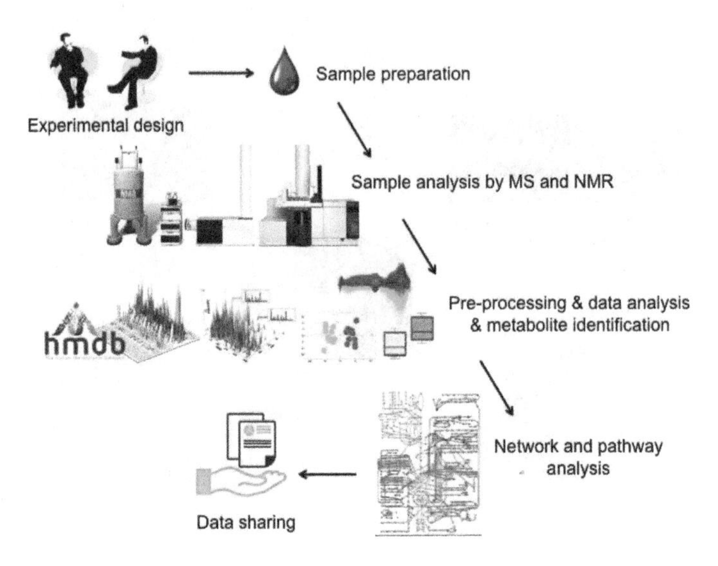

FIGURE 1.4
Untargeted metabolomics workflow. The key part is the experimental design, which determines the choice and appropriateness of sample collection, storage, extraction and the analytical techniques used to detect metabolites. Once data sets are generated, various statistical or chemometrics approaches are used to analyze the results leading to biological conclusions.

pathways (Figure 1.4). In contrast to targeted metabolomic results, untargeted metabolomic data sets are exceedingly large and complex, and present substantial challenges for interpreting metabolite profiling data. Major progress has been made in developing computational tools and chemometric approaches that can facilitate the analysis and the mining of such complex data sets, for example, see Mahieu and Patti (2017). A carefully constructed experimental design is key: one should *beforehand* define the goals of the experiment and the ways in which to obtain these goals (Dunn et al. 2012; Schripsema and Dagnino 2014). This includes defining treatments and treatment levels, available resources, defining the desired effect sizes, defining sample extraction and processing methods, assigning samples to batches, etcetera. Many things can be said about the definition of good and convincing experiments, true in metabolomics as in most other fields of science; some pointers are given in Chapter 2.

Untargeted metabolomics (or metabolite profiling), therefore, assesses the relative concentration of metabolites, providing a static readout of the biochemistry, physiological status and environmental exposure of biological systems. This approach may prove very valuable for biomarker discovery, due to the fact that metabolites have historically been proved to be clinical markers of human diseases such as heart disease (cholesterol) or diabetes (glucose). This idea goes back to Sir Archibald Garrod—the founder of modern clinical chemistry—who noted, already in 1908, that changes in metabolite concentrations often start before the onset of clinical symptoms (Garrod 1908). The very first clinical assay (and the first genetic disease test) was designed for detecting homogentisic acid among individuals with a rare inborn error of metabolism called alkaptonuria (Garrod 1908). This is a case where the gene defect directly causes accumulation of the marker in the blood and tissues; however, most metabolite biomarkers (and indeed combinations thereof) are only indirectly related to the disease usually because they are part of a mechanistic process or because they are produced uniquely by the disease. Despite the natural variation between individuals and within individuals over time, changes in metabolite concentrations have served as the basis to the development of >100 different chemicals tests, many of which are metabolite biomarkers that are commonly used today (for a range of available tests see, for example, https://labtestsonline.org).

Pathways

From a mechanistic view, changes in metabolite levels can be interpreted in order to generate testable hypotheses linking cellular pathways (Chapter 8) to biological mechanisms, which is shaping our understanding of cell biology and physiology (Yanes et al. 2010; Panopoulos et al. 2012; Patti et al. 2012; Samino et al. 2015). However, one must not confound metabolite levels (or concentrations) with metabolic fluxes (Cortassa et al. 2015); see Chapter 9. For example, it would be easy to make the erroneous assumption that an increase in the concentration of metabolite B, which is up-regulated in a disease, is due to increased production rates through the enzymatic reaction $A \Rightarrow B \rightarrow C$, though it would be an equally valid possibility that consumption of this metabolite is decreased in $A \rightarrow B \nrightarrow C$. This limitation may potentially constrain the number of direct leads to drug targets since in the first scenario the target protein that catalyzes the reaction from A to B may be overexpressed, whereas in the second scenario the protein that catalyzes the reaction from B to C may be downregulated or inhibited (see Chapter 8). To resolve this ambiguity in the interpretation of the results, flux analysis defines alterations in metabolic pathways in terms of upregulated or downregulated metabolic fluxes, a conceptual shift from the prevalent use of global metabolite profiling to a more dynamic exploration of cellular metabolism. By analogy to other omics-type analyses, the range of methods in experimental and computational biology that allow the determination of *in vivo* or *in vitro* fluxes in a metabolic network is termed "fluxomics" (Cascante and Marin 2008) (see Figure 1.5 and Chapter 9).

FIGURE 1.5
An analogy between metabolomics and a train network. The latter is continuously monitored to identify problems and abnormal situations. In metabolomics, we mainly take snapshots, providing often incomplete information, with limited dynamic information. The closest we can get is in the area of fluxomics, where alterations in metabolic pathway activities in terms of upregulation or downregulation are identified, similar to congestions observed in the train network.

Some fluxes define the overall rate of nutrients consumption or production, i.e., *exchange fluxes*, and can be easily measured (e.g., consumption of glucose or secretion of lactate). On the contrary, *intracellular fluxes* are more difficult to determine (Sauer 2006). In this sense, the use of metabolomic data from stable isotope labeling experiments is emerging as a powerful tool to determine the internal metabolic fluxes. These experiments are based on the use of labeled substrates such as glucose or glutamine enriched with ^{13}C—a non-radioactive and stable isotope of the carbon atom—introduced at specific carbon positions or uniformly enriched. The interconversion of substrates via intracellular enzyme-catalyzed reactions alters the labeling pattern of the carbon skeleton in metabolites (Stephanopoulos et al. 1998) (Figure 1.6a). The experimental measurements of ^{13}C enrichment of downstream metabolites, together with exchange fluxes and growth rates allow computational tools (Selivanov et al. 2004; Zamboni et al. 2009) to determine intracellular fluxes (Figure 1.6b). The isotopic enrichment of metabolites is represented with the concept of positional isomers (isotopomers) or mass isomers (isotopologues) that can be measured by NMR or MS techniques respectively.

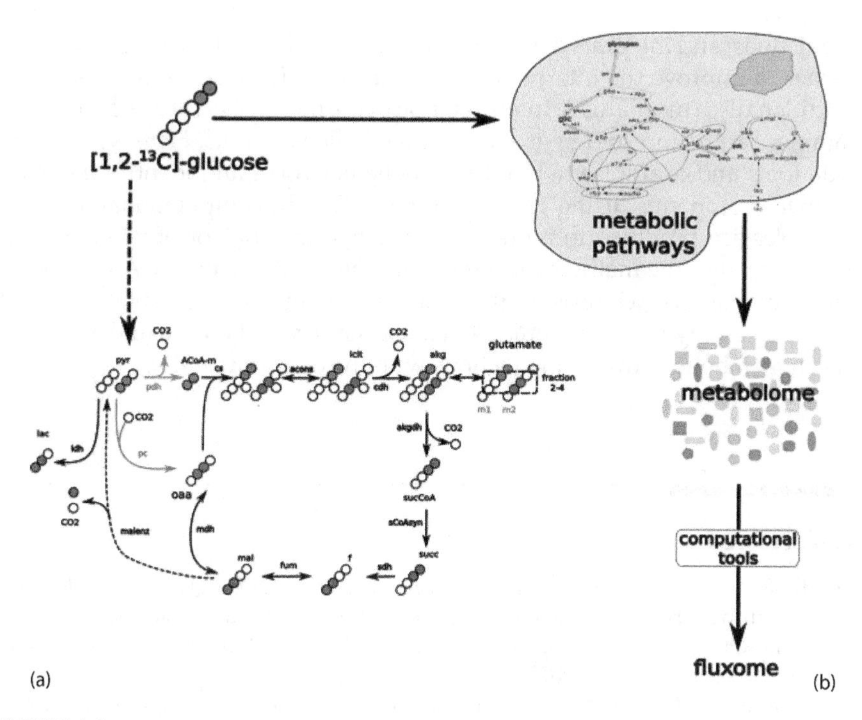

FIGURE 1.6
(a) Example of *stable isotope tracing.* Here is represented how the label of a glucose molecule enriched with ^{13}C (in the first and second carbon represented as orange circles) is propagated through the metabolic network. In this example, one glucose molecule produces two molecules of pyruvate via glycolysis, one non-labeled pyruvate and other labeled in the second and third carbon. Thus, if the labeled pyruvate is metabolized through pdh reaction, glutamate will be labeled in the first and second carbon (highlight in orange) while if the labeled pyruvate is metabolized through pc, the resulting glutamate will be labeled in the third and fourth carbon. Thus based on the labeling pattern of the measured metabolites one may infer the metabolic flux of the reactions involved in the process. Thus, we can analyze glutamate labeling using NMR or the fraction 2–4 using mass spectrometry techniques in order to determine the metabolic reactions involved in its metabolism. (b) Represents a workflow to determine metabolic fluxes from metabolomic data based on labeled substrates. Thus, using computational tools that integrate a previous knowledge of the metabolic pathways and tracer-based metabolic data we can infer the intracellular metabolic flux profile in different scenarios.

Conclusion

Overall, the potential applications of metabolomics are huge. However, converting raw spectrometric data into biological knowledge has become the major bottleneck for the field. This book is intended to facilitate this process by providing strategies, tools and examples to detect, identify and quantify metabolites from NMR, LC-MS and GC-MS spectra; visualize and explore the data using different statistical techniques; build statistical models that allow predictions; and integrate data from various instrumental setups or omic techniques to improve the interpretation about the differences between samples. Additionally, a mechanistic interpretation of the metabolic differences between samples can be inferred by exploring metabolic networks. Here we also provide tools and examples to build metabolic networks and identify metabolic pathways of interest using graph modeling. Finally, computational tools and examples are provided to represent the dynamic behavior of metabolic networks through the mathematical modeling of metabolic fluxes. Altogether, in the next chapters, we provide the basic knowledge and computational tools from processing raw spectrometric data to the comprehensive understanding of cell metabolism and its interactions with the environment.

References

Bondia-Pons, I., T. Barri, and K. Hanhineva. 2013. "UPLC-QTOF/MS Metabolic Profiling Unveils Urinary Changes in Humans after a Whole Grain Rye versus Refined Wheat Bread Intervention." *Molecular Nutrition & Food Research*. doi:10.1002/mnfr.201200571/full.

Cascante, M., and S. Marin. 2008. "Metabolomics and Fluxomics Approaches." *Essays in Biochemistry* 45: 67–82.

Cortassa, S., V. Caceres, L. N. Bell, B. O'Rourke, N. Paolocci, and M. A. Aon. 2015. "From Metabolomics to Fluxomics: A Computational Procedure to Translate Metabolite Profiles into Metabolic Fluxes." *Biophysical Journal* 108 (1): 163–172.

Dunn, W. B., I. D. Wilson, A. W. Nicholls, and D. Broadhurst. 2012. "The Importance of Experimental Design and QC Samples in Large-Scale and MS-Driven Untargeted Metabolomic Studies of Humans." *Bioanalysis* 4 (18): 2249–2264.

Garrod, A. E. 1908. "The Croonian Lectures on Inborn Errors of Metabolism." *The Lancet* 172 (4427): 1–7.

Kettunen, J. et al. 2012. "Genome-Wide Association Study Identifies Multiple Loci Influencing Human Serum Metabolite Levels." *Nature Genetics* 44 (3): 269–276.

Mahieu, N. G., and G. J. Patti. 2017. "Systems-Level Annotation of a Metabolomics Data Set Reduces 25,000 Features to Fewer than 1000 Unique Metabolites." *Analytical Chemistry* 89 (19): 10397–10406.

Mapstone, M. et al. 2014. "Plasma Phospholipids Identify Antecedent Memory Impairment in Older Adults." *Nature Medicine* 20 (4): 415–418.

Mentch, S. J. et al. 2015. "Histone Methylation Dynamics and Gene Regulation Occur through the Sensing of One-Carbon Metabolism." *Cell Metabolism* 22 (5): 861–873.

Panopoulos, A. D. et al. 2012. "The Metabolome of Induced Pluripotent Stem Cells Reveals Metabolic Changes Occurring in Somatic Cell Reprogramming." *Cell Research* 22 (1): 168–177.

Patti, G. J., O. Yanes, L. P. Shriver, J.-P. Courade, R. Tautenhahn, M. Manchester, and G. Siuzdak. 2012. "Metabolomics Implicates Altered Sphingolipids in Chronic Pain of Neuropathic Origin." *Nature Chemical Biology* 8 (3): 232–334.

Rajniak, J., B. Barco, N. K. Clay, and E. S. Sattely. 2015. "A New Cyanogenic Metabolite in Arabidopsis Required for Inducible Pathogen Defence." *Nature* 525 (7569): 376–379.

Rueedi, R. et al. 2014. "Genome-Wide Association Study of Metabolic Traits Reveals Novel Gene-Metabolite-Disease Links." *PLoS Genetics* 10 (2): e1004132.

Salek, R. M. et al. 2007. "A Metabolomic Comparison of Urinary Changes in Type 2 Diabetes in Mouse, Rat, and Human." *Physiological Genomics* 29 (2): 99–108.

Samino, S. et al. 2015. "Metabolomics Reveals Impaired Maturation of HDL Particles in Adolescents with Hyperinsulinaemic Androgen Excess." *Scientific Reports* 5: 11496.

Sauer, U. 2006. "Metabolic Networks in Motion: ^{13}C-based Flux Analysis." *Molecular Systems Biology* 2 (1): 62.

Schripsema, J., and D. Dagnino. 2014. "Metabolomics: Experimental Design, Methodology and Data Analysis." *Encyclopedia of Analytical Chemistry*, John Wiley & Sons, Ltd, pp. 1–17.

Selivanov, V. A., J. Puigjaner, A. Sillero, J. J. Centelles, A. Ramos-Montoya, P. W.-N. Lee, and M. Cascante. 2004. "An Optimized Algorithm for Flux Estimation from Isotopomer Distribution in Glucose Metabolites." *Bioinformatics* 20 (18): 3387–3397.

Shoaie, S. et al. 2015. "Quantifying Diet-Induced Metabolic Changes of the Human Gut Microbiome." *Cell Metabolism* 22 (2): 320–331.

Sreekumar, A. et al. 2009. "Metabolomic Profiles Delineate Potential Role for Sarcosine in Prostate Cancer Progression." *Nature* 457 (7231): 910–914.

Stephanopoulos, G., A. A. Aristidou, and J. Nielsen. 1998. *Metabolic Engineering: Principles and Methodologies*. Elsevier Science, San Diego, CA.

Strauch, R. C., E. Svedin, B. Dilkes, C. Chapple, and X. Li. 2015. "Discovery of a Novel Amino Acid Racemase through Exploration of Natural Variation in Arabidopsis Thaliana." *Proceedings of the National Academy of Sciences of the United States of America* 112 (37): 11726–11731.

Thorburn, A. N. et al. 2015. "Evidence That Asthma Is a Developmental Origin Disease Influenced by Maternal Diet and Bacterial Metabolites." *Nature Communications* 6: 7320.

Trushina, E., T. Dutta, X.-M. T. Persson, M. M. Mielke, and R. C. Petersen. 2013. "Identification of Altered Metabolic Pathways in Plasma and CSF in Mild Cognitive Impairment and Alzheimer's Disease Using Metabolomics." *PloS One* 8 (5): e63644.

Wang, T. J. et al. 2011. "Metabolite Profiles and the Risk of Developing Diabetes." *Nature Medicine* 17 (4): 448–453.

Yanes, O., J. Clark, D. M. Wong, G. J. Patti, A. Sánchez-Ruiz, H. P. Benton, S. A. Trauger, C. Desponts, S. Ding, and G. Siuzdak. 2010. "Metabolic Oxidation Regulates Embryonic Stem Cell Differentiation." *Nature Chemical Biology* 6 (6): 411–417.

Zamboni, N., S.-M. Fendt, M. Rühl, and U. Sauer. 2009. "^{13}C-Based Metabolic Flux Analysis." *Nature Protocols* 4 (6): 878–892.

2

Study Design and Preparation

Pietro Franceschi, Oscar Yanes and Ron Wehrens

CONTENTS

Introduction 15
Sampling Units 20
Experimental Design 21
Collection and Handling 25
Extraction and Pretreatment 26
Doing the Measurements 26
Analyzing the Data 28
  Preparations 28
  Reproducibility 30
References 32

Introduction

When designing a study there are several important considerations to be taken into account, irrespective of the measurement technology. The most important element is to precisely define the goals of the study. Even though these goals can be quite general, they must be more explicit than just expressing the desire to get to know the system under study. One easy rule of thumb is to think about the conditions under which the experiment can be called a success: "Finally, we are able to understand why this treatment leads to an upregulation of this particular metabolic pathway!" or "We have confirmed our hypothesis that compound X is not present in muscle tissue but is quite abundantly present in liver tissue." If you are able to define a successful experiment, your goals are testable. The more specific your research questions, the more likely it is that you will find a clear answer, based on your experimental results. The definition of the research question will also direct the choice of the most suitable experimental design and, consequently, will shape the data analysis plan, which is a critical part of metabolomics.

In that respect, it is informative to consider the difference between targeted and untargeted metabolomics, particularly when comparing two groups of samples (e.g., treated versus control samples). Untargeted metabolomics is usually employed to obtain global information, a fingerprint, often interpreted as a kind of metabolic phenotyping. In this approach, any feature showing a difference between the two sample classes is potentially a (tentative) biomarker of interest. The question is very general, but the results may not be easy to interpret. Since typically many variables are measured using a limited number of samples, and all statistical tests have an error rate associated with them, false negative, as well as false positive results, can be expected (Chapter 6). Often in untargeted metabolomics, we do not know the chemical nature of the biomarkers that are found in this way, and this makes it hard to assess their plausibility. Alternatively, if the research question is focusing on a limited number of metabolites, using a targeted approach will allow a more precise and sensitive measurement, easier for statistical comparison of the groups. Since the number of variables is much lower in this case and metabolite identities are usually known, immediate interpretation of the results is possible. Commonly, one often starts a new research project using untargeted metabolomics to find whether there are differences at all. Interesting candidates are then identified (e.g., using the statistical techniques from Chapter 6), and, if possible, validated using targeted approaches.

A common type of research question consists of comparing groups of samples, where a group can be defined as a set of biological samples with an important characteristic in common (genotype, tissue type, treatment, etc.). The variation within the group is caused by other factors, which are not taken into account in the grouping and which typically are of less interest. The question then is whether the differences between the groups are larger than the differences within the groups. Obviously, the more homogeneous the groups, the easier it will be to find differences of interest—that is why in many medical studies, subjects are stratified according to factors like age, BMI, gender or smoking behavior: within one group, one would expect individuals that are comparable.

Figure 2.1 shows typical cases where one, two or more homogeneous groups of samples are compared. In the case that we have only one group, one may be interested in assessing the overall variation within that group, perhaps with the goal to identify deviating samples (outliers). For example, in a statistical quality control setting, we look for samples that deviate from the expected behavior. This not only has applications in industry and manufacturing (products must agree to one or more specifications; process conditions should be in the "normal-operating range," otherwise something could be wrong), but also applies in areas like food authenticity: is this olive oil really a high-quality "extra virgin" or is it a cheaper imitation? If a test sample differs from the group of extra virgin oils, it is rejected. Such questions are typically assessed in a targeted approach because specifications or "normal operating conditions" have to be defined in a quantitative way.

The two-group case shown in Figure 2.1 is one of the classical situations in metabolomics, and also a typical example of a situation where untargeted analysis is most useful: typically, one already expects a difference between

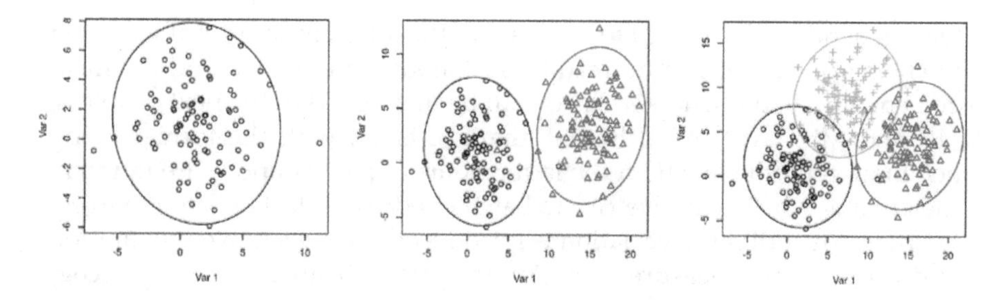

FIGURE 2.1
Examples of classification questions. The left plot, a one-group situation, allows one to determine how "typical" a new sample is, compared to a series of reference samples (in black). The middle plot shows a two-group situation, typically used to determine the most relevant differences between the groups. Here, Var 1 shows more pronounced differences between the groups than Var 2. The plot on the right shows a situation with more than three classes. The groups are still reasonably well separated, but both variables are needed to establish group memberships reliably. For simplicity, only two variables are shown; similar considerations hold for high-dimensional cases.

the groups but is not sure about the nature of the difference. Therefore it makes sense to test as many different variables as possible. Examples of this type of situation are abundant, e.g., case-control studies where each sample group—with each subject randomly chosen (but belonging to the same population)—gets a different treatment. Another example is the comparison between patients and healthy controls.

In the natural extension of this case, one is interested in comparing several groups simultaneously, e.g., when there is more than one treatment being compared to a control situation. Comparisons can be made to distinguish between one particular group from the rest (e.g., healthy people from people suffering from a variety of conditions) or, alternatively, to find the differences between all pairs of groups. In even more complex situations, the objective is to study the effects of one (or more) treatments on heterogeneous groups (typically found in human studies). In such cases, it can be advantageous to rely on crossover designs where each individual receives all the treatments at different times to take into account the natural differences among the individuals—basically, every subject is directly compared to himself only. In all the cases, it is necessary (before doing the experiment!) to decide what to measure, when, and if possible to give an estimate of the required sample size on the basis of the expected effect.

The aforementioned groups are defined a priori, e.g., using phenotypic measures (diseased/healthy), genetic information, or information from other sources. Obviously, there is also information that is not discrete but continuous, and one can also investigate relationships with such continuous measures, like age or weight in human studies. In such cases, we would be looking for variables, or groups of variables, that show, for example, large correlations with the variable of interest. Chapter 6 discusses several methods for doing so.

On a general level, we need to be aware of the different sources of variation influencing our experimental results. At the lowest level, we find the measurement variability (Chapter 3). As will be mentioned in the next chapters, NMR and MS have very different characteristics, with the former being much more reproducible than the latter. Some of the variations are random, some are not, which can give rise to batch effects, or effects related to injection order. An additional variation is introduced during sample preparation and during sample measurement. An analytical chemist can easily recognize these sources of variation and in many cases is able to control them, at least to some extent. The problem is that there are many other sources of variation, which cannot always be controlled. Biological variation often plays a significant part: it can contain random as well as systematic components. An apple growing in a position where it gets a lot of sunlight will develop differently than an apple that is in a more shady spot. So even within the same tree, we can see systematic differences. The same principle holds true for apples from within the same orchard, region, country or even continent. Not all of these differences are going to be relevant for our experiment, but we do need to be aware of it when we are formulating our conclusions and planning our experiments.

A schematic summary of these concepts is depicted in Figure 2.2. Statistically speaking, the total variance is the sum of the variances of the individual components:

$$s^2_{\text{total}} = s^2_{\text{location}} + s^2_{\text{biol}} + s^2_{\text{sampling}} + \cdots + s^2_{\text{data analysis}}$$

The last term, variation arising because of data analysis, may come as a surprise. Usually, we go through the whole process of data analysis only once—so what is this variability that we are talking about? Variation arises because of many decisions that are taken during the data analysis—if the analysis would be repeated, perhaps by another scientist, the results would probably not be exactly the same, especially if the point-and-click software is used. But when using analysis scripts (in principle perfectly reproducible), other data analysts may modify the settings to achieve optimal results. Importantly, one should realize that over all different stages of setting up an experiment, from sample collection, measurements and data analysis, variability is introduced. If one wants to decrease the overall level of variation, it is important to focus on the biggest contributors: improving technical reproducibility will not help much if the biological variation is dominant.

As the sensitivity of the measurement technologies for metabolomics increases, getting a good idea of the overall variability and the source of this variability becomes even more important. Regardless of its origin, unwanted variability limits our capacity to spot differences and identify trends. As is easy to imagine, subtler details will require more efforts on the experimental design, more control on the analytical conditions and more samples. Even though variability in itself does not decrease when measuring more samples,

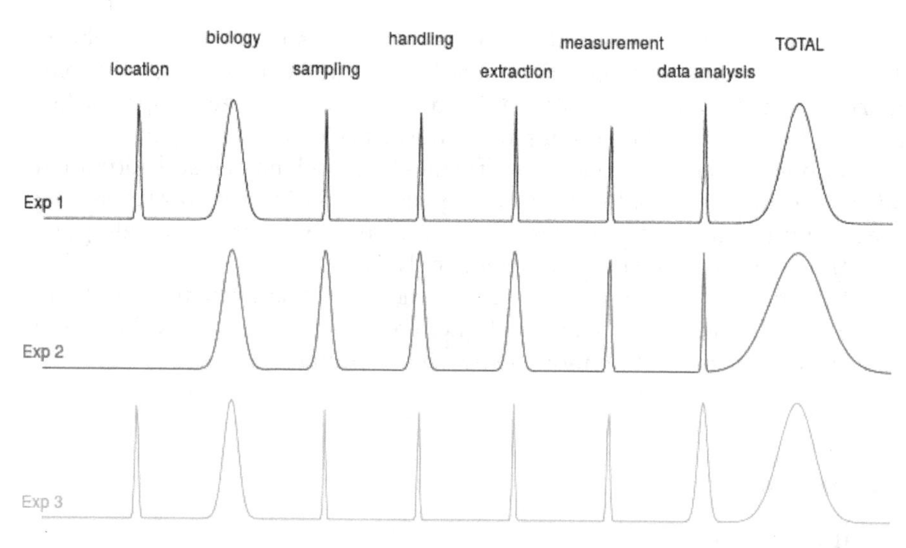

FIGURE 2.2
Sources of variation—schematic view. The width of the peaks indicate the size of the variation for that particular experimental stage—the rightmost peak is the total variation, i.e., the sum of the individual variance components. The top line shows a well-executed experiment, where the main variation component is the biological variation of the system of interest (i.e., differences between biological replicates). The second line shows an example where the overall variability is large even though only material from one location was used, resulting in absence of any environmental variation—the large variation in sampling, handling and extraction is at fault here. The final, bottom, line shows an example where non-reproducible data analysis leads to variation in outcomes. Note that in this drawing we only consider random variation and not systematic effects.

the variability of the mean does (with the square root of the number of samples), which is the effect of random errors cancelling each other out. This is the reason why statistical significance is easier achieved with larger numbers of samples (see Chapter 6 for more details).

The last point is particularly critical, and it is also one of the weaknesses of many large-scale metabolomics research projects for biomarker discovery. Estimating, a priori, how many samples should be analyzed to arrive at a particular level of certainty is often difficult if not impossible. For univariate cases, power analysis is providing a well-established framework to calculate the number of samples needed to be able to see a difference of a specific size in a given population (see Chapter 6). However, it is not at all obvious how to apply such an approach to untargeted metabolomics studies, where the size of the expected effect is often unknown, as is the population covariance structure.

Again, a multi-stage approach, separating the hypothesis generation phase from the validation phase, is strongly recommended. The first stage often consists of a pilot experiment, without stringent demands for statistical significance, and considering as many variables as possible; the second step is usually more focused and will be based on the results of the first phase. The outcome of the first experiment could not only be a list of putative biomarkers, but also an estimate of the size of the effect. Untargeted approaches are the ideal tools for the former part, while targeted analysis can be successfully applied in the latter. Such a multi-stage approach has several additional advantages: practice makes perfect, it gives a good indication of potential practical problems and provides the opportunity to test the complete pipeline, from experimental design to data analysis.

In this chapter, we will look at some of the data variation sources in more detail, and discuss ways to deal with them by designing our experiments in such a way that the information yield will be optimal.

Sampling Units

In science, we try to learn from nature. We measure things, make a hypothesis, use this to predict some future behavior and then observe whether we were right or wrong. What exactly it is that we measure, our "sample," to a large extent determines our conclusions about the unobserved population. Obviously, our sample needs to be *representative* of the population. However, this is not always easy to achieve. If we are interested in the metabolic differences between two apple varieties, simply buying two big bags of apples, one for each variety, from the supermarket is probably not good enough. Chances are that the apples (from one variety, at least) are from the same orchard, or even from the same tree. If we see differences between the apples of the two varieties, this can also be caused by the different locations of the apple trees. More generally, it is impossible to conclude anything about other apples from the same variety. Even the exact spot of the apple inside the tree is important: some areas are more shaded, some receive more sunlight. Shading may have profound effects on the metabolic profile of the fruit. In statistics, this problem has long been recognized, and readers will certainly have seen sentences like "... we assume that the values are independently and identically distributed...". This is not just a way to force real life in the straightjacket of a mathematical formula, but an essential statement, precisely defining one possible state of the system under study. By describing it so precisely, it becomes possible to detect deviations from the expected behavior. If your samples are not representative, their average may be quite different from the (unknown) population average, and you have a systematic error or a bias. No matter how many apples are in the bag, you will never be able to estimate this bias, and even worse: you will not even notice that there is a bias.

It, therefore, pays to think about how to collect your samples. In order to avoid systematic effects, the general advice is to use random sampling in order not to be biased in the choice of samples. Incidentally, there is a big difference between random sampling and haphazard sampling, which has an element of randomness, but unfortunately not enough of it. An example of haphazard sampling is walking in an orchard, and "randomly" picking apples from trees. How often, do you think, will the hard-to-reach apple at the top of the tree be picked in such a process? How do you know that the apple pickers are not bored after the second full basket and simply start picking apples from one and the same tree? How do you avoid picking apples from the one tree that has a particularly shaded position in the orchard, or is infected by a disease or a pest? A much better system is to a priori define what trees to be considered, and for each tree, define what apple to be picked. Essentially we need to make sure that every apple has an equal chance of being picked. People are notoriously bad at evaluating or generating random sequences, so it is best to leave this to a computer. Armed with a list of tree numbers and apple numbers, the pickers then can go to work. In this example, it may be essential (depending on the exact scientific goal in question) that for every apple in the tree, its position in the field, and the position of the apple in the tree and other relevant metadata (see Chapter 10) are captured, since all such parameters may influence the measurements and may need to be included in the analysis.

In particular, apples from one tree cannot be seen as independent samples—they are often called pseudoreplicates. Again, what is a real biological replicate depends on the "question in hand." In some cases, the tree is the sampling unit, in other cases the orchard, but there may also be cases where the apple is the sampling unit. An example of the latter situation is where we analyze differences between apples at different positions within one tree, for example, comparing apples from the top part of the tree with apples from the bottom part of the (same) tree. As with all other elements of experimental science, the sampling plan needs to be thoroughly documented so that everything can be repeated in exactly the same manner.

Experimental Design

Experimental Design is the field in statistics aiming to maximize the efficiency of experiments: as much information as possible is obtained with the least effort. It originated in agronomical sciences, early in the twentieth century, where fields, plots and subplots were used, e.g., to quantify the effects of specific treatments on different crop varieties. Many textbooks still present instructive examples from that field (e.g., Welham et al. 2014). The key point is that some variation cannot be avoided, but experiments can be set up in such a way that it does not interfere with the results. A well-known pedagogical example is a hypothetical experiment investigating "the wear and tear" in two types of shoe soles. One approach could be to distribute pairs of shoes with sole type A to one group of people, and pairs of shoes with sole type B to another, and after a month or so do an assessment on the wear in the two groups. This, however, would make it difficult to draw conclusions: obviously, different people will walk different amounts, and the spread within each group would be quite big. A better experiment would give each person one pair of shoes with different soles on both shoes so that the comparison can be made "within" each volunteer. This is an example of blocking: creating homogeneous experimental units within which the variation consists of the part that you are really interested in.

Blocking is (nearly) always advantageous: your analysis will focus on the relevant part of the variation, and is not distracted by other non-related "noise." Finding the right set up is not always easy. A typical example is the use of blocking in a design to account for batch effects. If we are comparing the response of human subjects to several treatments (e.g., diets), and we are forced to divide our measurements into several batches, then a good strategy is to put all measurements of one subject in the same batch. Comparing values within one subject is then possible without taking care of batch information. An example is shown in the left plot of Figure 2.3. If, on the other hand, we are in a situation where several populations are compared simultaneously, then the ideal experiment would be the one where each batch contains

I	II	III
x23	x41	x63
x14	x44	x51
x12	x32	x64
x24	x34	x54
x22	x33	x52
x21	x43	x62
x11	x42	x61
x13	x31	x53

I	II	III
T2	C	T1
T3	T3	C
C	T2	T1
T1	T1	C
T1	T2	T3
T2	C	T2
T3	T3	T2
C	T1	T3

FIGURE 2.3
Avoiding batch effects by blocking. In each case, a total of 24 measurements needs to be done, but only eight measurements can be done within one batch. The left plot shows a situation where four time points of two subjects (×12 signifies time point 2 of subject 1) are analysed per batch. Within each batch (I, II or III), the sample order is randomized. The relevant comparisons are between different time points per subject, which can be performed within a batch. The right plot shows a simpler experiment, where three treatments are compared to one control (indicated with T1-3 and C, respectively). Six biological replicates are available for each treatment—these are equally distributed over the batches so that the estimates within the batches are not affected by batch effects.

an equal number of biological replicates of all samples: the batches would be complete replicates of each other. This situation is depicted in the right part of Figure 2.3. Again, the relevant comparisons can be made within batches, eliminating any batch effects from the equation.

In cases where blocking cannot be done or is not useful, one should always randomize (note that also within blocks randomization is of crucial importance!). Just like in sample selection, one should use a computer to take care of the allocation of samples to treatments or to batches, to take care of injection sequence, etcetera. This is the only way to avoid subconscious influences of the researcher on the experiment, something that can have disastrous effects on the outcome. Apart from randomization in the allocation of subjects to treatment groups or batches (Figure 2.3), it should also be applied in defining the order in which samples are measured (see below). An overview of different randomization methods is given in Box 2.1 (after Suresh 2011).

Apart from these rather general guidelines, there are many different types of experimental designs that may or may not be appropriate for your situation. In Box 2.2 a couple of important ones are briefly described. More information can be found in the many textbooks that are available (e.g., George et al. 2005).

For metabolomics, the application of experimental design seems most straightforward where one knows a lot about the system under study, e.g., the level of variation that is expected. When concentrating on one metabolite, for instance, it is possible to estimate, at least roughly, how many samples need to be measured to achieve a certain level of statistical power (see Chapter 6). Conversely, if resources are available to analyze only a specific number of samples, one can calculate the power of the analysis or calculate the minimum effect that can be detected with a standard power level (e.g., 0.8). In a treatment-control situation, for example, one can calculate what would be the minimal size of a difference that one would expect to be able to find. For untargeted

BOX 2.1 RANDOMIZATION

- *Simple randomization:* Randomization based on a sequence of random assignments.
- *Block randomization:* A randomization method assigning subjects to groups so that one or more other characteristics are balanced across groups.
- *Stratified randomization:* A randomization method to address the need to control and balance the influence of covariates among groups (subjects' baseline characteristics).
- *Covariate adaptive randomization:* This method is particularly useful in the case of an imbalance of sample size among several covariates.

BOX 2.2 DESIGN OF EXPERIMENTS

- *Full factorial design:* In the case of an experiment testing the effects of two or more factors (e.g., treatment, diet, etc.), all possible combinations of these factors will be present. In this way, it will be possible to study the effects of all the factors and of their interactions on the response variables. As an example, considering three independent two-level factors (treated/untreated, male/female, low-BMI/high-BMI) will lead to $2^3 = 8$ test conditions. In the presence of a large number of factors or factor levels, full factorial designs can become prohibitively expensive (in terms of time and costs). In these cases is possible to rely on partial/fractional designs where only a subset of the design space is analyzed.
- *(Incomplete) block design:* A block is a group of samples measured in more or less homogeneous conditions, in this context typically one batch. Suppose we have four biological replicates for each sample, and we need four batches to measure them all, then the best option is to assign the samples in such a way that each batch corresponds to the analysis of one complete set of samples. Often, however, this cannot be arranged and then the batches do not all contain the same samples—in such a case an incomplete block design can be used.
- *Split-plot design:* This type of design is needed if it is more difficult to change the levels of a treatment during the experiment. In this way, the design has a nested structure. The factors difficult to change are randomly used on each block, while the others are fully represented within each block. As an example consider the effect of two types of fertilizers on the growth of four different types of seeds. A practical way to handle this question could be to treat two different fields with the two fertilizers and then to identify within each field four portions for the different seed types (randomly allocated!).
- *Crossover design:* In a crossover design, each sampling unit is exposed to all the different factors of the study over time; this makes these studies longitudinal. As an example consider the effects on patients of different treatments (A, B, C). With a crossover design, each patient will be exposed to all the possible treatments at different times, i.e., each patient crosses over from one treatment to the other during the course of a trial.

experiments, this is much harder. Not only is the number of metabolites that are measured quite large, also usually there is not much known about the samples. This basically makes it impossible to estimate required sample sizes or power. In these cases, the outcomes of pilot experiments can be used to estimate the order of magnitude of the expected effects. Usually, however, one cannot do much more than adhere to the principles of randomization and blocking, as mentioned in the previous paragraph. In addition, the outcome of such an experiment is typically interpreted in a multivariate analysis, where it is much harder to define very specific models and where instead, one usually focuses on global phenomena (again, see Chapter 6).

Recapitulating: the goal of untargeted experiments often is hypothesis generation. Targeted experiments, on the other hand, are often defined on the basis of prior knowledge (maybe from untargeted experiments!), and usually, are set up to confirm ideas. As a result, they should be more precise and have a lower variation. Note that taking the same samples analyzed with an untargeted method and putting them in a targeted analysis cannot be seen as an independent confirmation: one really needs independent samples for this.

Collection and Handling

The definition of an optimal strategy for sample collection and handling is an issue in metabolomics studies. This aspect becomes more and more important in studies, in particular when untargeted experiments are planned (Dunn et al. 2012). The key aspect in this phase is to avoid as much as possible the introduction of confounding effects originating from the non-uniform "mode of operation" of different researchers/technicians or from other preanalytical factors (time delay from sample collection and storage, the presence of contaminants, etc.). The impact of these differences will increase in big and complex experiments (think of multi-year, multi-country investigations).

As usual, it is impossible to completely eliminate the human factor, and the best that one can do is to try to control its impact. The first obvious strategy to do that is to rely as much as possible on automatic solutions for sample collection and handling and to standardize and document all the sample processing steps (collection, storage, quenching, etc.) defining a standard operating procedure (SOP) early in the study. The SOP can be optimized for designing a small scale pilot experiment to identify the more critical steps in the sample handling pipeline. On the base of this procedure, a specific training to the users involved in the study can be provided. It is important to point out that the first thing to do is to consider the possible presence of confounding effects in the phase of sample collection and handling during the design phase of the study.

As can be easily expected, the critical issues are very much dependent on the subject of the study and the reader can refer to the specific literature in the field (Yin et al. 2013, for example, discusses these aspects in blood metabolomics). As a general rule, it is always important to remember that the more complex a protocol is, the more it will be difficult to guarantee its reproducibility.

Extraction and Pretreatment

With the terms "sample extraction and pretreatment" we identify all the analytical steps which have to be applied to the samples before the actual injection in the analytical platform. Almost all the considerations valid for sample collection and handling hold also for these steps. In particular, for each project, it is necessary to strike an optimal balance between the complexity of the protocol, its coverage and its reproducibility, in particular for large multi-operator studies (De Vos et al. 2007).

In this phase, the use of internal standards could be of help to monitor and quantify sample extraction and pretreatment. On this respect, however, it is important to point out that, in the case of untargeted studies, it is not straightforward to draw general conclusions about the goodness of a specific method only considering a set of internal standards, because they are not able to capture the "chemical" complexity of the real samples (De Vos et al. 2007). Moreover, in these types of studies, it is important to realize that the optimal sample preparation method is a trade-off between the efficacy of the method (in terms of extraction efficiency) and the overall coverage of all the possible analytes and class of analytes. In this respect, a multipurpose method will most likely ensure good coverage of the "metabolome," but it will be less efficient than a specific method targeting a specific (class of) metabolites. A good practice is to identify the "optimal" conditions relying on the outcomes of a small scale pilot experiment designed for that purpose (Theodoridis et al. 2011).

Moving from these general considerations to more a more detailed treatment is beyond the scope of this book, and we invite the reader to refer to specific literature.

Doing the Measurements

The data acquisition step for MS and NMR is controlled by the instrument software. Commonly the samples are automatically analyzed with an order defined by a "sample list," which can be generated by the instrument software or imported from an external text file. This second possibility allows the use of automatic solutions for the setting up of long and complex experiments. No matter what technological platform is used, it is important to realize that the performance of the analytical pipeline will change with time and this can have a strong effect on the outcomes of the experiments. These changes become more evident for very long experiments, and the design of the analytical sequence has to ensure that analytical trends do not interfere with the objective of the study. In the case of a two-class problem, for example, it would be unwise to analyze all the samples belonging to class A at the beginning of the run and then move to class B, because a decrease in the efficiency of the analytical platform would result in features with higher average intensity in class A, which could be erroneously selected as putative biomarkers. A careful randomization of the samples is the first measure to be taken because it allows (as in the case of the experimental design) to disentangle the effects of the analytical variability from the ones resulting from the factors of the study. Also, in this case, the randomization should be performed by a computer to minimize the error-prone human intervention. For very long experiments, or for studies extending over many years, it is not possible to include all the samples in one single analytical run and it is necessary to define analytical batches. Not surprisingly, the analytical conditions will be different for each batch giving rise to so-called batch effects, which have to be carefully handled in the data analysis phase by means of specific algorithms for batch effect removal.

It is of fundamental importance to be able to monitor the variability of the analytical platform. To do this, it is necessary to design a suitable quality control strategy (QC). The QC strategy can include suitable internal standards to evaluate recovery and stability, but will, almost invariably, require the analysis of QC samples which are injected several times during the experiment, as illustrated in Figure 2.4. Monitoring the QC samples will allow to keep the stability of the system under control and check and correct for analytical drifts. What should be the nature of QC samples has been—and still is—the subject of much debate and research efforts (Godzien et al. 2014). However, there is a general agreement on the fact that they should give a reasonably good representation of the "chemical" space spanned by the samples. For this reason, "simple" mixtures of a pure chemical standard are not ideal, and pooled samples are often preferred. It is important to point out that the signal of QC samples can be monitored not only at the end of the analysis, but also during the course of the experiment in order to detect and correct problems in an early stage (Franceschi et al. 2014). This is of paramount importance to optimize the analytical resources and to avoid wasting high-value samples.

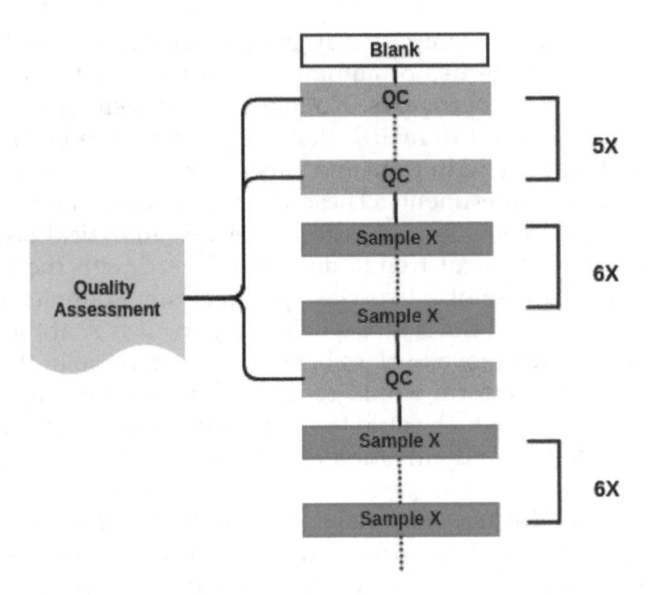

FIGURE 2.4
Typical injection sequence for untargeted metabolomics studies.

Analyzing the Data

Preparations

Considering all aspects we have been touching upon so far, one could be tempted to conclude that when the experiment reaches the "data analysis" phase all the complicated issues have been resolved. In reality, also getting out reliable and scientifically valid conclusions from metabolomics data requires care and specific skills. Again, a well-planned experimental design is the first element of a successful data analysis: an adequate number of samples and a reliable analytical pipeline facilitate data analysis and, eventually, data interpretation. It is always important to remember that also the most advanced data analysis strategy will be unable to turn lead into gold: "garbage in, garbage out."

In Chapter 6 some of the most commonly used statistical techniques will be discussed that allow the scientist to zoom in on the relevant part of the data, to distinguish between true and chance differences and to estimate quantitative effects. In the context of untargeted metabolomics, however, these issues are quite difficult: the number of variables usually exceeds the number of samples, often by one or more orders of magnitude; metabolite levels can be highly correlated; a large part (often the largest part) of the variables are unknowns, i.e., metabolites whose identity cannot be established yet. In order to be able to do any meaningful data analysis, it is imperative to clean up the data before submitting them to statistical procedures and to counter any characteristics that may hamper interpretation. The three most common cases are missing values, non-informative variables and batch effects.

Missing values arise when for some reason a particular variable cannot be measured for all samples. For NMR data this is usually not the case since the whole ppm axis is binned (perhaps with the exception of some values around the water peak), but in MS data this frequently occurs. Sometimes peaks are missed by the peak picking algorithm because they are in an unexpected position, have an unexpected shape or simply are too low in intensity to be detected. If mass peaks have been aggregated to allow the analysis to be done on the level of metabolite intensities, the mass spectrum at a particular time point in one sample may be too different from what is seen in other samples to be associated to the same metabolite. It is not uncommon to see 5%–30% of missing values in MS data sets, and obviously, it is impossible to manually check every single case. The simplest approach is to assume that peaks are missed because of low intensities. In the subsequent statistical analysis, it is sometimes possible to use methods for so-called censored data, i.e., data for which one knows the unobserved value is (in this case) below a certain limit. In other cases, one would rather replace the missing values with small random or non-random numbers, a process called imputation. Note that the imputed value may have a large effect on the result! Alternatively, an approach also implemented in XCMS (one of the most popular open source software solutions to analyze MS based untargeted metabolomics data sets), one could simply integrate the signal at that particular position to get an idea of the signal intensity, even though no real peak shape has been identified. In all other cases than low intensities, there is a real risk of making grave errors, since the missed peak may be quite large or important. When in doubt, there is no other way than going back to the raw data and doing a visual check.

Whereas missing values are predominantly associated with detection methods where each feature is a distinct and discrete phenomenon, like a mass peak in mass spectrometry, non-informative variables occur with all detection methods. In NMR, this may be a part of the spectrum where no peaks are present, or parts close to the water signal where every bit of information is drowned in the large background. In MS, we can label as non-informative variables peaks that have been found in only one or a few samples, peaks that can be unambiguously associated to column bleeding 1 or known contaminants (e.g., from the solvent) or peaks that basically do not change across all samples. If there are easy and automatic ways of detecting these cases, they should be used, and the corresponding variables should be removed. This makes the subsequent statistical analysis a lot easier, not only because the sample-to-variable ratio improves, but mostly because the statistical model is forced to concentrate on those variables that can be interpreted in a meaningful way. Sometimes it is easy to recognize uninformative variables (counting how often a mass peak has been found can be done automatically), and sometimes it is less easy (suppose one would consider the spread found in the intensity of a particular ppm value—what would be an appropriate cutoff? Sometimes small intensity differences are very important), but in any case, one should think about this and include it in the data analysis pipeline. Chapter 7 contains an example where only a small part of the variables is used for analysis. The selection is made on the basis of information content (i.e., variables with little variation are excluded from the analysis).

The final example of data cleaning, sometimes also called data preparation or data pre-processing (although the latter term is also used for choosing appropriate forms of scaling), is the batch correction. When the number of samples is too large to be prepared or measured in one batch there is a real risk that measurements in different batches will lead to different results. Even within batches one sometimes sees a gradual change. Batch effects can be due to the measurement system (mass spectrometers are more likely to lead to batch effects than NMR) but also due to the analytical handling (extraction, etc.). Whatever the reason, if not removed they will contribute to the noise in the data, and the worst thing is that this is not random noise, but noise with a particular structure. At the same time, this structure can be used to remove batch effects from the data: after all, we know the batch of each sample, we know the injection order, etcetera. Classical methods simply make sure—on a feature-by-feature basis—that the metabolite level in each batch has the same average, and that no trends are present (Hendriks et al. 2011). The latter part is achieved by simply estimating any trends, often using QC samples, and then subtracting them. Both linear models and nonlinear models can be used. QC samples are not explicitly needed, and one can also base a batch correction on the study samples (Rusilowicz et al. 2016; Wehrens et al. 2016)—this has the advantage that a few missing values in the QCs do not prevent batch correction to be performed. Since one can never be sure that batch correction is completely successful, it remains of paramount importance to randomize sample order: this is basically the only means we have to decrease the influence of any batch effects that have not been accounted for.

Reproducibility

An often disregarded aspect of data analysis is "reproducibility." It sounds obvious that it should be possible to reproduce any scientifically valid experiment (Mobley et al. 2013; Boulton 2016). The problem is that in metabolomics and in many other "big data" sciences, it is not always easy to reproduce the data processing steps which have been applied to achieve the final result.

Reproducibility of scientific results is becoming such a big issue that scientific journals are more and more requiring researchers to make the experimental data available as a part of the paper submission process. It is important to point out that this is done not only to allow the referees to check the data-analysis process but also to promote the incremental progress of science where the new evidence can be integrated with the established body of knowledge (see Chapter 10). The practical implementation of this concept is not at all easy. The first problem is that the details of the algorithms used for data processing are often not easily accessible by the user. As an example, consider the peak picking phase in the pre-processing of LC-MS untargeted data. In the majority of the cases, this task is performed by using the software provided with the instrument, which is applying a specific algorithm with tailored parameters. The users will never know (and often do not bother) about the details of that algorithm, which in many cases will be also undisclosed by the instrument producer. The peak list will be then the output of a specific software and to exactly reproduce it, it would be mandatory to use the very same software. In the logic of data sharing, this software should also be made publicly available and this is practically impossible.

The situation does not get better if we consider the subsequent phase of statistical analysis. With the commendable objective of making life easier for the users lacking a specific informatics background, software producers deliver complete point-and-click data analysis solutions. With these tools, it is possible to seamlessly use advanced statistical algorithms, but it is also almost impossible to keep track of the data analysis workflow, reproduce it and share it with the scientific community.

To circumvent all these problems, a possible solution is to convert the data into open source formats (see Chapters 4 and 5) and rely on data analysis pipelines, possibly developed with open source scripting languages like R or Python. This strategy is also ideal for multi-vendor metabolomics platforms because it facilitates cross-platform data sharing and integration.

A pipeline can include almost all the data analysis steps from pre-processing to statistical analysis, and the results of each "node" are "piped" inside the subsequent step of the analysis—an example is shown in Figure 2.5. This type of approach offers several advantages:

1. All the different nodes can be developed independently (also with different programming languages);
2. The pipeline script is a detailed description of the data analysis process, which can be used to document and reproduce the results;
3. The pipeline can be wrapped behind a high-level interface to facilitate routine analysis.

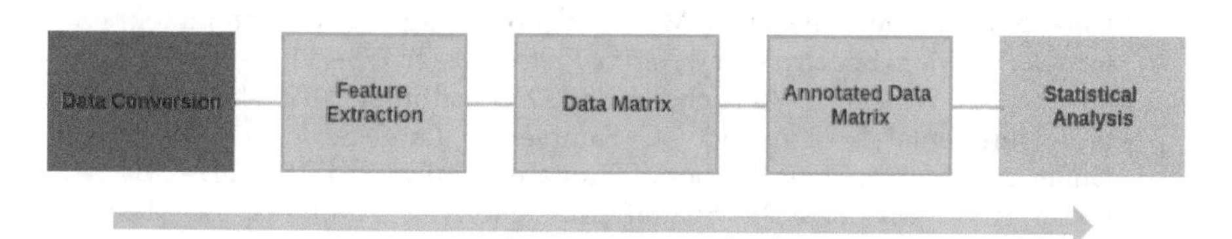

FIGURE 2.5
Block diagram of a data analysis pipeline for untargeted metabolomics studies.

A pipeline represents a convenient infrastructure for the whole data analysis phase, but how can one be sure of selecting the most appropriate data analysis strategy? As may be expected, there is no ultimate solution, and the optimal one should be selected depending on the specific data set at hand. On this topic, however, it is possible to make some general observations. There is always the temptation to go for the newest and most powerful algorithm published in the literature, with the idea that "more complex" is always a synonym of "more powerful." This can be the case, but it is always necessary to keep in mind that power has a price, which often is paid in terms of more stringent statistical assumptions, of more hypothesis on the structure of the data or, simply, having more parameters to optimize and tune. Well established, broadly taught and well-exercised methods usually still serve the purpose they were developed for, even though they may be old. In other words, to get the maximum from a powerful method it is often necessary to have more data of better quality, so we get back to the problem of experimental design, sample collection and data acquisition.

As a final remark, one should realize that only in very rare cases data analysis can be described by the linear process depicted above. In practice, nearly every single step features a feedback loop to earlier stages of the analysis. It is not uncommon, for example, to go back to the feature extraction after the annotation of the data matrix (terminology of Figure 2.5) to look for missed features. The same often happens after the statistical analysis: if on the basis of the general pattern found in the data one particular sample, or a feature in that sample, looks quite different from what would be expected, it is wise to go back to the raw data and check whether at some point in the chain errors have been made. If an error indeed is found, this—paradoxically—supports the sensibility of the analysis. Note that one should be very careful not to fall into the trap of only seeing what one is looking for: keep being critical!

References

Boulton, G. 2016. Reproducibility: International Accord on Open Data. *Nature* 530 (7590): 281.

De Vos, R. C. H., S. Moco, A. Lommen, J. J. B. Keurentjes, R. J. Bino, and R. D. Hall. 2007. Untargeted Large-Scale Plant Metabolomics Using Liquid Chromatography Coupled to Mass Spectrometry. *Nature Protocols* 2 (4): 778–791.

Dunn, W. B., I. D. Wilson, A. W. Nicholls, and D. Broadhurst. 2012. The Importance of Experimental Design and QC Samples in Large-Scale and MS-Driven Untargeted Metabolomic Studies of Humans. *Bioanalysis* 4 (18): 2249–2264.

Franceschi, P., R. Mylonas, N. Shahaf, M. Scholz, P. Arapitsas, D. Masuero, G. Weingart. 2014. MetaDB a Data Processing Workflow in Untargeted MS-Based Metabolomics Experiments. *Frontiers in Bioengineering and Biotechnology* 2: 72.

George, E. P. B., J. S. Hunter, and W. G. Hunter. 2005. *Statistics for Experimenters: Design, Innovation, and Discovery* (2nd ed.). Hoboken, NJ: John Wiley & Sons.

Godzien, J., V. Alonso-Herranz, C. Barbas, and E. G. Armitage, 2014. Controlling the Quality of Metabolomics Data: New Strategies to Get the Best Out of the QC Sample. *Metabolomics: Official Journal of the Metabolomic Society* 11 (3): 518–528.

Hendriks, M. M. W. B., F. A. van Eeuwijk, R. H. Jellema, J. A. Westerhuis, T. H. Reijmers, H. C. J. Hoefsloot, and A. K. Smilde. 2011. Data-Processing Strategies for Metabolomics Studies. *Trends in Analytical Chemistry* 30 (10): 1685–1698.

Mobley, A., S. K. Linder, R. Braeuer, L. M. Ellis, and L. Zwelling. 2013. A Survey on Data Reproducibility in Cancer Research Provides Insights into Our Limited Ability to Translate Findings From the Laboratory to the Clinic. *PloS One* 8 (5): e63221.

Rusilowicz, M., M. Dickinson, A. Charlton, S. O'Keefe, and J. Wilson. 2016. Erratum to: A Batch Correction Method for Liquid Chromatography–Mass Spectrometry Data That Does Not Depend on Quality Control Samples. *Metabolomics: Official Journal of the Metabolomic Society* 12 (11): 170.

Suresh, K. 2011. An overview of randomization techniques: An unbiased assessment of outcome in clinical research. *Journal of Human Reproductive Sciences* 4 (1): 8–11.

Theodoridis, G. et al. 2011. LC-MS Based Global Metabolite Profiling of Grapes: Solvent Extraction Protocol Optimisation. *Metabolomics: Official Journal of the Metabolomic Society* 8 (2): 175–185.

Wehrens, R. et al. 2016. Improved Batch Correction in Untargeted MS-Based Metabolomics. *Metabolomics: Official Journal of the Metabolomic Society* 12: 88.

Welham, S. J., S. A. Gezan, S. J. Clark, and A. Mead. 2014. *Statistical Methods in Biology: Design and Analysis of Experiments and Regression*. Boca Raton, FL: Taylor & Francis Group.

Yin, P., A. Peter, H. Franken, X. Zhao, S. S. Neukamm, L. Rosenbaum, and M. Lucio. 2013. Preanalytical Aspects and Sample Quality Assessment in Metabolomics Studies of Human Blood. *Clinical Chemistry* 59 (5): 833–845.

3

Measurement Technologies

Oscar Yanes, Katherine Hollywood, Roland Mumm, Maria Vinaixa, Naomi Rankin, Ron Wehrens and Reza Salek

CONTENTS

Introduction 36
Mass Spectrometry 36
Separation Techniques 36
GC-MS 38
LC-MS 39
CE-MS 39
Ionization 39
Detection 41
Data System 42
Tandem Mass Spectrometry (MS/MS) 44
Advantages and Disadvantages 47
Nuclear Magnetic Resonance Spectroscopy 47
Basic Concepts 47
Overview 47
Spin 52
The Proton NMR Spectrum 53
Signal Intensities 54
Relaxation Times 55
Hardware 55
The Magnet 55
The Probe 57
The Console and Accessories 57
Sample Preparation 58
Interpretation of ^{1}H NMR Signals 58
Parameters Affecting Spectral Data 59
LC—NMR Spectroscopy 59
Other Nuclei 60
Multidimensional NMR Spectroscopy 60
Advantages and Disadvantages 62
Vibrational Spectroscopy 62
Raman Spectroscopy 63
IR Spectroscopy 64

Advantages and Disadvantages....65
Applications of Vibrational Spectroscopy....66
Diode-Array Detection....66
Advantages and Disadvantages....68
References....69

Introduction

The main objective of a metabolomics experiment is to determine the composition of the set of small molecules in one or more biological samples, both with respect to quantity and chemical characteristics. Two detection platforms are most commonly used in metabolomics: mass spectrometry (MS) and nuclear magnetic resonance (NMR) spectroscopy. Since biological samples often consist of very complex matrices, containing several hundred to thousands of metabolites, these techniques are often coupled to different chromatographic separation techniques. In this chapter, we will treat the basic principles of the most commonly used chromatographic techniques and the very foundations of MS and NMR as major detection techniques in metabolomics. Numerous chromatographic techniques exist, but here we focus on three that are commonly used in metabolomics: gas chromatography, liquid chromatography and capillary electrophoresis. The focus is on concepts that are of particular importance for subsequent analysis and interpretation of the data. Less common detection methods based on vibrational and UV-Vis spectroscopy are treated briefly at the end of the chapter.

Mass Spectrometry

Here, we describe the foundations of MS, focusing on the production, separation and detection of ions. A mass spectrometer used in metabolomics typically contains the following elements: inlet, source, analyzer and ion detection module. All the modules are linked to a computer for processing the data, which eventually produces and visualizes the mass spectra. The relationship between these modules or elements is depicted in the diagram below (Figure 3.1) (de Hoffmann and Stroobant 2007).

Separation Techniques

Several types of chromatographic techniques can be used to introduce the compounds that are analyzed, such as liquid chromatography (leading to LC-MS), gas chromatography (GC-MS) and capillary electrophoresis (CE-MS). The small molecules of the sample mixture present in the mobile phase are

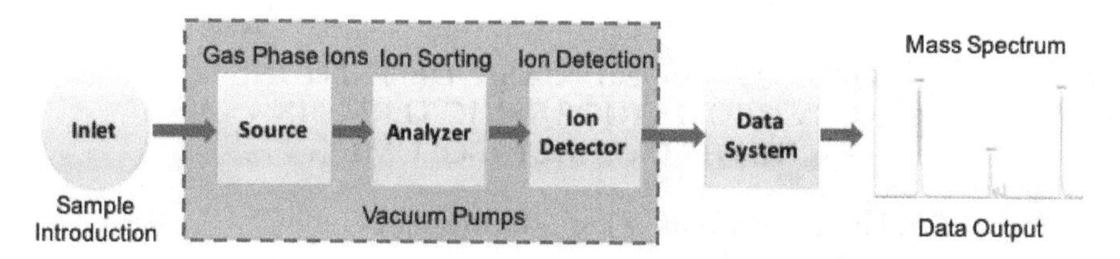

FIGURE 3.1
Diagram of a mass spectrometer. Once a sample is injected, metabolites are ionized in as gas phase (electron ionization, chemical ionization), liquid-phase (atmospheric pressure chemical ionization, electrospray ionization) or solid-phase (laser desorption ionization) in the "Source." The "Analyzer" subsequently separates the ions according to the type of Analyzer (Quadrupoles, Time-of-flight, Ion trap, Orbitrap, etc.). The ion detector translates the electrochemical signal into digital form and finally captures it in raw data files that can be read and visualized by appropriate software. Prior to injection, it is common to have a separation step. Depending on the type of sample, this separation can be based on, e.g., gas or liquid chromatography, capillary electrophoresis or solid-phase separation.

separated from each other due to their different degrees of interaction with the stationary phase. The chromatographic column introduces the sample to the MS for detection (Figure 3.1) either in the gaseous state, for example, GC-MS or in solution form, LC-MS and CE-MS. Typically, metabolites with the least affinity for the stationary phase will "travel" faster than those that have stronger interactions with the stationary phase (see Box 3.1).

BOX 3.1 BASIC DEFINITIONS OF CHROMATOGRAPHIC TERMS IN THE FIELD OF METABOLOMICS

- *Stationary phase:* The substance fixed in place for the chromatography procedure. Typically an immobilized phase on the support particles or on the inner wall of the column tubing.
- *Mobile phase:* In chromatography, a mobile phase transports the complex mixture of small molecules through a stationary phase. Upon contact, the sample interacts with the stationary phase resulting in separation of components dependent on the differential affinity with the stationary phase. The mobile phase may be a liquid (LC and CE) or a gas (GC). In the case of normal phase HPLC, the mobile phase consists of a non-polar solvent(s) such as hexane while in reverse phase chromatography polar solvents are used.
- *Retention time:* The characteristic time it takes for a particular analyte to pass through the system (from the column inlet to the detector) under set conditions.

(Continued)

BOX 3.1 (Continued) BASIC DEFINITIONS OF CHROMATOGRAPHIC TERMS IN THE FIELD OF METABOLOMICS

- *HPLC:* High-Performance Liquid Chromatography.
- *UHPLC or UPLC:* Ultra High-Performance Liquid Chromatography. The higher pressure capability of UHPLC allows the use of higher flow rates to maximize peak capacity.
- *Chromatographic resolution:* The resolution of an elution is a quantitative measure of how well two elution peaks can be differentiated in a chromatographic separation. It is defined as the difference in retention times between the two peaks, divided by the combined widths of the elution peaks.
- *Carrier gas:* Typically, inert or non-reactive gases, acting as a transporter of the volatile compounds in GC column.
- *Vapor pressure:* The pressure of the vapor resulting from evaporation of a liquid (or solid) above a sample of the liquid (or solid) in a closed container.
- *Temperature program:* The temperature that is set in a GC oven. Samples can be analyzed under isothermal conditions but typically a temperature gradient is applied covering the boiling points of the analytes.

GC-MS

Gas chromatography is a technique that can be used for the separation and analysis of gaseous, liquid solutions and volatile solid materials that are thermally stable (David et al. 2011; Poole 2012). Volatile samples are analyzed directly, while for non-volatile compounds derivatization or pyrolysis techniques are applied. Metabolites to be analyzed are vaporized in the injection system and transported by a gas stream to the analytical column. Typically, inert or non-reactive gases such as helium, nitrogen and sometimes also hydrogen are used as transporting (carrier) gases. The mobile phase containing the vaporized metabolites passes the stationary phase in the capillary column, typically consisting of a polymer adsorbent. The stationary phase is bound to the inner surface of a fused-silica tubing having a length of 10–100 m and an inner diameter of 0.18–0.53 mm. Compared to LC columns, GC columns are longer and have a smaller diameter, hence providing a higher separation capacity. This is due to the compressible nature of gases that can more easily be transported through a capillary. The distribution of the metabolites between the mobile and stationary phase is mainly determined by the metabolites' vapor pressure and partly by their polarity. Metabolites having a high vapor pressure will be transported faster through the column than compounds with a low vapor pressure. During the transport metabolites will adsorb/desorb repeatedly to/from the stationary phase resulting in different travel (retention) times through the column. Chromatographic resolution can be obtained by changing the length or selectivity of the stationary phase (column). Alternatively, comprehensive 2-dimensional GC (GC × GC) can be used, where two analytical columns with different selectivity are serially coupled. Ideally, the separate metabolites elute from the column individually and reach the detector at different times.

LC-MS

Several stationary phases are used for the separation of small molecules in LC-MS such as reversed phase (RP), ion (cation and anion) exchange and hydrophilic interaction liquid chromatography (HILIC) (Yanes et al. 2011). These phases separate small molecules based on hydrophobicity, positive and negative charges, and hydrophilicity. RP is most often used in metabolomics studies as RP has the highest chromatographic resolving power. The operational flow scale and pressures of the LC separation determine to a large extent the quality and characteristics of the LC-MS performance, and consequently, the properties of the data generated. Current LC-MS configurations can perform from high flow rates at 0.5–1 mL/min to nano flows at 20–200 nL/min, operating from 50 to 350 bar in HPLC systems to >1000 bar in UHPLC configurations. UHPLC provides better chromatographic resolution due to the smaller column particles size. In addition, UHPLC systems produce faster chromatographies and minimal retention time drifts as compared to HPLC.

CE-MS

Capillary electrophoresis is a well-suited technique for the separation of polar ionic and charged compounds (Klepárník 2015; Ramautar et al. 2015). These compounds are separated on the basis of their charge-to-size ratio in aqueous media. The main advantage of CE is that it consumes minimal amounts of sample and solvent (nL range), and the separation is fast. However, CE is less robust and reproducible as compared to GC and LC. The main problem is the large migration shift between runs that can hamper proper peak alignment, and consequently, quantification of metabolites. The most commonly used ionization technique for interfacing CE to MS is electrospray (ESI).

Ionization

The basic principle of mass spectrometry is to determine the exact mass of a number of molecules by converting them to a beam of charged gas phase ions and measure their mass to charge ratio (m/z). Ions can be produced by either adding or removing a charge from the neutral molecules. It is important to note that the ionization method refers to the mechanism of ionization while the ionization source is the mechanical device that allows ionization to occur (Rockwood 2004; Siuzdak 2004). The most important parameter is the internal energy transferred during the ionization process. Some ionization techniques are very energetic and cause extensive fragmentation. Other techniques are softer and mainly produce molecular species. This will have important implications for data processing (see Chapter 5). Three types of ion sources exist: (i) *gas-phase ion sources*, that include electron impact ionization (EI) and chemical ionization (CI), which are only suitable for gas-phase ionization and thus their use is limited to compounds that can be vaporized and which are thermally stable. Electron impact typically uses an energy of 70 eV. At this energy fragmentation patterns are comparably stable, making mass spectral databases and mass spectral matching using similarity algorithms routine in GC-MS-based metabolomics experiments (see Chapter 5 for additional details). However, a large number of compounds are thermally labile or do not have sufficient vapor pressure. Ions of these compounds must be extracted directly from the condensed to the gas phase. These ion sources exist as (ii) *liquid-phase ion sources*, such as electrospray (ESI), atmospheric pressure chemical ionization (APCI) and atmospheric pressure photoionization (APPI), and (iii) *solid-state ion sources*, including matrix-assisted laser desorption ionization (MALDI) or matrix-free modifications (LDI) (Northen et al. 2007; Peterson 2007), and secondary ion mass spectrometry (SIMS) (Benninghoven et al. 1987; Passarelli and Winograd 2011). The ion sources produce ions mainly by ionizing a neutral molecule through electron ejection, electron capture, protonation, deprotonation, adduct formation or the transfer of a charged species. This also has implications for MS signal processing (see Chapter 5) (Table 3.1).

TABLE 3.1

Commonly Used Ionization Sources in Metabolomics

Ionization Source	Main Event	Internal Energy	Coupled Chromatography
EI	Electron transfer	High: extensive fragmentation	GC-MS
CI	Proton transfer	Low/medium: little or some fragmentation[a]	GC-MS
ESI	Proton transfer, adduct formation	Low: little fragmentation	LC-MS & CE-MS
APCI	Proton transfer	Low: little fragmentation	GC-MS, LC-MS
MALDI	Proton transfer, adduct formation	Low: little fragmentation	Direct analysis or off-line separation

[a] Common reagent gases from high to less fragmentation: methane > ammonia > isobutane.

Detection

No matter which ionization technique or chromatographic method is used, three acquisition modes exist; scanning, selected-ion monitoring (SIM) and selected-reaction monitoring (SRM). Scanning mode is associated with mass profiling or untargeted metabolomics where the full spectrum range is monitored, whereas SIM and SRM modes are typical of targeted metabolomics. In SIM mode, only a limited mass-to-charge ratio range is transmitted/detected by the instrument. The SRM mode is used in tandem mass spectrometry (see Tandem Mass Spectrometry) to follow specific fragmentation reactions in which an ion of a particular mass, namely precursor ion, is selected in the first stage of a tandem mass spectrometer and an ion product of a fragmentation reaction of this precursor is selected in the second mass spectrometer stage for detection. These latter two modes of operation typically result in significantly increased sensitivity.

There are different mass analyzers that operate under these modes; SIM and SRM techniques are most effective on quadrupole mass analyzers and Fourier transform ion cyclotron resonance mass analyzers, whereas scanning mode is most effective on time of flight (TOF) and Orbitrap analyzers. Some mass analyzers separate ions in space while others separate ions by time. Ultimately, the performance of a mass analyzer can typically be defined by the following characteristics (Table 3.2): scan speed, transmission (for sensitivity), dynamic range, mass accuracy, mass resolution, mass range and tandem analysis capabilities (Hart-Smith and Blanksby 2011). All these characteristics will influence data processing tools (see Box 3.2).

BOX 3.2 BASIC DEFINITIONS IN MASS SPECTROMETRY

- *Mass resolution:* Resolution is the ability to distinguish between ions of different m/z ratios. The most common definition of resolution is given by the equation $R = M/\Delta M$, where M corresponds to m/z and ΔM represents the full width at half maximum (FWHM).
- *Mass accuracy:* Accuracy refers to the m/z measurement error—that is, the difference between the true m/z and the measured m/z of a given ion—divided by the true m/z of the ion. It is usually quoted in terms of parts per million (ppm).
- *Mass range:* The mass range is the range of m/z's over which a mass analyzer can operate to record a mass spectrum.
- *Dynamic range:* The linear dynamic range is the concentration range over which the ion signal is directly proportional to the

(*Continued*)

BOX 3.2 (Continued) BASIC DEFINITIONS IN MASS SPECTROMETRY

analyte concentration. This measure of performance is of importance to the interpretation of mass spectral abundance readings.

- *Abundance sensitivity:* Abundance sensitivity refers to the ratio of the maximum ion current recorded at an m/z of M to the signal level arising from the background at an adjacent m/z of (M + 1). This is closely related to dynamic range: the ratio of the maximum usable signal to the minimum usable signal (the detection limit).
- *Scan speed:* Refers to the rate at which the analyzer scans a mass spectrum over a particular mass range. It is expressed in mass units per second or in mass units per millisecond.
- *Tandem mass spectrometry:* Involves multiple steps of m/z selection, with some form of fragmentation occurring in between the stages of m/z separation.

TABLE 3.2

Mass Analyzers Most Commonly Used in Metabolomics

Mass Analyzer	Resolution	Accuracy	Mass Range	Dynamic Range	Sensitivity
Quadrupole	100–10^3	100 ppm	4000	10^7	10^4–10^6
Ion Trap	10^3–10^4	50–100 ppm	4000	10^2–10^3	10^2–10^3
TOF	10^3–10^4	5–50 ppm	>100,000	10^6	10^6
FT-ICR	10^4–10^6	<1–5 ppm	>10,000	10^3–10^4	10^3–10^4
Orbitrap	10^4–10^5	1–5 ppm	6000	10^3–10^4	10^4

Since the chromatograph is typically connected directly to the mass spectrometer, also known as online configuration, the mass spectra are acquired while the compounds of the mixture are eluting. The ion beam passes through the mass analyzer and then is detected and transformed into a usable signal by the detector.

Data System

The spectrometer provides two series of data as a function of time: the number of detected ions (i.e., counts) and, simultaneously, a physical value indicating the mass of these ions. The ions of every mass appear with a certain distribution over a time period. The area under the curve is proportional to the number of detected ions, whereas the value at the centroid of the peak is an indicator of the ion mass. MS data collected off an instrument is presented as either profile or centroid mode. Shown below are two mass spectra illustrating an ion cluster for profile data and a centroid mass spectrum created from the profile data (Figure 3.2). In chapter profile mode, a peak is represented by a collection of signals over several scans. The advantage of profile data is it is easier to classify a signal as a true peak from the noise of the instrument. In centroid mode, the signals are displayed as discrete m/z with zero line widths based on the average m/z value weighted by the intensity and assigned m/z values based on a calibration file. The advantage of centroid data is the file size is significantly smaller as there is less information describing a signal.

In order to achieve this, the acquisition processor accumulates the signal related to the number of ions during a certain period of time. For example, a spectrometer can cover a range of 500 mass units within 1s, i.e., one mass within 2 ms. During that time, the processor must carry out about eight measurements of the number of ions; 0.25 ms is allotted for every sample. In other words, it must measure 4000 scans per second. Its sampling frequency is said to be 4 kHz. To convert it into a bar graph, a suitable algorithm (see Chapter 4, *the section on peak detection*) allows the processor to determine the limits of the peak and its centroid. The sum of the values read

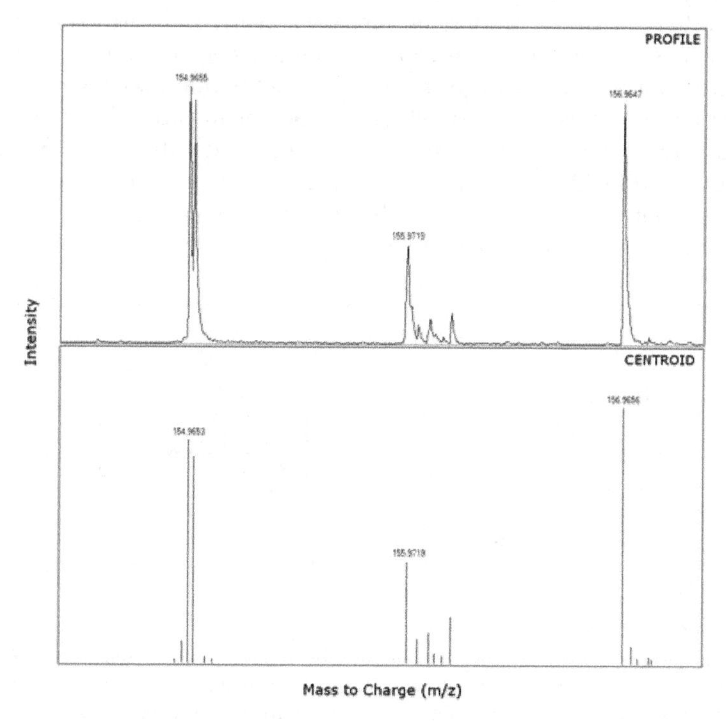

FIGURE 3.2
Comparison of profile versus centroid data. The two mass spectra illustrating an ion cluster for profile data and a centroid mass spectrum created from the same profile data.

within these limits during the condenser discharges is proportional to the number of ions, whereas the interpolation of the indicator value at the centroid yields the ion mass. If a larger mass range is scanned per unit time, or if a higher mass resolution is used, the sampling speed must be increased because the number of data points per unit time increases. Another important characteristic of the acquisition processor is its dynamic range, which is linked in part to the signal digitization possibilities. For example, if an ion detector can typically detect between one and one million ions reaching the detector simultaneously, its own dynamic range—the ratio of the largest to the smallest detectable signal—is thus equal to 10^6.

Tandem Mass Spectrometry (MS/MS)

Tandem mass spectrometry, abbreviated MS/MS, involves at least two stages of mass analysis in conjunction with a dissociation process or a chemical reaction that causes a change in the mass or charge of an ion (de Hoffmann and Stroobant 2007; Wiley Series on Mass Spectrometry 2013). In the most common MS/MS experiment a first analyzer is used to isolate a precursor ion, which undergoes fragmentation to yield product ions and neutral fragments that are directly or indirectly (neutral loss) detected with a second mass analyzer (Figure 3.3). It is possible to increase the number of steps to yield an MS/MS/MS or MS^n experiment, where *n* refers to the number of generations of ions being analyzed.

There are two main categories of instruments that allow MS/MS experiments: tandem mass spectrometers in space or in time. Common in-space instruments have two mass analyzers, typically quadrupoles such as QqQ or qTOF. The QqQ configuration indicates an instrument with three quadrupoles where the second one, indicated by a lower-case q, is the collision cell where fragmentation reaction takes place. Besides this spatial separation

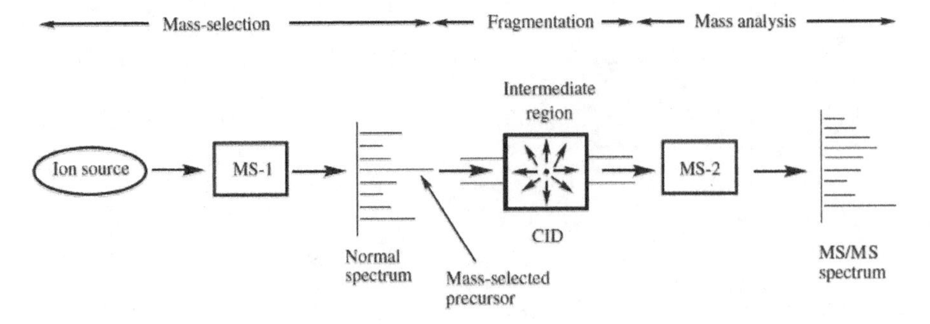

FIGURE 3.3
The principle of MS/MS: an ion M_1 is selected by the first spectrometer MS-1, fragmented through collision, and the fragments are analyzed by the second spectrometer MS-2. Thus ions with a selected m/z value, observed in a standard source spectrum, can be chosen and fragmented to obtain their product ion spectrum. (From Sindona, G. and Taver, D., Modern methodologies to assess the olive oil quality, in Innocenzo Muzzalupo (Ed.), *Olive Germplasm—The Olive Cultivation, Table Olive and Olive Oil Industry in Italy*, InTech, 2012.)

method using successive mass analyzers, MS/MS can be achieved also through time separation with ion traps, and ion cyclotron resonance (ICR) or Fourier transform mass spectrometry (FTMS), so that the different steps are carried out successively in the same mass analyzer (see Box 3.3).

BOX 3.3 THE NOMENCLATURE USED IN MS/MS

- *Precursor ion (also known as parent ion):* Any ion undergoing either a decomposition or a charge change.
- *Product ion (also known as daughter ion):* Ion resulting from the above reaction.
- *Fragment ion:* Ion resulting from the fragmentation of a precursor ion.
- *Neutral loss:* Fragment lost as a neutral species.
- *CID:* Collision-induced dissociation, occurs when an inert gas is present in the collision cell.
- *Product ion scan:* Consists of selecting a precursor ion (or parent ion) of chosen mass-to-charge (m/z) ratio and determining all of the product ions (daughter ions) resulting from collision-induced dissociation (CID) (Figure 3.4).
- *Precursor ion scan (parent scan):* Consists of choosing a product ion (or daughter ion) and determining the precursor ions (or parent ions). In this method, the precursor ions are identified. This scan mode cannot be performed with time-based mass spectrometers. All of the precursor ions that produce ions with the selected m/z through reactions of fragmentations thus are detected (Figure 3.4).
- *Neutral loss scan:* Consists of selecting a neutral fragment and detecting all the fragmentations leading to the loss of that neutral fragment. This scan mode cannot be performed with time-based mass spectrometers. For a mass difference of a, when an ion of mass m goes through the first mass analyzer, detection occurs if this ion has produced a fragment ion of mass m-a when it leaves the collision cell (Figure 3.4).
- *Selected reaction monitoring (SRM):* Consists of selecting a fragmentation reaction. For this scan, both the first and second mass analyzers are focused on selected masses. There is no scan, the ions selected by the first mass analyzer are only detected if they produce a given fragment by a selected reaction. The absence of scanning allows one to focus on the precursor and fragment ions over longer times, increasing sensitivity and selectivity (Figure 3.4).

(*Continued*)

BOX 3.3 (Continued) THE NOMENCLATURE USED IN MS/MS

Note that precursor or neutral loss scans are not possible in time separation analyzers. The product ion scan is the only one available directly with these mass spectrometers. However, with time-based instruments, the process can be repeated easily over several ion generations, allowing an MS^n product ion spectrum to be obtained.

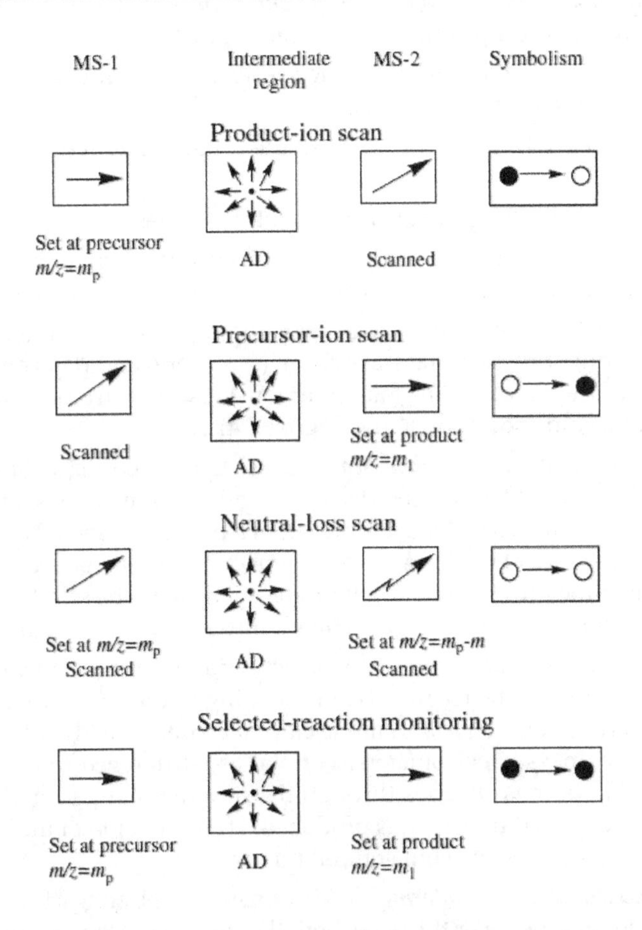

FIGURE 3.4
Main processes in tandem mass spectrometry (MS/MS). The AD refers to ion activation and dissociation. The filled and open circles stand for fixed and scanning mass analyzers, respectively. (From Sindona, G. and Taver, D., Modern methodologies to assess the olive oil quality, in Innocenzo Muzzalupo (Ed.), *Olive Germplasm—The Olive Cultivation, Table Olive and Olive Oil Industry in Italy*, InTech, 2012.)

Advantages and Disadvantages

Mass spectrometry is an almost universal detection method offering very high sensitivity and specificity. Especially when coupled to a chromatographic separation system, the method allows one to detect many compounds, also at very low concentrations, in one experiment, a major reason why this technique has become so important in metabolomics. The only compounds that are hard to detect are compounds that do not ionize well, are in very low abundance or are unstable hence degenerate. This versatility does come at a price: data processing is extremely complex. Results from different data processing pipelines, although internally consistent, rarely are directly comparable to each other. In addition, the detection system is quite sensitive to environmental influences and may require frequent calibration or batch correction.

Nuclear Magnetic Resonance Spectroscopy

In this section, we will go through a brief introduction to the basics, theory and concepts of nuclear magnetic resonance (NMR) spectroscopy. There are several books already available on this topic (Ziessow 1987; Keeler 2010; Levitt 2013); we do not aim to replicate what already is out there, but rather provide an overview on the application of NMR in metabolomics. First, we outline the basic concepts of NMR spectroscopy; describe how the information available from the proton NMR spectrum aids metabolite identification; describe the basic hardware making up the NMR spectrometer; briefly cover LC-NMR spectroscopy and multidimensional NMR spectroscopy and finish by outlining the advantages and disadvantages of NMR spectroscopy in comparison to MS.

Basic Concepts

Overview

NMR spectroscopy is a powerful analytical technology for structural elucidation of small molecules or metabolites in solution. In a typical NMR experiment, a sample is placed in an NMR tube and inserted into a probe that is housed in a strong and homogeneous magnetic field (Figure 3.5a). NMR active nuclei (such as ^{1}H, ^{2}H, ^{13}C ^{15}N or ^{31}P) in the sample align themselves with (or against) the magnetic field direction eventually reaching an equilibrium. A specific radiofrequency pulse is applied to the sample, causing the NMR-active nuclei to move out of alignment with the magnetic field,

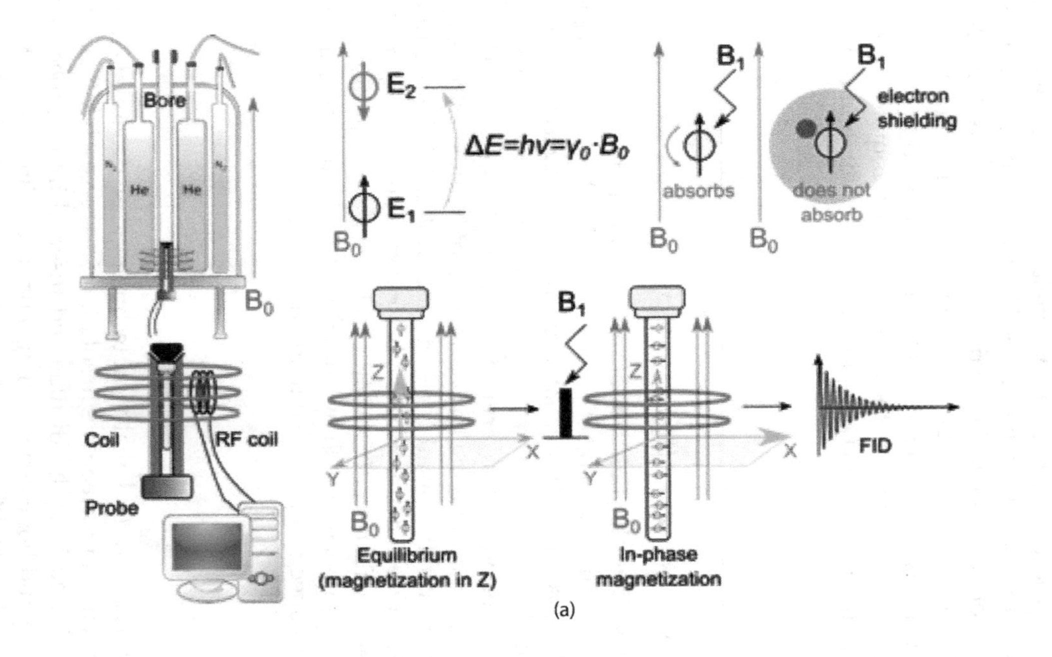

FIGURE 3.5
(a) Diagram of an NMR spectrometer. On the left-hand side, you can see a schema of a cross-section of an NMR magnet. A set of superconducting coils usually made of (NbTaTi)3Sn embedded in copper is wound into multi-turn solenoid coils. The coils are immersed in a liquid helium (at about 4.2 K) dewar itself surrounded by liquid nitrogen in a large dewar (77.4 K), which acts as a thermal buffer. Below the magnet, you can see a schema of a sample probe. The probe holds the sample, generates the B_1 excitation pulses and contains the signal receiver coils, temperature control circuits and gradient coils. When a particular nucleus is immersed into an external static and homogeneous magnetic field referred to as B_0 and subjected to a second oscillating magnetic field (B_1).

(Continued)

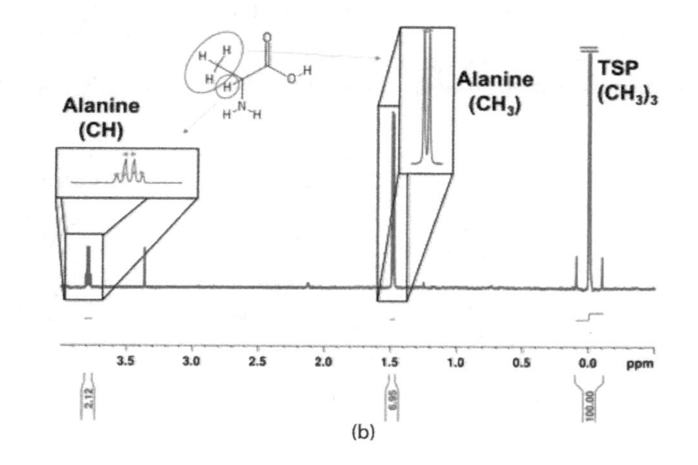

FIGURE 3.5 (Continued)
(b) 1D ^{1}H NMR spectrum of alanine in PBS buffer containing TSP. The 9 protons from deuterated TSP give a strong singlet at a chemical shift of 0 ppm (integral set to 100 units, hole peak not shown). The signals from the methine (CH) group are further downfield (3.8 ppm) due to their close proximity to the electronegative N atom than the methyl (CH_3) group protons (1.5 ppm). Protons on N and O are in a fast exchange and not observed under these conditions. Note the integral for the methyl (CH_3: 6.95) protons is approximately three times greater than the methine proton (CH: 2.12). In theory this should be exactly three times; however, the delay between pulses has been optimized to balance run-time and signal-to-noise ratio. The intensities of the peaks are dependent on the number of protons in that molecular environment, i.e., the concentration of alanine (1 mM). Note the signal from the methyl (CH_3) protons has been split into a doublet (1:1 ratio) due to the presence of a proton neighbor, and the signal from the methine (CH) protons has been split into a quartet (1:3:3:1 ratio) due to the presence of three proton neighbors (the n + 1 rule). The coupling constant (J) for all is 7.3 Hz (orange arrows).

(Continued)

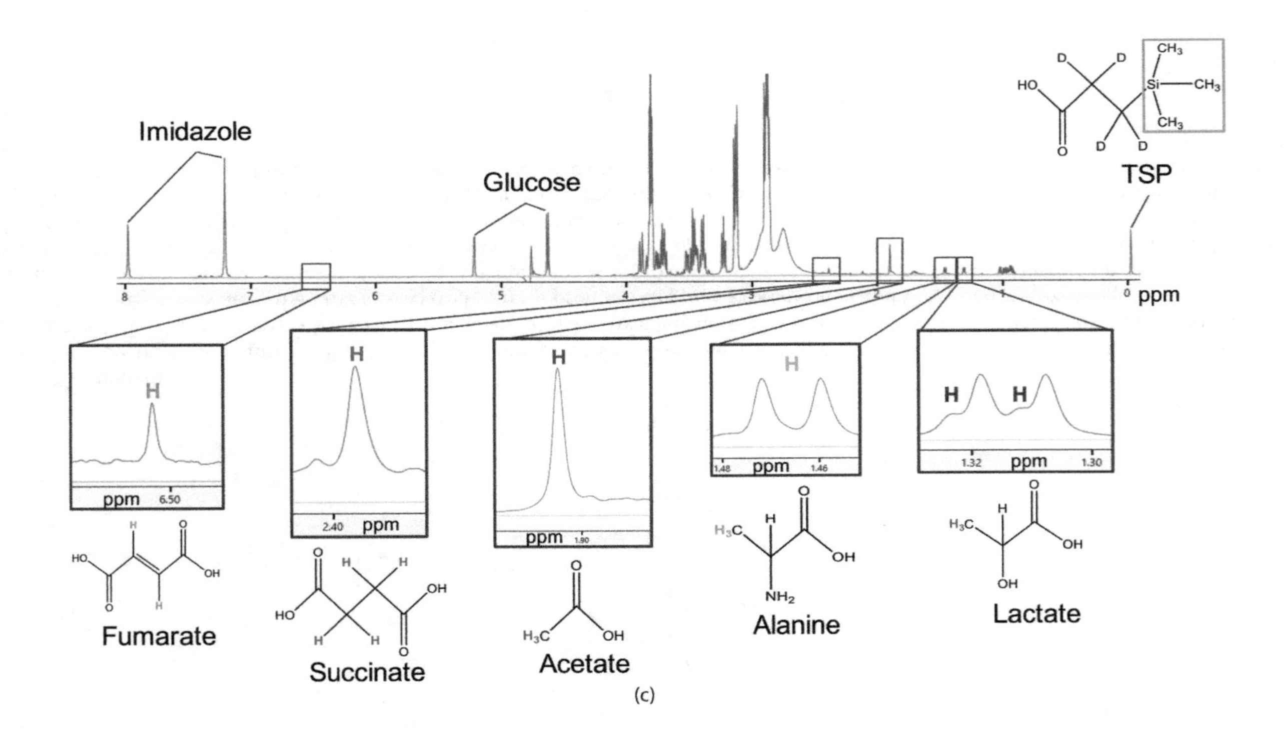

FIGURE 3.5 (Continued)
(c) 1D ^{1}H NMR spectrum of Leishmania Mexicana media. The 9 protons from deuterated TSP give a strong singlet at a chemical shift of 0 ppm. The signals from the metabolites are downfield (between approximately 1 and 8 ppm). Protons from the same metabolite but from different parts of the metabolite (different molecular environments) give signals in different parts of the spectrum (e.g., glucose). Note the signal from alanine at 1.5 ppm is attributed to the methyl (CH_3) protons; this signal has been split into a doublet (1:1 ratio) due to the presence of a proton neighbor. Protons on N and O (NH_2 and OH) are in fast-equilibrium and generally do not give good signals. The intensities of the peaks are dependent on the number of protons in that molecular environment and the concentration of the metabolite in the biofluid.

with subsequent relaxation of these excited nuclei back to their original state. These oscillating nuclear magnetizations induce an oscillating electric current in the receiver coils of the NMR probe, giving the signal known as the free induction decay (FID). Fourier transformation of the time-domain FID signal gives rise to the frequency domain NMR spectrum (see Chapter 5).

Every NMR spectrum is characterized by the chemical shifts, coupling constants and relaxation times of the nuclei of the metabolites under observation, and for complex mixtures (e.g., biofluids) by the relative abundance of the components within the mixture (see Box 3.4). An understanding of these fundamental properties of the nuclei is essential for the understanding and the interpretation of NMR spectra.

BOX 3.4 BASIC DEFINITIONS IN NMR

- *Chemical shift (δ):* The Larmor precession frequencies of nuclei in different parts of a metabolite or in different metabolites are different. When the signals are plotted on an NMR spectrum, their position on the x-axis (chemical shift) will be dependent on the chemical environment. ^{1}H and ^{13}C NMR chemical shifts are affected by proximity to electronegative atoms (O, N) and unsaturated groups (C=C, aromatic), which de-shield the nuclei by reducing the electron density that surrounds it. The signal is shifted to the left (downfield). The NMR chemical shift is expressed in dimensionless units of parts per million (ppm) to allow spectra from different NMR magnets (e.g., 600 vs 750 MHz) to be easily compared.
- In an aqueous biofluid, TSP (sodium 3-trimethylsilyl)propionate-2,2,3,3-d4 is generally used as a standard to calibrate chemical shifts: δTSP = 0ppm.
- *Spin-spin (J) coupling:* Means that the NMR signal is split by one or more "neighbor" protons which act like tiny magnets adding to or detracting from the local magnetic field. The multiplicities follow the n + 1 rule (i.e., one neighbor will split the signal into a doublet).
- The signal splitting, a distance between peaks measured in Hz, is termed the coupling constant. The splitting pattern can give information about the number of the neighboring protons. J-coupling is mediated through bonds via the electrons and is generally observed over 2–4 bonds. Magnetically equivalent protons do not give rise to spin-spin splitting e.g., methyl group protons.

Spin

NMR spectroscopy exploits a phenomenon occurring in nuclei known as spin to generate the signals detected. Subatomic particles (i.e., electrons, protons, neutrons, etc.) have an intrinsic quantum property called spin, a form of angular momentum. Only certain nuclei are NMR spin active: for those with an odd mass number, their nuclear spin is a half-integer (I = ½; e.g., ^1H and ^{13}C); for those with an even mass number but an odd atomic number, their nuclear spin is an integer (I = 1; e.g., ^2H); for those with an even mass number and an even atomic number their integer is zero (I = 0; e.g.,^{12}C and ^{16}O) i.e., the spin of protons and neutrons cancel each other out. Such nuclei are not NMR spin active and are "NMR invisible" (Bothwell and Griffin 2011; Levitt 2013). When a particular nucleus with a non-zero spin quantum number (I = ½ or 1) is immersed into an external static and homogeneous magnetic field referred to as B_0 (the large superconducting magnet in Figure 3.5a), the nuclear spins will align with or against B_0 in almost equal numbers. As with all charged particles, the spinning nucleus generates a magnetic field. The magnetic moment (μ) is related to the angular momentum of the nucleus by:

$$\mu = \gamma l$$

where (γ) is the gyromagnetic ratio, a proportionality constant unique to each nucleus (de Graaf 2013). In contrast to the behavior of macroscopic magnets, the nuclei do not align themselves perfectly with B_0, for I = 1/2 nuclei, the magnetic moments align themselves at an angle of 54.73° with B_0 and, in analogy to spinning tops in a gravitational field, the force trying to align the nuclear magnetic moments with B_0 makes these nuclei precess (like a gyroscope) about B_0 at a characteristic rate of precession (frequency) known as the *Larmor frequency* (ω_o) (de Graaf 2013).

$$\omega_o = \gamma B_0$$

This spin property is quantized, i.e., it has *only* two directions, either an up or down state (Figure 3.5a) (Bothwell and Griffin 2011). When the sample is placed in a magnetic field B_0, the spin state of the nucleus can align along the direction of this magnetic field (z) (stable state, hence lower energy E_1) or oppose the magnetic field (unstable state, higher energy E_2) (Figure 3.5a). This distribution is governed according to the Boltzmann equation (de Graaf 2013) where N_{low} is the number of spins in the low energy state, N_{high} is the number of spins in the high energy state, ΔE is the energy difference between the two states, K is the Boltzmann constant and T is the temperature in Kelvin.

$$\frac{N_{low}}{N_{high}} = e^{\frac{\Delta E}{KT}}$$

The low sensitivity of NMR is due to the fact that the energy gap between the two states is very small and the excess of nuclei in the lower energy state (population difference) is typically only around 0.01% (Bothwell and Griffin 2011). It is possible to make the nuclear spin state flip from the more stable alignment to the less stable one by the input of radiofrequency energy (B_1) of the correct frequency. The energy needed to make this flip (E) increases linearly with the strength of the external magnetic field used. The energy difference E is such that radio frequency waves applied orthogonally to **z** (direction of the magnetic field) for a defined period of time will perform this switch at frequencies of about 400–800 MHz (B_1-pulse) for magnetic field strengths of 9.4–18.8 Tesla, as typically used in modern metabolic profiling experiments. This flipping of protons from one magnetic alignment to the other position is termed nuclear magnetic resonance. The precession about the applied magnetic field of the excited nuclear magnetizations induces tiny oscillating currents at specific frequencies that are detected in the receiver coils of the NMR probe as the FID. Fourier transformation converts the time domain into the frequency domain resulting in the NMR spectrum (Figure 3.5a) (Bothwell and Griffin 2011).

The Proton NMR Spectrum

The lines (peaks) in the NMR spectrum represent nuclei in different parts of the molecule. The nuclei experience different local magnetic fields according to their local "molecular environment," which is determined by the molecular structure. Chemically distinct nuclei will be *shielded* from the applied magnetic field (B_0) to different extents by local electrons (Figure 3.5a). They, therefore, have different frequencies at which they absorb energy and resonate. This difference is expressed as the chemical shift (δ) (Keeler 2010).

Proton NMR spectroscopy (^{1}H-NMR) is the most common form of NMR spectroscopy. In ^{1}H, NMR it is the environment of the hydrogen atom in the chemical structure of the metabolite that determines the chemical shift in the spectrum. A commonly used internal standard in NMR spectroscopy is tetramethylsilane (TMS), a silicon group surrounded by 4 methyl groups, with 12 equivalent protons resulting in a single peak assigned as d 0 ppm (Figure 3.5b). Protons on other naturally occurring metabolites are deshielded in comparison to TMS, since other nuclei are not as electron donating as silicon, and their signals are observed downfield (to the left) of the TMS signal. This means the opposition to the magnetic field resulting from the spin of the electrons (which themselves have a magnetic vector). Electronegative atoms such as oxygen, nitrogen and chloride further de-shield the protons in the local environment and shift those signals further downfield. Double bonds and aromatic structures also reduce the electron cloud and result in signals which are downfield.

Consider a ^{1}H-NMR spectrum of a urine sample. Each chemically distinct proton within each metabolite in the sample will resonate at its own characteristic chemical shift position in the spectrum. There is a close and well-understood relationship between the chemical shift of the proton and its (molecular) structural environment: hence the mere observation of a signal immediately gives information on the structural environment. This makes NMR spectroscopy a powerful tool in the identification of unknown metabolites (structural elucidation).

Signal Intensities

Within a metabolite, the area (or integral) of the ^{1}H NMR signal is directly proportional to the number of protons accounting for that signal. For example, a methylene (CH_2) or methyl (CH_3) group will be exactly 2 or 3 times larger, respectively, than that from a methine (CH) group, provided that certain precautions are observed by the spectroscopist (Figure 3.5b). Consequently, the area of each ^{1}H signal will be directly proportional to the molar concentration of the molecule (e.g., in the urine) responsible for the signal, again given certain provisos. The inclusion of a single internal standard (e.g., TMS) at a known concentration allows the absolute concentration of the metabolites in the sample to be calculated. Hence NMR metabolomics can provide absolute quantification of metabolites within a complex mixture (unlike in MS where generally only relative concentrations are obtained). NMR spectroscopy can be completely quantitative with no requirement for reference standards (using ERETIC—Electronic REference To access In vivo Concentrations) methods (Akoka et al. 1999).

Spin-Spin (J) Coupling

Not all signals in the urine spectrum will be single lines, or singlets as they are known. Many signals will be split into 2, 3, 4 or more lines, known as doublets, triplets, quartets or multiplets (Figure 3.5b). This is due to the presence of non-equivalent nuclei (within a different molecular environment) which themselves act as weak magnets. This is the phenomenon of spin-spin (or J) coupling (Bothwell and Griffin 2011). This is a through-bond, magnetic interaction, mediated via bonding electrons, occurring between nuclei which are 2–4 bonds apart (generally) in a molecule. The splitting pattern follows the $n + 1$ rule, i.e., if a proton has one proton neighbor it will be split into a doublet (with a 1:1 ratio) and if it has two proton neighbors it will be split into a triplet (1:2:1 ratio). The relative intensities of each line are given by the coefficients of Pascal's triangle (Dona et al. 2016). If a proton has one neighboring proton, with different chemical environment, its signal will be split in two (a doublet) since 50% of the neighboring protons will slightly add to the magnetic field experienced by the nucleus (shifting the signal slightly downfield) and 50% of the neighboring protons will slightly detract from the magnetic field experienced by the nucleus (shifting the signal slightly upfield). If a proton has two neighboring protons, with a different chemical environment, its signal will be split in three (triplet) since 25% of the time, *both* neighboring protons will slightly add to the magnetic field experienced by the nucleus (shifting the signal slightly downfield) and 25% of the time *both* neighboring protons will slightly detract from the magnetic field experienced by the nucleus (shifting the signal slightly upfield) and 50% of the time the two proton neighbors will cancel out (one adding and one detracting from the local magnetic field). If a proton has three proton neighbors, its signal will be split into a quartet. For example, in the metabolite lactate (CH_3-CH(OH)-CO_2^-) the three methyls (CH_3) protons split the single methine (CH) signal into a quartet (3 + 1 = 4) with a 1:3:3:1 ratio.

The distance between the peaks, measured in Hz, is the coupling constant. This is independent of the magnetic field as it is caused by the magnetic field of another nucleus, not the magnet (B_0) itself. Again, there is a close and well-understood relationship between the size of the signal splitting (the spin-spin coupling constant or J-coupling) and molecular structure. The observation of mutual splitting between two or more nuclei demonstrates that they are connected by chemical bonds in a molecule. Thus spin-spin coupling provides a mechanism for piecing together the structures of unknown molecules from the signals that their nuclei exhibit in the NMR spectrum.

Relaxation Times

Finally, each distinct proton in each different molecule in the sample will be associated with its own characteristic relaxation times (Emwas et al. 2016). These relaxation times govern the width of the signals and also limit the rate at which NMR spectra can be acquired. Here again, there are close relationships between the molecular environment that a particular proton inhabits and its relaxation times, and further structural and motional information can be obtained.

Hardware

In this section, we turn from the fundamentals of NMR theory to the practicalities of obtaining an NMR spectrum on a modern, high-field, NMR spectrometer. We shall look in turn at the spectrometer and the way in which the NMR signals are generated and then processed to provide the NMR spectrum. For newcomers to NMR spectroscopy, we provide general guidance on the magnet, digital resolution, dynamic range, with other issues related to data processing addressed in Chapter 5. Here we describe the main parts of every NMR spectrometer, the magnet, the probe and the console.

The Magnet

All high field NMR spectrometers (B_0 ca. 4.7–23.5 T, equivalent to ^{1}H NMR at 200–1000 MHz) employ superconducting magnets in which the cylindrical solenoids are wound from a niobium alloy such as niobium/titanium or niobium/tin. These solenoids are immersed in liquid helium at ca. 4 K (–269°C)

BOX 3.5 NMR DATA ACQUISITION TERMINOLOGIES

- *FID:* NMR signal is different in form to the absorption or emission signals which we are familiar with from optical and vibrational spectroscopy. The FID is best understood as being derived from an electrical signal induced in the receiver coil of the NMR probe by the coherent precession of the magnetic nuclear spins and which decays over time as those spins relax.
- *Dynamic range:* In biofluid NMR spectroscopy, the signals from many different low concentration (mM or sub mM) biofluid components are measured in the presence of very high concentrations of water (ca. 100 M in protons). The measurement of a very small signal in the presence of a very large signal (ca. 1 part in 10 ^ 5) poses a problem of dynamic range. The problem arises chiefly due to the limited resolution of the ADC. Prior to the start of an acquisition, the receiver gain is adjusted so that the signal will not overload the ADC, but almost reaches the maximum voltage.
- *The resolution:* The NMR spectrum signal resolution will have contributions from:
- The natural linewidth which relates inversely to the transverse relaxation time (homogeneous broadening) and increases with a more efficient relaxation of the spins or fast transverse relaxation caused by field inhomogeneity (inhomogeneous broadening).
- The digital resolution. If the spectrum is insufficiently digitized, the digital resolution may be less than necessary to accurately define the signals and their frequencies.

at which temperature the wire loses all electrical resistance (Bothwell and Griffin 2011). After energizing the solenoid, the external power supply can be disconnected, and the superconducting current will remain stable and persistent, and produce a correspondingly stable and persistent magnetic field, for many years. For high resolution (see Box 3.5) work on small liquid samples such as biofluids, the magnets are designed with a vertical room temperature bore of ca. 50 mm containing the probe. Wide bore vertical magnets with bores of ca. 90 mm are available for work on slightly larger samples (greater than ca. 20 mm diameter) and can accommodate isolated perfused organs from small animals as well as the small animals themselves, albeit with rather a tight fit. Magnets (generally 3–7 T) with horizontal bores from ca. 15 to 100 cm are employed for non-invasive *in vivo* spectroscopy (known as Magnetic Resonance Imaging (MRI) when signals from water in solids versus liquids are used to create maps of the body, or Magnetic Resonance Spectroscopy (MRS) when high concentration metabolites such as glucose are measured) on larger laboratory animals and humans (Bothwell and Griffin 2011). In all cases, the superconducting magnet is carefully temperature shielded by means of vacuum and liquid nitrogen jackets, so that liquid helium loss is minimized (Bothwell and Griffin 2011). The liquid helium and liquid nitrogen in these magnets must be replenished, with a frequency which can vary from weeks to months depending on design.

The Probe

The probe accommodates the sample, which for biofluid NMR is normally contained within a precision-made glass tube of 1.7, 3.0 or 5 mm outside-diameter typically, although flow probes are also available and work well. The tube is mounted in an air turbine spinner to allow spinning of the sample during the experiment (required only in older magnets) (Bothwell and Griffin 2011). The probe contains the coils for both transmitting radiofrequency pulses to the sample and for receiving the NMR signals, in addition to coils tunes to the frequency of deuterium (^{2}H) for the purpose of "locking" the field/frequency ratio of the spectrometer to ensure stability and also allow optimization of magnetic field homogeneity (Bothwell and Griffin 2011). A single radiofrequency coil can be used for both transmitting and receiving for a particular nucleus or range of nuclei: this coil is perpendicular to B_0. Many different types of probes are available including: (i) probes dedicated to the observation of one particular nucleus, (ii) dual probes which allow detection of two types of nuclei, (iii) inverse dual probes in which the ^{1}H coil is on the inside and the X-nucleus observation/decoupling coil is on the outside and (iv) broadband probes of either normal or inverse geometry, in which the X nucleus coil can be tuned to observe a wide range of X nuclei.

The Console and Accessories

The NMR console contains the computer(s), the electronics necessary to generate and control the timing, phase, frequency and shape of the radiofrequency excitation pulses and the detectors and the analogue to digital converter(s) (ADC) to convert analogue signals into digital form. The preamplified signals are sent to a detector prior to digitization and at this stage, the excitation frequency is subtracted from the incoming radio frequency signals. This subtraction converts the NMR signals from a series of oscillations at radio frequencies to a narrow band of audio frequencies which are easier to digitize. For a ^{1}H NMR spectrum at 400 MHz, the frequency range will be 4 kHz for a spectral width of 10 ppm. (1 ppm is equivalent to 400 Hz on a 400 MHz spectrometer.) Note that the conversion from radiofrequency to audio frequency has no effect on the relative frequency separation of the signals in the NMR spectra, and chemical shifts are unaffected, as they are defined relative to the resonance frequency of a standard. The signal which is detected following excitation of the spins using an RF pulse arises from transient free precession of the nuclei and is called a free induction decay (FID). It is converted into the familiar NMR spectrum by the NMR software, using a Fourier transformation (Chapter 5).

Sample Preparation

Generally, the sample preparation required for NMR spectroscopy is simple: the biofluid is mixed with an NMR buffer and transferred to a glass NMR tube (typically 5 mm) (Salek et al. 2011). The buffer contains a known concentration of an internal standard; for aqueous samples, this is often deuterated sodium 3-trimethylsilyl propionate-2,2,3,3-d4 (TSP) or deuterated 4,4-dimethyl-4-silapentane-1-sulfonic acid (DSS), as these are more water soluble than TMS. The use of deuterated internal standards reduces the introduction of additional signals in the ^1H NMR spectrum since ^2H are invisible in ^1H spectra. The buffer also contains deuterated water (D_2O/^2H_2O) to provide a lock signal: the ^2H is monitored on another channel and any shift in ^2H due to fluctuations in the magnetic field is compensated for in the ^1H spectrum (Bothwell and Griffin 2011). The NMR buffer usually contains a pH buffer to minimize sample to sample variation in pH (a particular problem with urine samples) which can affect the chemical shift of pH-sensitive NMR signals. Sometimes a molecule sensitive to pH change, such as imidazole, is added to allow the pH to be monitored between samples. For samples with a high protein concentration (such as serum or plasma), protein precipitation, solid phase extraction, liquid-liquid extraction or spin filtering may be required (Barding et al. 2012; Viant et al. 2007).

Interpretation of ^1H NMR Signals

A typical one-dimensional (1D) ^1H-NMR spectrum from a biological specimen (i.e., cell extracts, blood plasma, urine, etc.) contains thousands of spectral lines representing resonances from hundreds of—mainly—low molecular weight compounds. As described above, the chemical shifts of each spectral band (position in the spectra), the relative intensities between signals and the splitting pattern observed give information on the molecular composition of the compound (i.e., its structure). Integration of the signals also provides absolute quantification of the metabolites in the sample.

The majority of samples and solvents used in metabolomics contain a large amount of water. For example, serum and urine are approximately 95% water (Psychogios et al. 2011). The difference in proton concentration between water (110 M) and metabolites of interest (generally mM) is of several orders of magnitude. Unless the water peak is suppressed, the water peak would cause radiation damping or dominate the spectrum and the metabolites of interest may not be visible (Beckonert et al. 2007; Bothwell and Griffin 2011). Therefore, 1D ^{1}H NMR experiments generally include a water suppression scheme (Barding et al. 2012; Viant et al. 2007). It should be noted that when using pre-saturation based water suppression methods, protons in a fast chemical exchange with water protons will also suffer signal suppression (Viant et al. 2007). Care must be taken in the quantitation of peaks close to solvent peaks, as their resonances may have been suppressed as part of the solvent suppression too or there may be baseline distortions resulting from the suppression (Barding et al. 2012). One of the most commonly used pulse sequences in NMR metabolomics is the 1D-^{1}H-NOESY Nuclear Overhauser Effect SpectroscopY (NOESY) pulse with pre-saturation of the water signal (Noesy-Presat). This is a simple sequence, with few optimization requirements or hardware restrictions, and good quality water suppression is generally easily obtained (Mckay 2011).

The Carr-Purcell-Meiboom-Gill (CPMG) pulse sequence is used to identify small molecules in the presence of large proteins and lipoproteins, as seen in native serum and plasma samples (Barding et al. 2012). Essentially, there is a delay before the FID is recorded so only the FIDs of small molecules are detected, not the FIDs from large molecules, in a method known as spectral editing or filtering (Viant et al. 2007).

Parameters Affecting Spectral Data

A basic understanding of a few key aspects of NMR spectroscopy is necessary in order to assure that you get accurate, realistic spectra for NMR metabolomics purposes. In order to overcome the basic insensitivity of NMR spectroscopy, signal averaging, over multiple pulses/scans, is widely used. This involves an acquisition of the FID into the same region of computer memory in a repetitive fashion. The basic concept is simple. The signal component of the FID is constant and will build up in strength linearly with the number (n) of FIDs accumulated, but the noise component is random and will build up only as the square root of n. Thus, the signal-to-noise ratio in an NMR spectrum improves as n ^ 1/2. If four scans are accumulated, the signal-to-noise ratio will be doubled, but 16 scans will be required in order to affect a further doubling. Eventually, the benefits of further signal averaging become insignificant relative to the time investment required and alternative methods of sensitivity enhancement should be used. Other methods include increasing the concentration of the sample, increasing the diameter of the sample tube and using mathematical manipulation of the FID.

LC—NMR Spectroscopy

The on-line coupling of high-performance liquid chromatography with NMR spectroscopy gives the analyst the prospect of combining one of the most efficient separation technologies with a detector that is non-selective (albeit rather insensitive compared with UV or MS) and provides more structural information. LC-NMR has been widely applied in chemical and biological areas, capable of identifying and profiling both endogenous and xenobiotic materials in biofluids. For the analysis of complex mixtures like biofluids, the combination of HPLC and NMR is extremely powerful and adds another dimension to the array of NMR techniques that can be employed. One attraction of the technique is that the results of the LC-NMR experiment can be compared with those from LC-MS experiments if run under identical conditions. This combined NMR and MS information gives a tremendous power to solve the structures of endogenous and xenobiotic biofluid components, without any need to isolate and purify the components of interest. On-line, hyphenated LC-NMR-MS represents the ultimate way to extract retention time, NMR and MS data from the components in a biological fluid (Burton et al. 1997). However, this can be technically demanding, and off-line LC-NMR spectroscopy may be required, where fractions from LC are collected for analysis by NMR spectroscopy; often these can be pooled, concentrated or extracted before analysis as required (Bothwell and Griffin 2011).

Other Nuclei

As mentioned previously other nuclei are NMR active, such as ^{13}C, ^{31}P and ^{15}N. In metabolomics, 1D ^{13}C NMR spectroscopy is useful for aiding the identification of metabolites due to its larger chemical shift range (approximately 200 ppm compared to approximately 10 ppm for ^{1}H NMR spectroscopy) (Dona et al. 2016; Kumar Bharti and Roy 2014). It is, therefore, more sensitive to differences in molecular structure, so even in complex mixtures most chemical shifts will be unique (Dona et al. 2016). However, the sensitivity of ^{13}C NMR spectroscopy is low, so samples generally need to be labeled with ^{13}C, for example by the incorporation of ^{13}C glucose. This is highly useful in tracer studies where the flow of carbon during metabolism is monitored. The study of ATP and other phosphor-metabolites can be achieved using ^{31}P NMR spectroscopy (Bothwell and Griffin 2011). A combination of ^{13}C and ^{15}N NMR spectroscopy was used to monitor fluxes of ^{13}C labeled glucose and ^{13}C, ^{15}N labeled glutamine in lung cancer cells (Fan and Lane 2011).

Multidimensional NMR Spectroscopy

Two-dimensional (2D) NMR spectroscopy is very important in metabolomics. It helps determine which peaks are connected (part of the same spin system on the same molecule) when multiple metabolites are present, in order to help determine metabolite identity (Chapter 5). It is also useful when there is significant spectral overlap in 1D ^{1}H NMR as it increases signal dispersion.

To perform a basic 2D NMR experiment, hundreds of 1D spectra are recorded, and the delay in the pulse sequence is increased with each acquisition. A typical 2D NMR experiment consists of four stages: preparation, evolution (t1), mixing and detection. In the preparation phase, a pulse is used to excite the nuclei. An evolution period (t1) is the delay; this is increased by a fixed amount of time for each pulse. During the mixing time, the magnetization is transferred from the excited nuclei to other coupled nuclei within the same spin system via j-coupling or nuclear Overhauser effects. In the acquisition phase the FIDs produced by the precessing nuclei are detected (Tal and Frydman 2010). The resulting spectra are stacked with the lowest t1 value at the bottom. FIDs of each acquisition are Fourier transformed as normal to obtain the F2 (direct) frequency spectra. A new FID will be formed across the stacks, in the indirect time domain (t2), which is then also Fourier transformed to obtain the orthogonal F1 (indirect) spectrum. The indirect time domain is a result of the FIDs being sampled at discrete time intervals. When F1 is plotted against F2 in a topographic (contour) plot, cross peaks are observed when the nuclei are coupled (Tal and Frydman 2010). The intensity of each cross-peak is dependent on the efficiency of the transfer of magnetization from an excited nucleus to another, nearby nucleus. The 2D spectrum, therefore, represents paired nuclei which can be assigned to specific positions on metabolites (Dona et al. 2016; Emwas 2015; Emwas et al. 2015).

There are two main types of 2D NMR: homonuclear, where magnetization is transferred from one nucleus to another nucleus of the *same* type (usually ^{1}H to ^{1}H) or heteronuclear, where magnetization is transferred from one nucleus to another nucleus of a *different* type (usually ^{1}H to ^{13}C) (Dona et al. 2016; Emwas 2015; Emwas et al. 2015). There are many different types of homonuclear and heteronuclear 2D NMR experiments. These mainly differ in the pulses and delays used during the mixing time to transfer the magnetization.

The simplest 2D NMR spectroscopy method is J-resolved NMR spectroscopy. It is also the fastest with acquisition times of about 20 minutes (Ludwig et al. 2012). It is used to create a projected 1D ^{1}H NMR spectrum without j-coupling, effectively a broadband ^{1}H-decoupled spectrum. This greatly reduces spectral overlap allowing signals to be resolved, therefore improving identification (Dona et al. 2016; Fonville et al. 2010; Ludwig et al. 2012; Viant et al. 2007).

^{1}H-^{1}H Correlation spectroscopy (COSY) is one of the most useful 2D NMR methods in metabolite identification. It identifies which protons are spin-spin coupled (Emwas 2015). ^{1}H-^{1}H Total Correlation spectroscopy TOCSY provides information on which ^{1}Hs are directly coupled, as in COSY; however, it also provides information of which ^{1}Hs are indirectly coupled (Dona et al. 2016).

^{1}H-^{13}C Heteronuclear single quantum coherence spectroscopy (HSQC) is also extremely useful in metabolite identification. It provides information on which protons are coupled to which carbons (Dona et al. 2016). It also exploits the larger chemical shift range of the ^{13}C spectrum to further resolve cross-peaks (Dona et al. 2016; Everett 2015). ^{1}H-^{13}C Heteronuclear multiple bond correlation spectroscopy (HMBC) can be very useful in metabolite identification as it allows connectivities between parts of the metabolite structure separated by heteroatoms or quaternary carbons to be identified (Dona et al. 2016).

The major disadvantage of 2D NMR spectroscopy is that it is very slow, typically taking several hours to obtain the spectrum. However, this would only be required for a handful of samples in a study: once the metabolites are known, they can generally be identified from 1D ^{1}H NMR spectroscopy (Dona et al. 2016; Giraudeau and Frydman 2014).

Advantages and Disadvantages

The major strengths of NMR spectroscopy for metabolomics are that it is fast, quantitative and reproducible, in comparison to MS. The sample preparation is simple and NMR spectroscopy is now fully automated and high-throughput, allowing the analysis of over 100 samples a day (depending on the pulse sequences chosen). The high-throughput nature of the analysis is associated with a reduction in cost per sample. As discussed above, absolute (rather than relative) quantification of metabolites is possible in NMR spectroscopy. For clinical metabolomics cohorts, this is essential (Emwas 2015). The spectra are highly reproducible, making it suitable for large-scale metabolomics studies with no batch effects. Another important feature of NMR spectroscopy in metabolomics is that it can be used for structural elucidation of novel metabolites (Emwas 2015). Major limitations of NMR spectroscopy for metabolomics are lack of sensitivity and signal dispersion. This means that only 40–200 metabolites can be detected, compared to the several hundred metabolites that can be detected by MS, meaning that a less detailed view of the changes in the metabolome can be achieved (Emwas 2015). As with mass spectrometry, data analysis remains the major bottleneck for metabolomics NMR applications. NMR spectra of biological matrices such as cells, tissues (dissolved) or biofluids contain hundreds of resonances. It is of paramount importance to accurately process such spectra to obtain biologically relevant information from NMR-based metabolomics experiments.

Vibrational Spectroscopy

Raman and infrared (IR) spectroscopy offer an alternative detection method to mass spectrometry or NMR for metabolomics applications. Raman and IR are the most common types of vibrational spectroscopy used and can provide a wealth of information regarding the molecular composition of samples. Both of these techniques arise as a result of the interaction of light and matter. If we consider light as a collection of multiple particles termed photons; upon interaction with physical matter these may be absorbed, scattered, reflected or may simply pass straight through. In Raman spectroscopy, we are recording the scattering of light while with IR we are monitoring the absorption of light. Together these techniques are complementary in the molecular information that they provide.

Raman Spectroscopy

As stated, for Raman spectroscopy we are observing the scattering of photons from a monochromatic laser source. The interaction of a scattered photon promotes the molecule under interrogation to a virtual energy state. The promotion is short-lived, as a consequence of the instability of the virtual state the molecule returns almost instantaneously to its ground state. The most common example of scattering is Rayleigh scattering,[1] in which the molecule returns to the exact ground state from which promotion occurred. In Rayleigh scattering, there is no net change in energy and the process is overall more thermodynamically likely—in fact, Rayleigh scattering accounts for 99.99999% of all scattering events. Rayleigh scattering is an example of elastic scattering. Raman scattering is observed when this theory is not obeyed, i.e., when the molecule returns to a higher or lower energy state than its origin and thus emits a photon with a different energy to the incident photon. This type of scatter is referred to as inelastic scattering. There are two types of Raman scattering, termed Stokes and anti-Stokes. Stokes occurs when a molecule is promoted from its ground state to a higher excited state by the overall absorption of energy. Anti-Stokes occurs when a molecule is already excited and is demoted to the ground state by a transfer of energy from the molecule to a photon. A schematic representation of these scattering events is shown in Figure 3.6.

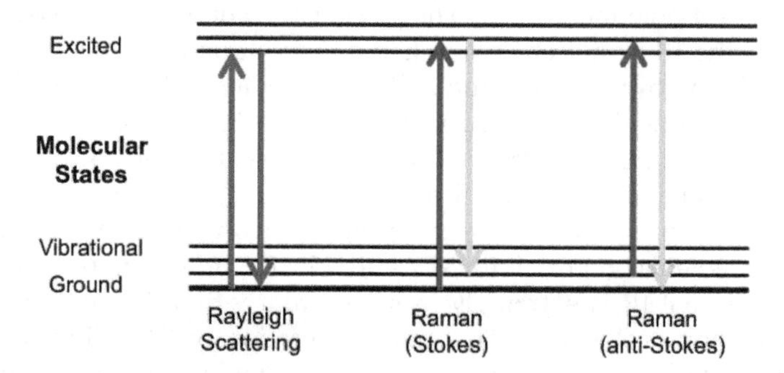

FIGURE 3.6
An energy level diagram illustrating the transitions occurring during Rayleigh and Raman scattering events. The most common example of scattering is Rayleigh scattering and this is depicted by the purple illustration, whereby the molecule returns to the original ground state from which promotion has occurred. Raman scattering occurs when there is a net change in energy and these are illustrated by the blue and yellow arrows. Stokes occurs when a molecule is promoted from its ground state to a higher excited state by the overall absorption of energy. Anti-Stokes occurs when a molecule is already excited and is demoted to the ground state by a transfer in energy from the molecule to a photon.

[1] "Rayleigh scattering—Wikipedia." https://en.wikipedia.org/wiki/Rayleigh_scattering. Accessed October 19, 2017.

For a molecule to be Raman-active there must be a change in the polarization of the molecule. The degree of polarization incurred is directly proportional to the intensity of the Raman scatter observed. Specific shifts in energy can be directly correlated to functional group chemistry (Figure 3.6).

The major advantage of Raman spectroscopy is that water is a weak Raman scatterer and thus does not mask biological information. Raman is a non-destructive and minimally invasive technique that allows for sample recovery after data acquisition. The technique is rapid and can be applied to solids, liquids and gases. The major caveat with the application of Raman spectroscopy is that the Raman effect is inherently weak and as a consequence detection limits are substantially higher than those obtainable by MS or NMR. There are a number of enhancement variants of Raman spectroscopy that can begin to overcome this sensitivity issue including resonance Raman (RR) spectroscopy and surface-enhanced Raman scattering (SERS). An additional limitation is the occurrence of fluorescent interference, observed as a large, broad signal that will greatly mask any sample specific spectral features.

IR Spectroscopy

Infrared is a technique regularly used for sample characterization. It is an absorption technique in which we are observing the promotion of molecules to an excited vibrational energy state upon interrogation with an infrared light source. A molecule is only promoted to this excited state if there is a change in its dipole moment, μ. The greater the change in dipole moment, the greater the IR signal observed. Upon absorption of light, certain functional groups within the molecule will vibrate in a number of possible manners, i.e., stretching, bending or rocking. These vibrations are characteristic of certain biochemical features and can thus be correlated to specific functional group behavior and classification.

Akin to Raman spectroscopy, infrared spectroscopy is not as sensitive or specific when compared to MS and NMR. However, the associated advantage is that IR is extremely rapid, regularly high-throughput and in some cases non-destructive. The major limitation of IR is that water is a very strong absorber and as a consequence, samples are often dried prior to data acquisition, thus leading to a low recovery of the sample. Liquid samples can be analyzed using attenuated total reflectance (ATR); however, this approach is more manual and thus limits the throughput and speed of the technique.

The spectral outputs from Raman and IR are very similar in form. A Raman spectrum contains information in the structure of Raman shift (cm^{-1}) versus intensity while IR spectra consist of wavenumber (cm^{-1}) versus absorbance or transmittance (% or total). The chemical information is visible in the form of peaks that can be compared and verified by using well-established peak ID tables or indeed by running standards. These techniques will not provide specific metabolite identification but

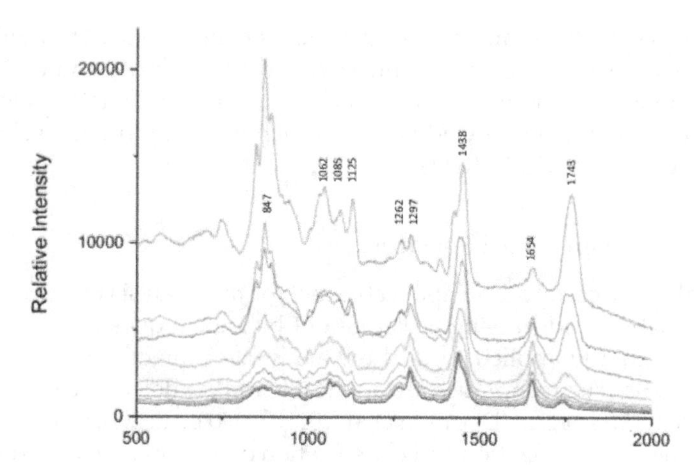

FIGURE 3.7
Example Raman spectra collected from a porcine skin section treated with a topical API, using 785 nm excitation. Tentative band assignments are illustrated which can be attributed to different functional group chemistry associated with the tissue matrix and the API. These include COC vibration (847), C-C stretches (1062–1125), CH_2 twist (1297), CH_2 deformation (1438), C=O stretch (1743).

will illustrate changes within specific functional groups, i.e., lipid CH_2 or CH_3 stretches or protein stretches (Figure 3.7).

Some degree of data pre-processing is routinely undertaken for both approaches. Baseline correction algorithms are applied to remove baseline drift over the spectral range due to scattering, reflection and contributions from temperature, sample and instrument effects. The spectra are also customarily scaled or normalized to reduce sample-to-sample variation when undertaking the high-throughput analysis. A commonly used scaling method is extended multiplicative scatter correction (EMSC). Filtering can also be performed for both sets of data. Raman spectra regularly contain contributions from cosmic rays (random, high intensity, sharp spectral bands resulting from high energy particles from outer space) that can be removed using zapping or filtering approaches. In respect to IR, IR spectra suffer from CO_2 interferences and thus this contribution is always subtracted. After sufficient spectral pre-processing, data is interpreted by multivariate analysis or by a simple comparison of peak areas and peak positioning.

Advantages and Disadvantages

Raman and IR spectroscopy cannot rival the performance and wealth of information provided by MS and NMR approaches; however, they are beneficial tools present within the suite of metabolomics analytics. At present these techniques provide robust, rapid, non-destructive screening options before undertaking costly (time and money) MS/NMR experiments. They will provide complex spectral fingerprints illustrating the complexity and variability of the metabolome. There are many variants of instrumentation available and the technology is rapidly advancing which will only further substantiate these techniques within the field.

Applications of Vibrational Spectroscopy

Vibrational spectroscopy is an applicable tool for the analysis of a wide range of sample matrices addressing a plethora of biological questions. The techniques are regularly applied toward the analysis of intact mammalian and eukaryotic cells. The techniques are applicable for the detection and classification of bacterial species (Dettman et al. 2015; Garip et al. 2009; Oliveira et al. 2012), and the characterization and classification of mammalian cells (Hughes et al. 2012; McManus et al. 2012; Pascut et al. 2013; Zwielly et al. 2009).

The arena is now readily focused on using these techniques for disease detection from clinical samples, i.e., serum, plasma and tissue biopsies with the aim to develop non-invasive methods for the rapid characterization of disease status (Hands et al. 2014; Kwak et al. 2015; Nallala et al. 2016; Stevens et al. 2016). Beyond the clinical arena, vibrational spectroscopy is also being applied to the detection of food fraud (Ellis et al. 2015) and food classification (Jandrić et al. 2015).

Diode-Array Detection

Diode array (DA) data contain much less information than MS or NMR data, not only because only a subset of all metabolites is active in the typical wavelength range of 200–600 nm, but also because the spectra do not contain much information on the identity of the metabolites. Yet, the technique has a number of advantages that warrant a place in the toolbox of metabolomics. For one thing, DA detection is extremely stable and quantitative. This provides a useful complementary characteristic to MS, which could be exploited in a more rigorous fashion. Moreover, with LCMS data the DA data are very often recorded in a separate trace, so acquiring them does not constitute any additional effort. The more sparse nature of the response (i.e., fewer metabolites give a signal) makes it easier to check alignments, or can be used to suggest appropriate alignments for the MS data in the case of long measurement sequences. And finally, some compounds such as carotenoids are not easily analyzed by MS but do have a distinct and clear signal in the UV-Vis wavelength range (Wehrens et al. 2013).

Analysis of DA data in the context of metabolomics is mostly done by hand, concentrating on one wavelength, or a small set of wavelengths known to be important for several molecules of interest. In such a case, the analysis has the character of a targeted analysis. In the field of chemometrics, multivariate methods have been developed specifically for this type of data, known as multivariate curve resolution (MCR) techniques (Juan and Tauler 2006, 2007). They not only constitute a full-spectrum approach, much less sensitive to the choice of wavelength, and able to distinguish also components with very similar spectra, but they also are able to identify unknowns, bringing the processing of this type of data firmly in the realm of untargeted analysis. In principle, these multivariate techniques can be applied to MS-based metabolomics data, too, but this is not often done in practice since it takes quite a lot of time and computing power (see e.g., Mullen et al. 2009). Basically, abstract components are defined adhering to certain constraints such as non-negativity of elution profiles and spectra. Each component corresponds to one metabolite or a group of metabolites with equal or highly similar UV-Vis spectra.

The very flexible form of Beer's Law allows us to fit LCDA data quite easily. However, one does need to take care of several issues that are hard to automate (Wehrens et al. 2014). Pre-processing is very important: baseline correction and possibly smoothing of the data. Then the number of components is crucial. Furthermore, one should be aware that the system is essentially ill-defined, and those constraints, like the aforementioned non-negativity constraint (but also others, e.g., unimodality) may be essential to arrive at meaningful conclusions. This is one technique where more samples lead to better results, so it pays to also include pure standards of compounds that are expected in the samples during the model building.

For components with very different spectra, resolution in the retention time dimension is not necessary, and the MCR components can be quantified easily. Conversely, even if two metabolites have exactly the same UV-Vis spectrum but are well separated in retention time, quantification is also not a problem. This means that in the vast majority of cases the conversion of MCR components to a data matrix containing peak areas or peak heights is easy and fast. In case that metabolites overlap both in the spectral and chromatographic domains, they will not be identified as separate species, and the MCR result is simply one variable.

An example of the application of MCR, taken from a stability study of grape extracts, is shown in Figure 3.8: in a window of slightly more than two minutes, it is possible to identify three separate compounds, two of which have very similar spectra. These compounds can be quantified very easily using the elution profiles—note that for one of these compounds two major peaks are visible, indicating the presence of two distinct chemical species with the same chromophore.

Annotation can be achieved by comparison to pure standards, measured in the same sequence or (less certainly) by the usual combination of matching retention times and spectral characteristics. The good news is that spectra are extremely reproducible; the bad news is that spectra are usually quite

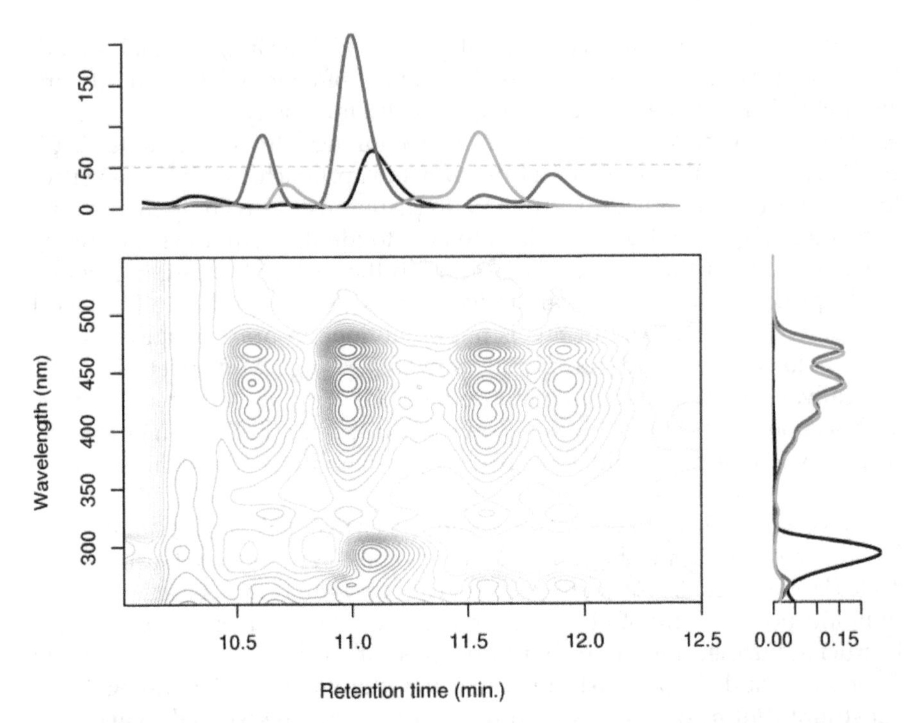

FIGURE 3.8
MCR deconvolution of LC-DA data into three components. The contour plot shows the raw UV-Vis data; at the top and the right, the MCR components are visualized. The top shows the estimated elution profiles for the three components; the right the spectra. Two of the spectra correspond to carotenoid species, the third to alpha-tocopherol.

similar. Measures like simple correlations are often unable to distinguish between the subtle differences: these can consist of shifts in peak maxima of only a few nanometers.

Advantages and Disadvantages

At first sight, DA data mainly have disadvantages: many compounds do not give rise to a spectrum at all, and the spectra that can be measured are not very informative. In addition, sensitivity is lower than a mass analyzer (though usually higher than NMR). And yet, the technique has some use, especially when analyzing compounds with characteristic colors, like pigments. One of the earlier drawbacks, the necessity for manual and laborious data analysis, is being remedied by novel software pipelines. The fact that DA data are often recorded "for free" when performing LC-MS analyses makes it an attractive additional layer of information.

References

Akoka, S., L. Barantin, and M. Trierweiler. 1999. "Concentration Measurement by Proton NMR Using the ERETIC Method." *Analytical Chemistry* 71 (13): 2554–2557.

Barding, G. A., Jr, R. Salditos, and C. K. Larive. 2012. "Quantitative NMR for Bioanalysis and Metabolomics." *Analytical and Bioanalytical Chemistry* 404 (4): 1165–1179.

Beckonert, O., H. C. Keun, T. M. D. Ebbels, J. Bundy, E. Holmes, J. C. Lindon, and J. K. Nicholson. 2007. "Metabolic Profiling, Metabolomic and Metabonomic Procedures for NMR Spectroscopy of Urine, Plasma, Serum and Tissue Extracts." *Nature Protocols* 2 (11): 2692–2703.

Benninghoven, A., F. G. Rudenauer, and H. W. Werner. 1987. *Secondary Ion Mass Spectrometry: Basic Concepts, Instrumental Aspects, Applications and Trends*, January. http://www.osti.gov/scitech/biblio/6092161.

Bothwell, J. H. F., and J. L. Griffin. 2011. "An Introduction to Biological Nuclear Magnetic Resonance Spectroscopy." *Biological Reviews of the Cambridge Philosophical Society* 86 (2): 493–510.

Burton, K. I., J. R. Everett, M. J. Newman, F. S. Pullen, D. S. Richards, and A. G. Swanson. 1997. "On-Line Liquid Chromatography Coupled with High Field NMR and Mass Spectrometry (LC-NMR-MS): A New Technique for Drug Metabolite Structure Elucidation." *Journal of Pharmaceutical and Biomedical Analysis* 15 (12): 1903–1912.

David Sparkman, O., Z. Penton, and F. G. Kitson. 2011. *Gas Chromatography and Mass Spectrometry: A Practical Guide*. Academic Press, Burlington, MA.

de Hoffmann, E., and V. Stroobant. 2007. *Mass Spectrometry: Principles and Applications*. New York: John Wiley & Sons.

Dettman, J. R., J. M. Goss, C. J. Ehrhardt, K. A. Scott, J. D. Bannan, and J. M. Robertson. 2015. "Forensic Differentiation of Bacillus Cereus Spores Grown Using Different Culture Media Using Raman Spectroscopy." *Analytical and Bioanalytical Chemistry* 407 (16): 4757–4766.

Dona, A. C., M. Kyriakides, F. Scott, E. A. Shephard, D. Varshavi, K. Veselkov, and J. R. Everett. 2016. "A Guide to the Identification of Metabolites in NMR-Based Metabonomics/Metabolomics Experiments." *Computational and Structural Biotechnology Journal* 14: 135–153.

Ellis, D. I., H. Muhamadali, S. A. Haughey, C. T. Elliott, and R. Goodacre. 2015. "Point-and-Shoot: Rapid Quantitative Detection Methods for on-Site Food Fraud Analysis—Moving out of the Laboratory and into the Food Supply Chain." *Analytical Methods* 7 (22): 9401–9414.

Emwas, A.-H. et al. 2015. "Standardizing the Experimental Conditions for Using Urine in NMR-Based Metabolomic Studies with a Particular Focus on Diagnostic Studies: A Review." *Metabolomics: Official Journal of the Metabolomic Society* 11 (4): 872–894.

Emwas, A.-H. M. 2015. "The Strengths and Weaknesses of NMR Spectroscopy and Mass Spectrometry with Particular Focus on Metabolomics Research." *Methods in Molecular Biology* 1277: 161–193.

Emwas, A.-H. et al. 2016. "Recommendations and Standardization of Biomarker Quantification Using NMR-Based Metabolomics with Particular Focus on Urinary Analysis." *Journal of Proteome Research.* doi:10.1021/acs.jproteome.5b00885.

Everett, J. R. 2015. "A New Paradigm for Known Metabolite Identification in Metabonomics/Metabolomics: Metabolite Identification Efficiency." *Computational and Structural Biotechnology Journal* 13: 131–144.

Fan, T. W.-M., and A. N. Lane. 2011. "NMR-Based Stable Isotope Resolved Metabolomics in Systems Biochemistry." *Journal of Biomolecular NMR* 49 (3–4): 267–280.

Fonville, J. M., A. D. Maher, M. Coen, E. Holmes, J. C. Lindon, and J. K. Nicholson. 2010. "Evaluation of Full-Resolution J-Resolved 1H NMR Projections of Biofluids for Metabonomics Information Retrieval and Biomarker Identification." *Analytical Chemistry* 82 (5): 1811–1821.

Garip, S., A. Cetin Gozen, and F. Severcan. 2009. "Use of Fourier Transform Infrared Spectroscopy for Rapid Comparative Analysis of Bacillus and Micrococcus Isolates." *Food Chemistry* 113 (4): 1301–1307.

Giraudeau, P., and L. Frydman. 2014. "Ultrafast 2D NMR: An Emerging Tool in Analytical Spectroscopy." *Annual Review of Analytical Chemistry* 7: 129–161.

de Graaf, R. A. 2013. *In Vivo NMR Spectroscopy: Principles and Techniques.* John Wiley & Sons, Chichester, UK.

Hands, J. R. et al. 2014. "Attenuated Total Reflection Fourier Transform Infrared (ATR-FTIR) Spectral Discrimination of Brain Tumour Severity from Serum Samples." *Journal of Biophotonics* 7 (3–4): 189–199.

Hart-Smith, G., and S. J. Blanksby. 2011. "Mass Analysis." In *Mass Spectrometry in Polymer Chemistry*, pp. 5–32. Wiley-VCH Verlag GmbH & Co. KGaA, Weinheim, Germany.

Hughes, C., M. D. Brown, N. W. Clarke, K. R. Flower, and P. Gardner. 2012. "Investigating Cellular Responses to Novel Chemotherapeutics in Renal Cell Carcinoma Using SR-FTIR Spectroscopy." *The Analyst* 137 (20): 4720–4726.

Jandrić, Z., S. A. Haughey, R. D. Frew, K. McComb, P. Galvin-King, C. T. Elliott, and A. Cannavan. 2015. "Discrimination of Honey of Different Floral Origins by a Combination of Various Chemical Parameters." *Food Chemistry* 189: 52–59.

de Juan, A., and R. Tauler. 2006. "MCR from 2000: Progress in Concepts and Applications." *Critical Reviews in Analytical Chemistry* 36: 163–176.

de Juan, A., and R. Tauler. 2007. "Factor Analysis of Hyphenated Chromatographic Data Exploration, Resolution and Quantification of Multicomponent Systems." *Journal of Chromatography A* 1158 (1–2): 184–195.

Keeler, J. 2010. *Understanding NMR Spectroscopy.* 2nd edition. John Wiley & Sons.

Klepárník, K. 2015. "Recent Advances in Combination of Capillary Electrophoresis with Mass Spectrometry: Methodology and Theory." *Electrophoresis* 36 (1): 159–178.

Kumar Bharti, S., and R. Roy. 2014. "Metabolite Identification in NMR-Based Metabolomics." *Current Metabolomics* 2 (3): 163–173.

Kwak, J. T., A. Kajdacsy-Balla, V. Macias, M. Walsh, S. Sinha, and R. Bhargava. 2015. "Improving Prediction of Prostate Cancer Recurrence Using Chemical Imaging." *Scientific Reports* 5: 8758.

Levitt, M. H. 2013. *Spin Dynamics: Basics of Nuclear Magnetic Resonance.* Chichester, UK: John Wiley & Sons.

Ludwig, C. et al. 2012. "Birmingham Metabolite Library: A Publicly Accessible Database of 1-D 1H and 2-D ^{1}H J-Resolved NMR Spectra of Authentic Metabolite Standards (BML-NMR)." *Metabolomics: Official Journal of the Metabolomic Society* 8 (1): 8–18.

Mckay, R. T. 2011. "How the 1D-NOESY Suppresses Solvent Signal in Metabonomics NMR Spectroscopy: An Examination of the Pulse Sequence Components and Evolution." *Concepts in Magnetic Resonance* 38A (5): 197–220.

McManus, L. L., F. Bonnier, G. A. Burke, B. J. Meenan, A. R. Boyd, and H. J. Byrne. 2012. "Assessment of an Osteoblast-like Cell Line as a Model for Human Primary Osteoblasts Using Raman Spectroscopy." *The Analyst* 137 (7): 1559–1569.

Mullen, K. M., I. H. M. van Stokkum, and V. V. Mihaleva. 2009. "Global Analysis of Multiple Gas Chromatography-Mass Spectrometry (GC/MS) Data Sets: A Method for Resolution of Co-Eluting Components with Comparison to MCR-ALS." *Chemometrics and Intelligent Laboratory Systems* 95: 150–163.

Nallala, J., G. R. Lloyd, N. Shepherd, and N. Stone. 2016. "High-Resolution FTIR Imaging of Colon Tissues for Elucidation of Individual Cellular and Histopathological Features." *The Analyst* 141 (2): 630–639.

Northen, T. R., O. Yanes, M. T. Northen, D. Marrinucci, W. Uritboonthai, J. Apon, S. L. Golledge, A. Nordström, and G. Siuzdak. 2007. "Clathrate Nanostructures for Mass Spectrometry." *Nature* 449 (7165): 1033–1036.

de Siqueira Oliveira, F. S., H. E. Giana, and L. Silveira. 2012. "Discrimination of Selected Species of Pathogenic Bacteria Using near-Infrared Raman Spectroscopy and Principal Components Analysis." *Journal of Biomedical Optics* 17 (10): 107004–107004.

Pascut, F. C., S. Kalra, V. George, N. Welch, C. Denning, and I. Notingher. 2013. "Non-Invasive Label-Free Monitoring the Cardiac Differentiation of Human Embryonic Stem Cells *In-Vitro* by Raman Spectroscopy." *Biochimica et Biophysica Acta* 1830 (6): 3517–3524.

Passarelli, M. K., and N. Winograd. 2011. "Lipid Imaging with Time-of-Flight Secondary Ion Mass Spectrometry (ToF-SIMS)." *Biochimica et Biophysica Acta* 1811 (11): 976–990.

Peterson, D. S. 2007. "Matrix-Free Methods for Laser Desorption/Ionization Mass Spectrometry." *Mass Spectrometry Reviews* 26 (1): 19–34.

Poole, C. F. 2012. *Gas Chromatography*. Elsevier, Burlington, NJ.

Psychogios, N. et al. 2011. "The Human Serum Metabolome." *PloS One* 6 (2): e16957.

Ramautar, R., G. W. Somsen, and G. J. de Jong. 2015. "CE-MS for Metabolomics: Developments and Applications in the Period 2012–2014." *Electrophoresis*. doi:10.1002/elps.201400388/full.

Rockwood, A. L. 2004. "The Expanding Role of Mass Spectrometry in Biotechnology.." *Clinical Chemistry* 50 (6): 1108–1109.

Salek, R., K.-K. Cheng, and J. Griffin. 2011. "The Study of Mammalian Metabolism through NMR-Based Metabolomics." *Methods in Enzymology* 500: 337–351.

Sindona, G., and D. Taver. 2012. "Modern Methodologies to Assess the Olive Oil Quality." In *Olive Germplasm—The Olive Cultivation, Table Olive and Olive Oil Industry in Italy*, edited by Innocenzo Muzzalupo. InTech, London, UK.

Siuzdak, G. 2004. "An Introduction to Mass Spectrometry Ionization: An Excerpt from The Expanding Role of Mass Spectrometry in Biotechnology, 2nd ed.; San Diego, CA: MCC Press, 2005." *Journal of the Association for Laboratory Automation* 9 (2): 50–63.

Stevens, O., I. E. Iping Petterson, J. C. C. Day, and N. Stone. 2016. "Developing Fibre Optic Raman Probes for Applications in Clinical Spectroscopy." *Chemical Society Reviews* 45 (7): 1919–1934.

Tal, A., and L. Frydman. 2010. "Single-Scan Multidimensional Magnetic Resonance." *Progress in Nuclear Magnetic Resonance Spectroscopy* 57 (3): 241–292.

Viant, M. R., C. Ludwig, and U. L. Günther. 2007. "Chapter 2:1D and 2D NMR Spectroscopy: From Metabolic Fingerprinting to Profiling." In *Metabolomics, Metabonomics and Metabolite Profiling*, edited by William J Griffiths, pp. 44–70. Royal Society of Chemistry, London, UK.

Wehrens, R., E. Carvalho, and P. D. Fraser. 2014. "Metabolite Profiling in LC–DAD Using Multivariate Curve Resolution: The Alsace Package for R." *Metabolomics: Official Journal of the Metabolomic Society* 11 (1): 143–154.

Wehrens, R., E. Carvalho, D. Masuero, A. de Juan, and S. Martens. 2013. "High-Throughput Carotenoid Profiling Using Multivariate Curve Resolution." *Analytical and Bioanalytical Chemistry* 405 (15): 5075–5086.

"Wiley Series on Mass Spectrometry." 2013. *Cluster Secondary Ion Mass Spectrometry*, pp. 349–50. John Wiley & Sons.

Yanes, O., R. Tautenhahn, G. J. Patti, and G. Siuzdak. 2011. "Expanding Coverage of the Metabolome for Global Metabolite Profiling." *Analytical Chemistry* 83 (6): 2152–2161.

Ziessow, D. 1987. "Book Review: Principles of Nuclear Magnetic Resonance in One and Two Dimensions (International Series of Monographs on Chemistry 14). R. E. Ernst, G. Bodenhausen, and A. Wokaun." *Angewandte Chemie* 26 (11): 1192–1195.

Zwielly, A., J. Gopas, G. Brkic, and S. Mordechai. 2009. "Discrimination between Drug-Resistant and Non-Resistant Human Melanoma Cell Lines by FTIR Spectroscopy." *The Analyst* 134 (2): 294–300.

4

Mass Spectrometry Data Processing

Steffen Neumann, Oscar Yanes, Roland Mumm and Pietro Franceschi

CONTENTS

Introduction 73
From Vendor Formats to mzML........ 74
From Raw Data to Features 77
Alignment: From Features to Data Matrices........ 80
Parameter Optimization and Performance Metrics 83
From Data Matrices to Annotated Data Matrices........ 84
From (Annotated) Mass Spectra to (Tentatively) Identified Metabolites........ 88
Mass Spectral Databases Containing Pure Standards 89
Identification with *In Silico* Tools........ 91
Comparison with Theoretical Mass Spectra........ 91
Approximate Matches........ 92
Prediction of Retention Times and Retention Indices........ 93
Incorporating Network Information 94
References 94

Introduction

This chapter focuses on the mass spectrometry data processing workflow. The first step consists of processing the raw MS data, leading to a feature matrix amenable for statistical analysis. In the second stage, features of interest are identified, i.e., annotated with names of metabolites, or compound classes. Finally, different network-based approaches may be used to infer molecular pathways and components via integrative analysis of metabolite features. More often than not, one then is going back to data from an earlier stage (sometimes even to the raw data) to double-check key elements in the data and resolve ambiguities. This complicated process asks for flexible and interactive software tools. Instrument manufacturers typically provide generic software packages with their spectrometers, often containing intuitive graphical user interfaces. This allows also experimental scientists with little or no bioinformatic background to perform data processing and quality control of the measurements. In addition, the metabolomics community has

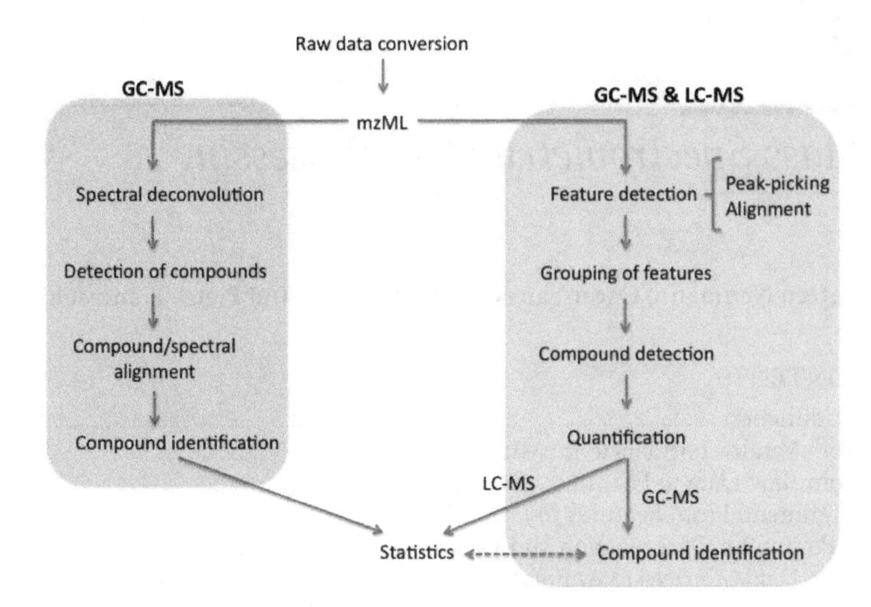

FIGURE 4.1
Flowchart for data processing in LC-MS and GC-MS mass spectrometry.

developed open source software capable of processing large-scale data commonly occurring in metabolomics studies, often even before such capabilities are available in vendor software.

In this chapter, we will introduce the general concepts of the individual steps and give examples for typical workflows. A general overview is shown in Figure 4.1.

From Vendor Formats to mzML

Mass spectrometry raw data as stored and managed by the vendor software usually is not directly usable by third-party software. Each vendor has developed its own proprietary file format, often binary, and not compatible with others. In general, it requires a vendor-supplied software library (typically a Windows™ Dynamic-Link Library, DLL) to open such files and to extract the actual data for further processing. Because each of these DLLs has its own Application Programming Interface (API), it would be impractical especially for academic software, to be programmed to access all individual MS vendor formats. Instead, vendor-neutral and open data formats have emerged, such as netCDF, mzData, mzXML and mzML, which can be imported by most third party software packages. As an added bonus, reading these open formats is not tied to Windows™ and they can be accessed within different operating systems, like macOS or

Linux. The early netCDF format is limited to simple GC-MS or LC-MS data; it cannot contain a mix of MS and MS/MS data, and has almost no metadata about the instrument and spectrum settings, such as collision energy or precursor information. To overcome these limitations, mzXML and mzData formats were introduced. Their individual strengths have been merged into the currently recommended mzML file format, which not only can store spectral information but also chromatograms from, e.g., SRM-style experiments.

Usually, mzML files can be produced by the following two routes. The first approach is to use vendor provided converters and exporters. Sometimes they are easy to find as part of the main data analysis software (e.g., File→Export as…→mzML); sometimes they consist of dedicated command line tools, which usually have the advantage of being capable of doing batch conversion for many files at once. The second approach to perform file conversion is to rely on community projects, which have developed independent converters. The Proteowizard (http://proteowizard.sourceforge.net/) team have obtained permission to redistribute most of the required vendor DLLs. The Proteowizard msconvert tool can convert from most of the LC-MS data, as well as some of the GC-MS formats, to mzML. The msconvert tool also allows a set of operations (like simple processing or filtering) on the input data. A graphical user interface (msconvertGUI) is also available.

Note that in many cases the proprietary formats contain additional (and sometimes important) data that are not necessarily required for metabolomics data analysis. Examples are temperature (which might come handy to "debug" mass calibration issues due to malfunctioning laboratory air conditioning), pressure or details of the chromatographic method. This additional data is typically not included in the converted mzML data. As an example, Box 4.1 shows the resulting mzML for one spectrum from a Thermo Orbitrap XL instrument after conversion.

BOX 4.1 METADATA FROM AN EXAMPLE mzML FILE FROM A THERMO ORBITRAP XL INSTRUMENT AFTER CONVERSION

Excerpt from an mzML file, with some of the (meta)data highlighted in boldface. This is an MS/MS spectrum with 90 eV collision energy of the precursor with 212.12 m/z, eluting at 0.49 minutes. The actual spectrum is base64 encoded in a binary data array, and not human-readable. The encoding has the benefit of fully preserving the numerical accuracy. For readability, non-informative XML attributes like cvRef="MS", or unitCvRef="UO" and value="" have been omitted.


```
<spectrum index="53" id="controllerType=0
  controllerNumber=1 scan=54" defaultArrayLength="26">
 <cvParam accession="MS:1000580" name="MSn spectrum" />
```


(*Continued*)

BOX 4.1 (Continued) METADATA FROM AN EXAMPLE mzML FILE FROM A THERMO ORBITRAP XL INSTRUMENT AFTER CONVERSION

```
<cvParam accession="MS:1000511" name="ms level" value="2" />
<cvParam accession="MS:1000130" name="positive scan" />
<cvParam accession="MS:1000285" name="total ion current"
  value="99364.1171875"/>
<scanList count="1">
<cvParam accession="MS:1000016" name="scan start time"
  value="0.492676666667" unitAccession="UO:0000031"
  unitName="minute"/>
<cvParam accession="MS:1000512" name="filter string"
  value="FTMS + p ESI Full ms2 212.12@hcd90.00
  [50.00-250.00]"/>
</scan>
</scanList>
<precursorList count="1">
<precursor spectrumRef="controllerType=0
  controllerNumber=1 scan=51">
<isolationWindow>
<cvParam accession="MS:1000827" name="isolation window
  target m/z" value="212.12" unitAccession="MS:1000040"
  unitName="m/z"/>
</isolationWindow>
<selectedIonList count="1">
<selectedIon>
<cvParam accession="MS:1000744" name="selected ion m/z"
  value="212.12" unitAccession="MS:1000040"
  unitName="m/z"/>
<cvParam accession="MS:1000042" name="peak intensity"
  value="4533.17236328125" unitAccession="MS:1000131"
  unitName="number of counts"/>
</selectedIon>
</selectedIonList>
<activation>
<cvParam accession="MS:1000422" name="high-energy
  collision-induced dissociation" value=""/>
<cvParam accession="MS:1000045" name="collision energy"
  value="90.0" unitAccession="UO:0000266"
  unitName="electronvolt"/>
</activation>
</precursor>
</precursorList>
<binaryDataArrayList count="2">
```

(*Continued*)

BOX 4.1 (Continued) METADATA FROM AN EXAMPLE mzML FILE FROM A THERMO ORBITRAP XL INSTRUMENT AFTER CONVERSION

```
  &lt;binaryDataArray encodedLength="280"&gt;
  <cvParam accession="MS:1000523" name="64-bit float"
    value=""/>
  <cvParam accession="MS:1000514" name="m/z array"
    unitAccession="MS:1000040" unitName="m/z"/>
  &lt;binary&gt;AAAAg...AcBSkAAAAC&lt;/binary&gt;
  &lt;/binaryDataArray&gt;
  &lt;binaryDataArray encodedLength="280"&gt;
  <cvParam accession="MS:1000523" name="64-bit float"
    value=""/>
  <cvParam accession="MS:1000515" name="intensity array"
    unitAccession="MS:1000131" unitName="number of counts"/>
  &lt;binary&gt;AAAAQGWYo...0AAAAAAH9CnQAAAAKAb&lt;/binary&gt;
  &lt;/binaryDataArray&gt;
  &lt;/binaryDataArrayList&gt;
&lt;/spectrum&gt;
```

From Raw Data to Features

The raw MS files contain the list of spectra acquired during the run, and each spectrum contains many (m/z, intensity) pairs. In direct infusion MS, the sample is introduced into the MS instrument directly and one or more spectra are acquired—the raw data for one sample consists of repeated spectral measurements; see Box 4.2. In chromatography-separated data (i.e., CE-MS, LC-MS and GC-MS) the metabolites present in the sample are separated along the chromatographic gradient, resulting in different retention times. Note that in general, one compound appears in several consecutive spectra because a chromatographic peak is covered by several scans of the MS. Additionally, in complex samples more than one compound can co-elute at the same retention time, giving rise to a mixed spectrum. In practice, depending on the mass range, scan rate and mass resolution of the mass spectrometer, millions of raw data points could be recorded.

To reduce this complexity, several feature detection algorithms have been developed. We refer to all raw data points that originate from one particular

BOX 4.2 DIRECT INFUSION MASS SPECTROMETRY DATA

In the field of metabolite fingerprinting, Direct Infusion Mass Spectrometry (DIMS) and Flow Injection Mass Spectrometry (FIMS) can be used to complement chromatography-based MS techniques. In both cases, the sample is directly injected in to the spectrometer without an upstream chromatographic separation, thus allowing the high throughput analysis of up to several thousands of samples (Fuhrer and Zamboni 2015; Kirwan et al. 2014). From the analytical point of view, matrix effects constitute the major drawback of these acquisition types, because without separation, all the analytes reach the ion source at the same time and the ion population can be only a partial representation of the chemical complexity of the sample. This limitation, however, is less severe for discovery and hypothesis generation studies.

Without separation, all the isobaric and isomeric metabolites will be detected as ions with the same nominal mass, but if isobaric molecules can be still discriminated by using a mass spectrometer with an adequate resolution due to their *different exact mass* (e.g., N_2, CO, C_2H_4), the separation of isomeric compounds (like Leucine and Isoleucine) cannot be done on the base is of an m/z value. Relying on mass spectrometry, the only possibility to solve this ambiguity is to use ion-mobility devices which could do the trick if the ionic structures of the isomers are different. To at least mitigate these drawbacks, high resolution spectrometers are preferred in DIMS and FIMS applications.

Due to the absence of the chromatographic dimension, DIMS data are less complex and, in the basic setting, each feature is a represented by a tuple (m/z, intensity). The data can become again three dimensional in case of mass spectrometers equipped with ion mobility. As in the case of LC/GC-MS data, feature detection is normally performed sample wise and the different lists of features are then "clustered/grouped" together to guarantee that the same feature has the same meaning across the samples. This process of clustering is facilitated by the use of detectors with high mass accuracy. Peak detection (i.e., the identification of the peaks in the m/z dimension) is often performed on the "average" mass spectrum of each sample, and it is a critical step. Due to the presence of co-eluting compounds, the algorithm has to be able to manage rich spectra with high dynamic range. To optimize such a step it is often necessary to bypass the vendor algorithm and directly access to the raw profile data.

DIMS is preferentially used in high throughput fingerprinting, so the use of QC strategies which allow the monitoring of the analytical run is extremely important. In this respect it is possible to implement many of the strategies already discussed. The same consideration is true when considering the process of annotation of the list of aligned features.

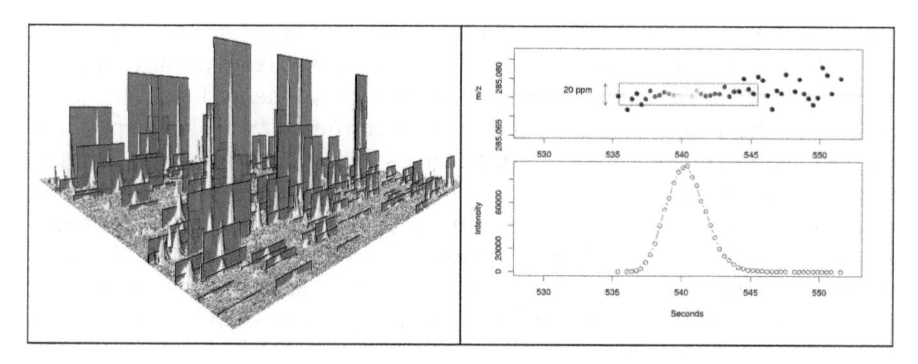

FIGURE 4.2
Left: A three dimensional view of MS raw data in grey and the superimposed bounding boxes of detected features. Right: A closeup on one of the peaks.

ion species as a *"feature."* A *feature* is characterized by a retention time, an m/z value and an intensity (RT, mz, intensity). Figure 4.2 shows a set of features as red boxes, superimposed onto the underlying raw data in grey. Feature detection algorithms are sometimes also referred to as "peak pickers." The three-dimensional structure of features may be extended by an additional dimension if two-dimensional chromatographic systems are used. For example, in comprehensive GC × GC analysis, retention times of the first and second column are obtained, or in ion-mobility MS a spectrum is also characterized by the "drift time."

In the following sections, we will describe some of the general ideas behind the processing steps, which are common to many different commercial and open source solutions. The data that are used (throughout the book) are available under the accession number MTBLS18 from EMBL-EBI MetaboLights repository (Haug et al. 2012). More information on this data set can be found in the Appendix. Feature detection is usually performed on individual input files (samples) and can easily be parallelized. The general question, to detect a signal among potentially noisy data, can be tackled with concepts commonly used in signal processing applications. One of the most intuitive feature detection algorithms is the filter approach, where the raw data are processed through filter functions that resemble the shape of the feature to be detected: where the signal will show a good similarity with the shape of the filter, a feature will be detected. Examples of typical filters are the Mexican hat filter function, or more general wavelet transformations. Due to the large number of raw data points, the algorithms often select regions of interest from the raw data with fast algorithms first, before the (computationally expensive) filter functions are applied. The last step in the feature detection is the calculation of the m/z, RT and intensity characterizing each feature. The actual m/z can be derived, e.g., as an intensity-weighted average, the retention time is the position of the maximum intensity (the so-called peak apex) and the intensity could be calculated as maximum count (peak height) or the area under the curve of the feature (peak area).

One of the most important characteristics of a feature detection algorithm is the ability to distinguish features from the noise. Typical noise in mass spectrometry consists of *electronic* and *chemical* components. Electronic noise is an inherent characteristic of the mass spectrometer and is—depending on the mass trace—relatively constant across the chromatography. Chemical noise originates mostly from the chromatographic system, and is generated by phenomena such as column bleeding or by the presence of contaminants within the mobile phase. Chemical noise is typically dependent on the mass trace and contributes to a noisy baseline of a chromatogram. Depending on the software package, several quite different approaches can be used for feature detection (Busch 2002).

Alignment: From Features to Data Matrices

In microarray data analysis it is relatively straightforward to compare the expression of some genes in one sample against the expression in a second one, because the RNA is hybridized to a particular "known" probe sequence. In contrast, due to the nature of hyphenated MS data, it is normal to find slight analytical differences between the instrumental measurements resulting in small deviations of the m/z values or slight shifts in the retention times. As a consequence, the "same" feature will be characterized by slightly different m/z and RT values in different samples. For this reason, in order to be sure of comparing the same feature across the different samples, it is necessary to perform feature grouping (including optionally a retention time correction step) to go from individual feature lists to a rectangular data matrix as shown in Figure 4.3. This matrix is required for subsequent

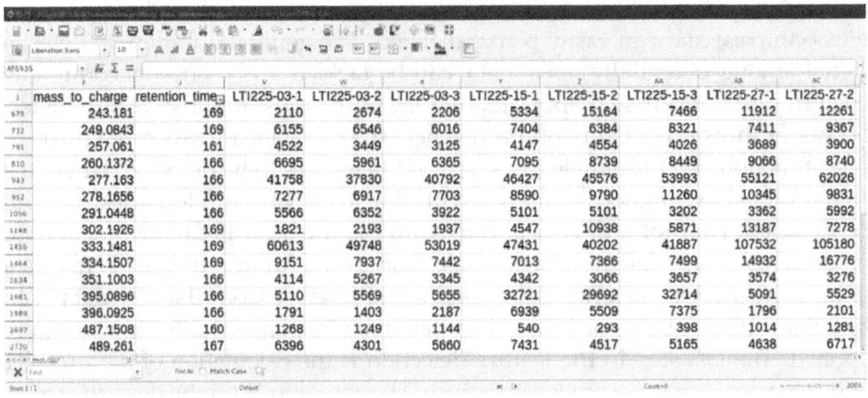

mass_to_charge	retention_time	LTI225-03-1	LTI225-03-2	LTI225-03-3	LTI225-15-1	LTI225-15-2	LTI225-15-3	LTI225-27-1	LTI225-27-2
243.181	169	2110	2674	2206	5334	15164	7466	11912	12261
249.0843	169	6155	6546	6016	7404	6384	8321	7411	9367
257.061	161	4522	3449	3125	4147	4554	4026	3689	3900
260.1372	166	6695	5961	6365	7095	8739	8449	9066	8740
277.163	166	41758	37830	40792	46427	45576	53993	55121	62026
278.1656	166	7277	6917	7703	8590	9790	11260	10345	9831
291.0448	166	5566	6352	3922	5101	5101	3202	3362	5992
302.1926	169	1821	2193	1937	4547	10938	5871	13187	7278
333.1481	169	60613	49748	53019	47431	40202	41887	107532	105180
334.1507	169	9151	7937	7442	7013	7366	7499	14932	16776
351.1003	166	4114	5267	3345	4342	3066	3657	3574	3276
395.0896	166	5110	5569	5655	32721	29692	32714	5091	5529
396.0925	166	1791	1403	2187	6939	5509	7375	1796	2101
487.1508	160	1268	1249	1144	540	293	398	1014	1281
489.261	167	6396	4301	5660	7431	4517	5165	4638	6717

FIGURE 4.3
After grouping, the features can be written as a matrix. The screenshot shows an excerpt of the feature matrix for MTBLS18.

statistical analysis (Chapter 6), or to perform *ab initio* reconstruction of metabolic networks as shown in Chapter 8. A matrix like this can also be saved in different formats, such as mzTab-M (Hoffmann et al. 2019) or mwTab for the submission to the NIH Metabolomics Workbench (Sud et al. 2016), or a Metabolite Assignment File (MAF) for submission to MetaboLights (Haug et al. 2012; Salek et al. 2013); see also Chapter 10.

To determine which features in a sample should be linked to corresponding features in other samples, several approaches have been proposed, implemented and compared (Eva Lange et al. 2008). Most of the approaches are variations of some sort of clustering based on m/z and retention time. This is implemented in e.g., OpenMS (Lange et al. 2005) as FeatureLinker using the "unlabeled_qt" algorithm (Weisser et al. 2013). A different approach is implemented in XCMS (Smith et al. 2006), which uses extracted ion chromatogram (EIC): all features from all files are mapped to m/z slices. Then it is possible to calculate the density of features along the retention time and group all features which have "the same" retention time. Here "same" retention time depends on the stability of the chromatography. The important parameters are the width of the EIC-like slices and the amount of smoothing in the density estimation. Once the groups of features are collected, we can perform filtering. The most important parameter is the absolute or relative minimum occurrence of features across the samples, to avoid groups with just one or very few features. If the experiment contains different sample classes, it might be that for instance a metabolite is present in the wild-type, while it is absent in the mutant. So the minimum number of occurrences should be calculated on a per-class level.

As mentioned above, the retention time for the same metabolite can vary across different runs. To correct for this, one can try to align or warp all measurements, so that the retention deviations are minimized. Several approaches are possible. The first approach operates directly on the raw spectra: given two sequences of spectra, the pairs of retention times that maximize the similarity of the corresponding spectra are found. Then, a global alignment is created from the individual pairwise alignments. The second class of algorithms requires the extracted feature lists and uses so-called landmark peaks, i.e., features that can be reliably found across (almost) all samples. Then a smoothed curve can be fitted to adjust the retention time in each sample to minimize this retention deviation. A third approach uses optimization to find a deformation (called a warping) of the time axis that leads to maximal overlap of EICs, without first defining spectra or landmark peaks. An example of the latter is Parametric Time Warping (Eilers 2004), implemented in the R package ptw (Bloemberg et al. 2010; Wehrens et al. 2015). This can be used to align both EICs and peak lists.

The main difficulty in alignment is to prevent overfitting: one should allow enough freedom to get the right signals aligned but not so much that also non-related signals are apparently connected. In the XCMS setting, a recommended strategy with rather large expected retention time deviations is to perform a grouping with quite relaxed retention time windows to obtain enough landmark features for the correction, or alternatively start right away with the raw spectra alignment. In the next iteration, a second grouping with much stricter retention time requirements can lead to an improved feature matrix. If retention time differences are small enough, say, just a few seconds in case of short (e.g., 20 minutes) gradients on modern UHPLC systems, then an actual correction step might not be necessary. The alignment plots can still provide diagnostics about the retention time stability of all measurements; see Figure 4.4. Such plots can not only expose outlier runs for "misbehaving" analytical equipment, but also can give evidence for overcorrection, where the retention time correction is far from smooth.

One common problem after the grouping step is that we need to deal with missing values (or NA) in a number of samples. NAs can cause issues with the downstream statistical analysis, and in many cases we need to somehow obtain a suitable numeric value instead of NA, as already discussed in Chapter 2. Missing values can happen due to several reasons: for example, features can be below the detection limit (Limit of Detection; LOD) or otherwise not reliably detected in a sample. If the feature detection has missed a peak in one of the samples, there is no intensity value in the matrix, but the raw

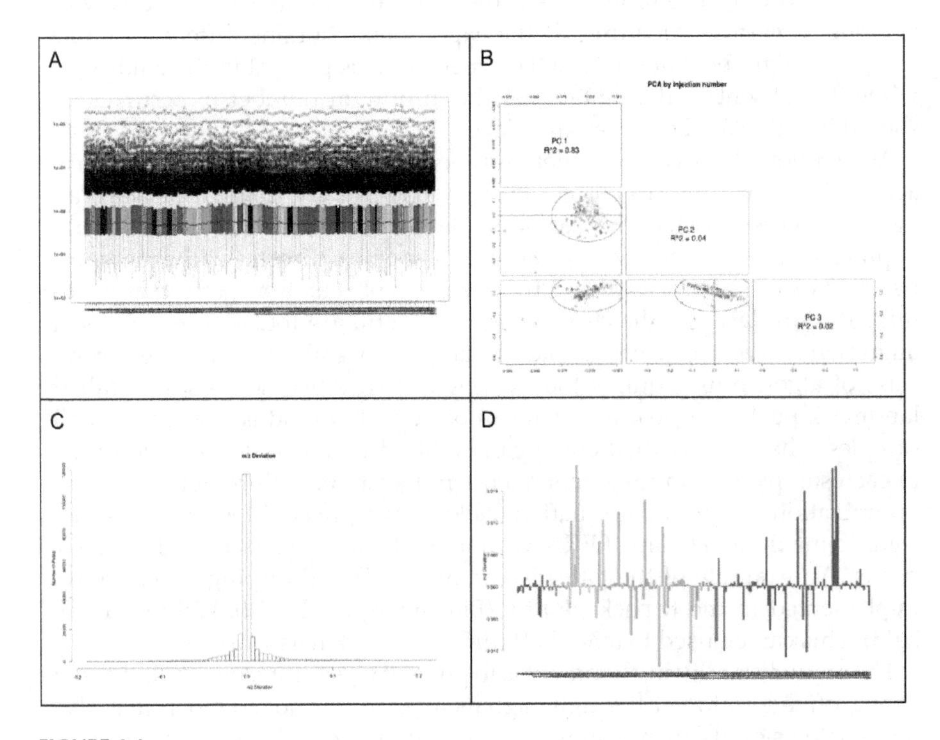

FIGURE 4.4
Several metrics can be used to calculate and visualize the performance of the LC-MS data.

data might still have recorded intensities. So one approach is to use a consensus m/z and retention time window, based on the values from other samples, and blindly integrate the raw intensities in that window. In XCMS, this is performed by the fill Chrom Peaks() function. Another possible cause for NA values, even in cases where a feature was detected across all samples, is that the grouping step has "split" a feature, placing the values for the same compound in different samples in different rows. The latter can happen if the m/z or RT values differ too much between samples to be collected into the same row of the matrix. Other (raw data-independent) possibilities for missing value imputation are methods which "borrow" information about the missing numeric value from neighboring samples (Stacklies et al. 2007), but these methods do rely on strong assumptions which are often difficult to verify.

Parameter Optimization and Performance Metrics

So far, we have applied several processing steps in the faith that everything went well in previous steps, and used settings and parameters that were hopefully chosen well. We now retrace our steps and add diagnostic visualizations and quality metrics, discussing how the choices of parameter values could be optimized for a particular analytical setting. Several of the concepts mentioned here have already been touched upon in Chapter 2, in a quite general way. Here we focus on those elements that are pertinent to mass spectrometry.

A global overview of the data can be seen in the boxplot in Figure 4.4a, which shows the intensity distribution of the detected features for each sample. If the samples are displayed following the order of acquisition, this boxplot can reveal run order or batch artifacts, like a decrease in intensity over time. Run-order effects can also be detected in PCA plots if the samples are colored in e.g., a rainbow color gradient (following the injection order) as shown in Figure 4.4b. Also variation between batches, defined as a set of samples that have been extracted in one go and have been analyzed in one uninterrupted sequence, can often be detected through a PCA by labeling the different extraction or analysis batches which are commonly required in large scale metabolomics studies. This illustrates the necessity and importance to store as many meta information of the samples as possible.

In case an accurate-mass spectrometer has been used, it is also important to determine the mass accuracy in the measurements. If internal standards are present, one can simply calculate the m/z error for these known compounds. Another option is to determine the relative mass deviations between the samples. One can imagine a matrix similar to the one in Figure 4.3, where instead of the intensities, each cell contains the deviation of the m/z value of a feature in one sample to the median within that row. Figure 4.4c shows the histogram of deviations for MTBLS18. It is also possible to calculate the median mass deviation for each sample, as shown in Figure 4.4d. The observed relative mass deviations can depend on the prior data processing steps: small m/z tolerances in the grouping step can be expected to also lead to lower observed mass deviations, because features from samples where the instrument was very de-calibrated would be in a different row of the matrix, and hence the large mass difference to other samples would go unnoticed.

Finding "the best" choice of the algorithm parameters is not easy for most software, and often requires experience (or gut feeling) and some trial-and-error, including checking of the obtained results. Actually, even the "best" settings for an analytical setup and sample type are only a compromise, as the features of different compound classes do have different elution characteristics, but none of the common algorithms take these explicitly into account.

Several approaches have been described to achieve less subjective parameter choices. Eliasson et al. (2012) described a strategy for optimizing LC-MS data processing, where a dilution series with several different concentrations of a representative sample (e.g., the pooled QC sample) is measured. The assumption is that the parameters are chosen well if the intensities of as many as possible of the resulting features follow the dilution series. Later, the R package IPO (Isotopologue Parameter Optimisation) was developed (Libiseller et al. 2015), which uses a similar approach, but instead of requiring a dilution series, IPO aims to maximize the number of detected isotopes in the obtained list of features (consequently, it will not work with ^{13}C labeled samples). Also, don't assume that the optimization can magically salvage poor data, but it might help finding good settings for your analytical setup.

From Data Matrices to Annotated Data Matrices

Up to here, we treated each feature individually and equally, which leads to a data matrix like Figure 4.3 that can be directly subjected to different types of univariate and multivariate statistical analysis. But most features are redundant both in LC-MS and GC-MS data sets, generating multitude of features for a single metabolite due to naturally occurring isotopes, adducts, cluster ions, multiply charged ions and fragment ions, and of course combinations of these. The extent of the redundancy may vary between metabolites and analytical platforms, depending on, e.g., metabolite abundance, chromatographic mobile phase, the ionization mode, other acquisition parameters and of course the chemical nature of the metabolite. The actual number of likely ion species as combinations of adducts, cluster ions and in source fragments depends on the analytical setup. The most influential parameters are the geometry and voltages in the ion source or the collision gas (like nitrogen or argon), the sample preparation and ionization-enhancing additives like formic acid, ammonium acetate, acetic acid or ammonium fluoride (Yanes et al. 2011). This results in an unequal representation of metabolites regarding the number of features (rows) in the data matrix. In electrospray ionization (ESI), there are between a few and up to tens of features per metabolite (or chemical entity). This number is in general higher in GC-MS due to extensive fragmentation of molecular ions by hard electron impact ionization (EI), leading to large and complex data sets with many features per metabolite.

In general, the statistical analysis and the interpretation of the results will be much easier when the number of columns in the data matrix is limited, and when the correlations between the columns are not very high (see Chapter 6). Therefore, if it is possible to aggregate the individual features into groups corresponding to metabolites, one should definitely do so. In the following, we show some approaches how to group features into compound spectra, and annotate the ion species in order to reduce redundancy and facilitate the identification of metabolites. Reconstructing GC-MS profile data into identified and quantified metabolites across multiple samples can be tackled using multiple computational approaches, like peak extraction by curve resolution and spectral deconvolution techniques as shown in Box 4.3.

The first question is, which ions are coming from the same metabolite? The first hint is that all ions derived from one metabolite have the same retention time. However, due to uncertainties in the feature detection,[1] in practice a certain retention time window has to be applied. A second hint for the common origin of ions is the chromatographic peak shape. While above we said that for example the metabolite "creatinine" has a retention time in a chromatographic run, this is just the mean of the retention time distribution of all individual creatinine molecules leaving the chromatographic column, related to a particular elution profile. Thus, a metabolite has a chromatographic peak at a certain retention time, with a certain width and a peak shape as shown in Figure 4.2 (lower right). Often, that shape is close to a gaussian curve, but depending on the metabolite's physico-chemical properties and the interaction with the analytical column, the shape might be asymmetric and exhibit so-called fronting or tailing. It requires a high-enough scan rate in relation to the expected peak widths to resolve the shape. In these cases, other distributions could be used to describe the peak shapes. More than 90 (!) different mathematical functions (Di Marco and Bombi 2001) have been suggested over the years to model peak shapes.

It is only in the mass spectrometers' ion source that the molecules are turned into one or more ion species before they hit the detector. Under the assumption that the ratios of their intensities are constant, the correlation of the raw intensities across the retention time can give a hint whether two or more ion species originate from the same metabolite.[2] Under some assumptions, the ion counts can even be modeled via probability distributions

[1] Especially for noisy data, it can be challenging to determine the apex and hence the exact RT of a feature, and get exactly the same RT for e.g., the different isotopes. Shifts of a few spectra to the left or to the right are possible.

[2] There are some practical limitations where the assumption does not hold true. First, very high abundance metabolites can cause the detector to saturate. Secondly, very low abundances can cause the intensities to drop below the limit of detection.

BOX 4.3 PEAK EXTRACTION BY CURVE RESOLUTION AND SPECTRAL DECONVOLUTION TECHNIQUES

While peak picking approaches are based on extracting individual ion features, other techniques focus on the compound as the analysis entity. Particularly in GC-MS, co-elution and in-source fragmentation of compounds cause the resulting raw mass spectra to consist of mass peaks from all of the co-eluting metabolites. Co-eluting compounds are quantified and identified based on a multivariate deconvolution process that extracts and constructs pure compound spectra from complex raw data. For each extracted ion chromatogram (EIC) that results from two or more compounds, deconvolution calculates the contribution of each compound to the EIC. Although deconvolution has been developed for some time (Biller and Biemann 1974; Colby 1992; Dromey et al. 1976), it became especially popular with the launch of the software package AMDIS (Automated Mass spectrometry Deconvolution and Identification System) (Stein 1999/8; "AMDIS" n.d.). Deconvolution using AMDIS typically consists of a number of steps: (i) noise analysis, (ii) compound determination and (iii) spectral deconvolution (Du and Zeisel 2013; Stein 1999/8). AMDIS makes use of the NIST library to identify compounds, but it does not include spectral alignment. Other tools like BinBase (Kind et al. 2009; Skogerson et al. 2011) use the spectral deconvolution provided by a proprietary algorithm in the commercial software ChromaTOF (LECO Corporation) to align compounds across samples, and provides compound quantification and identification based on self-constructed libraries. More representative tools using deconvolution include eRah (Domingo-Almenara et al. 2016), MetaboliteDetector (Hiller et al. 2009), AnalyzerPro (SpectralWorks ["AnalyzerPro | SpectralWorks" n.d.]), TNO-DECO (Jellema et al. 2010) and ADAP-GC (Ni et al. 2016). Although many of the deconvolution tools use similar approaches, for certain steps like AMDIS other strategies also are implemented. For detailed comparative analyses of the different approaches, the reader is referred to some previously published articles, e.g., Du and Zeisel 2013; Lu et al. 2008/3.

(Ipsen et al. 2010). In practice, the assumptions do not always hold, and e.g., the gain control to improve dynamic range in modern mass spectrometers can thwart the modeling, and under detector saturation even the simple peak correlation is lost.

Similarly, the intensity of ions originating from the same compound correlates not only across the neighboring scans within an LC-MS run, but also across the different samples measured. So if there is a large enough number of samples, this correlation can also be calculated for pairs of detected features within a given retention time window across all the samples. The number of pairwise correlations to be calculated per metabolite can range from a few for soft-ionization techniques typically used in LC-MS, to more than 100 in hard-ionization techniques as used in GC-MS. Feature pairs with a correlation above a user-defined threshold and significance are then considered to originate from the same metabolite. Again, watch out for external effects that could influence the intensity ratio across samples. In the case of labeling experiments designed to follow the metabolite fluxes, for example, isotope ratios will change in the different samples by design. Another obvious example is the ratio between the different adduct types $[M + H]^+$ versus $[M + Na]^+$ in a study on the influence of *salt* stress on plants at different concentrations of NaCl. In this case it is indeed impossible to assume that ratios of adduct intensities are constant. However, the peak shape correlation *within* a sample described above is not influenced by these experimental setups.

A final hint which features belong to the same metabolite can be very specific mass differences, such as ~1.003 for isotope peaks (Tikunov et al. 2012), or the m/z distance of 21.982 between the adduct ions $[M + H]^+$ and $[M + Na]^+$. Especially the presence of several isotope peaks is easily recognizable and immediately indicates that one or more metabolites are eluting at this retention time, rather than an isolated chance occurrence of a feature that is recognized as a peak.

Based on these hints, one can merge the ions belonging to the same metabolite into a reconstructed compound spectrum. Several techniques including clustering have been described, such as the software tools MSClust (Tikunov et al. 2012), CAMERA (Kuhl et al. 2012), MetaMS (Wehrens et al. 2014), TagFinder (Luedemann et al. 2008), MetaboliteDetector (Hiller et al. 2009) and PyMS (O'Callaghan et al. 2012), allowing the putative identification of compounds by comparing their mass spectra with reference MS libraries as described later in this chapter.

In case of high-resolution MS combined with ESI, ion species in the compound spectrum can be annotated. The feature annotation of ion species facilitates the calculation of the molecular mass, which is essential to calculate the elemental formula or to query compound libraries. Again, easiest to annotate are isotope clusters, since their occurrence, mass differences and intensity ratios do not depend on the analytical setup, but only on the molecular formula and charge *z*. The number of annotated ions can be increased taking into account known mass differences between known ion species, such as the above mentioned 21.982 between the $[M + H]^+$ and $[M + Na]^+$. With a large list of possible adducts and common neutral losses, it is in many cases possible to annotate the ion species.

Modern mass spectrometers allow advanced acquisition modes with names like MS^E, All-Ion, broad-band CID (bbCID) or Sequential Windowed Acquisition of All Theoretical Fragment Ion Mass Spectra (SWATH). These are combinations of normal MS1 scans, followed by one or more scans with a higher collision energy to also obtain fragmentation information that can be used for metabolite identification. Since these acquisition schemes use no or only very broad precursor isolation, the task is to assign fragment ions to their likely precursor ion. Again, peak shape and correlation across many samples (Broeckling et al. 2014; Tsugawa et al. 2015) can be used to create "pure" compound spectra.

From (Annotated) Mass Spectra to (Tentatively) Identified Metabolites

It is evident that different analytical techniques have different resolving power, not only in the sense of the technical term introduced in Chapter 3, but also when it comes to providing evidence to decide if two features are the same or different, and to identify the molecular formula or eventually the molecular structure among all possibilities. A scoring system for the analytical power of different analytical platforms is given by Sumner et al. (2014), but although these scores are numeric and can easily be compared, much depends on the experimental context and compound class in question. The bottom line is that it is important to report the analytical equipment and conditions (an aspect of the MSI paper, which is somewhat neglected), the analytical evidence leading to an identification, as well as the uncertainty (like positional or even stereoisomers) that is left (Chapter 10).

Spectral libraries vary with respect to the number and quality of the compound spectra, depending on how well they are curated. Currently, databases are still far from containing experimental data from all known metabolites, mostly because of the comparably small number of compounds commercially available as pure metabolite standards. Nevertheless, the use of spectral libraries is one of the best approaches to annotate the structure of metabolites and, consequently, convert raw spectral data into biological knowledge. Most of the annotations based on mass spectral similarity are combined with other physico-chemical properties (e.g., chromatographic retention time or retention index, see Box 4.4) in order to increase the discriminative power and to reduce the number false positive candidates. In the following we will give an overview of some well-curated representative databases. A more detailed overview about the most commonly used mass spectral databases for GC-MS and LC-MS-based untargeted metabolomics can be found in Vinaixa et al. (2016/4). An interesting question is the overlap and complementarity of metabolites covered among different spectral databases, and overlap with metabolic networks. This can be analyzed using the InChIkeys of the covered compounds; see (Vinaixa et al. 2016/4, Clément 2018) for an in-depth analysis. We see that there is a lot to gain if more than just one reference library is consulted for identification. If no spectra for pure standards are available, then we have to resort to *in silico* approaches, which are covered later in this chapter.

BOX 4.4 RETENTION INDICES AS TOOLS FOR IDENTIFYING COMPOUNDS IN GC-MS METABOLOMICS

The main parameters used in GC-MS metabolomics studies to identify metabolites are similarity to the mass spectra with known metabolites as well as the comparison of retention time behavior on the analytical column. In the ideal situation candidate compounds can be verified by comparing the mass spectra and retention time of a pure reference compound using the same analytical conditions as were applied for the experimental study, i.e., the same GC-MS is used with the same settings (see also MSI level 1). If a pure reference standard is not available, the same parameters should be used from literature data. However, when referring to retention times in the literature one needs to keep in mind that the retention time of a particular metabolite strongly depends on several analytical factors including column length, column type or temperature gradient applied in the GC oven, just to name a few. Therefore, retention times published in the literature become less reliable for identification purposes unless the exact same GC-MS settings are applied, which is very difficult to achieve.

Therefore, to overcome this obvious drawback, the concept of the retention index (RI) has been developed where the retention time is converted to a system independent value. Often the RI is also referred to as Kovats index (KI) as the concept has been first published by the Swiss chemist Ervin Kováts (1958). The retention index of a certain compound is its retention time normalized to the retention times of mostly adjacently eluting n-alkanes. The n-alkanes have by convention a fixed RI that corresponds to the number carbon atoms multiplied by 100, e.g., n-hexane (C_6H_{14}) has a RI of 600 while decane ($C_{10}H_{22}$) has a RI of 1000 and so on. Retention indices for other compounds are obtained by simple linear or polynomial interpolation. While the KI was originally developed for isothermal GC runs, for temperature programmed GC a simple way to calculate the RI was developed by van Den Dool and Kratz (1963).

Since RIs were introduced by Kovats, RIs have been progressively included in mass spectral databases commonly used for GC-MS metabolomics and provide an essential tool for the annotation of metabolites especially valuable to distinguish structurally similar metabolites that have similar spectra but differ in their retention behavior.

Data on retention indices are, however, available for only a small fraction of the compounds that are detected in metabolomics studies.

Mass Spectral Databases Containing Pure Standards

Tandem mass spectrometry (introduced in Chapter 3) provides valuable hints for metabolite structure annotation. The MS/MS spectrum of a compound

TABLE 4.1

Mass-Spectrometric Databases for Metabolomics

Database	Website	LC-MS	GC-MS	General Comments
Human Metabolome Database	http://www.hmdb.ca/	Yes	Yes	• Public and downloadable • Real and *in silico* generated spectra
METLIN	http://metlin.scripps.edu/	Yes	No	• Public • Real and *in silico* generated spectra
MassBank	http://www.massbank.jp	Yes	Yes	• Public and downloadable • Real spectra
NIST	http://www.sisweb.com/software/ms/nist.htm	Yes	Yes	• Commercial license • Real spectra
Golm Metabolome Database	http://gmd.mpimp-golm.mpg.de	No	Yes	• Public • Real spectra
Fiehn library	http://fiehnlab.ucdavis.edu/projects/fiehnlib	No	Yes	• Commercial license. • Real spectra
LipidBlast	http://fiehnlab.ucdavis.edu/projects/lipidblast	Yes	No	• *In silico* generated library
Lipid Maps	http://www.lipidmaps.org	Yes	No	• Public and downloadable • *In silico* generated spectra
mzCloud	https://www.mzcloud.org	Yes	No	• Public • Only Orbitrap spectra
GNPS	https://gnps.ucsd.edu/ProteoSAFe/static/gnps-splash.jsp	Yes	No	• Public and downloadable

can be considered its fingerprint, and just like in a crime scene investigation, the unknown compound can be identified if the fingerprint has been stored in a database previously. Several public and/or commercial spectral libraries have been created (see Table 4.1), containing mass spectra of dozens, hundreds or even thousands of compounds.

The comparison of spectra can be performed with different similarity measures. Common choices include the cosine-similarity used, e.g., in MassBank, or the fit, reverse fit and purity scores in HMDB. Probability-based measures like X-Rank (Mylonas et al. 2009) are relatively new, and can be expected to improve the reliability and power of reference libraries.

Several databases focus more on LC-MS data, such as the Human Metabolome Database; others contain mostly GC-MS data. Mass spectral databases most commonly used in GC-MS metabolomics consist of spectra that were created by electron impact at 70 eV. EI is a highly reproducible ionization process across many different GC-MS platforms. Hence, GC-MS experimental spectra collected in different labs can be easily compared to recorded EI database samples for metabolite annotation. In turn, the high reproducibility mass spectra enabled the development of extensive databases consisting of hundreds up to several hundred thousand spectra. Databases most often used in GC-MS metabolomics experiments are the one of the US National Institute of Science and Technology (NIST), the Golm Metabolome Database and the Fiehn/Binbase library. The latter two mainly contain mass spectra of TMS-derivatized metabolites.

Identification with *In Silico* Tools

Even if all available spectral libraries were combined, we would not have reference spectra for all compounds we are interested in. Sometimes authentic standards are extremely expensive, sometimes they simply are not available at all. In the case that reference spectra are not available, spectral databases will still return a best match, which then corresponds to the wrong structure. If the scoring function gives a low match for this, we may be inclined not to believe it, but in other cases the match may seem reasonably good, resulting in an incorrect assignment. This is definitely something to avoid, and until robust approaches to calculate a false discovery rate (FDR) in metabolomics arrive (Scheubert et al. 2017), results still have to be validated manually in the end.

As an alternative to comparing experimental data to pure-compounds spectral databases, the MS/MS data of the unknowns can be annotated with *in silico* tools. Several different strategies exist: some tools compare experimental mass spectra to theoretical spectra; other tools give approximate matches that can be refined further, focus on predictions of retention behavior or use knowledge about metabolic networks to guide the search for the best fitting structure.

Comparison with Theoretical Mass Spectra

One common approach is to obtain a list of candidates from generic compound databases containing molecular structures. These are typically obtained via an accurate mass search or a molecular formula search. Each candidate is then processed, and a score is calculated so that candidates can be ranked by that score. Software tools like MetFrag (Ruttkies et al. 2016; Wolf et al. 2010) or MAGMa (Ridder et al. 2012, 2014) use combinatorial fragmentation: they create all possible fragments and compare their mass against the measured spectrum. For specialized compound classes like lipids, rule-based methods like Lipid-Blast (Kind et al. 2013) can predict the fragmentation.

Recently, and with the availability of more training spectra, machine learning methods have been very successful (Figure 4.5). The CSI:FingerID (Dührkop et al. 2015) approach generates structural fingerprints from the MS/MS

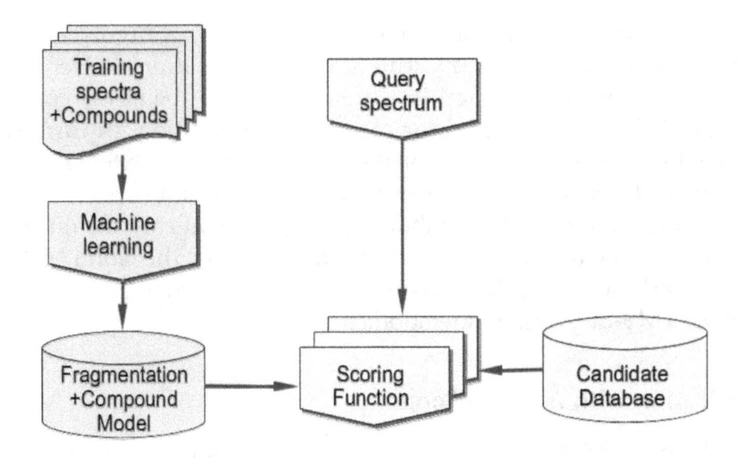

FIGURE 4.5
Workflow for *in silico* tools. Candidate structures are selected from a database, possibly with additional filter criteria. Some approaches use machine learning to obtain an optimized scoring function that is derived from a set of training data.

spectra, which in turn can be matched against a compound database. The CFM-ID (Allen et al. 2015) method also uses machine learning, and is one of the few approaches that can actually predict spectra for many compound classes, so that the actual identification and ranking is a standard Jaccard similarity coefficient.[3]

The choice of the database is important. While the large PubChem (Kim et al. 2016) and ChemSpider (Pence and Williams 2010) databases contain huge numbers of compounds, most of them are of synthetic origin and not biologically relevant. In contrast, KEGG COMPOUND (Kanehisa et al. 2014) contains almost exclusively biologically relevant compounds. Sadly, it covers only a fraction of known natural compounds, particularly for non-model species, plants or fungi. The situation is different if you have environmental samples, where, e.g., wastewater has a considerable amount of corrosion inhibitors, flame retardants, surfactants, pharmaceuticals, sweeteners and many more man-made compounds—in these cases PubChem and ChemSpider may be very useful indeed.

Approximate Matches

In some circumstances, the compound structure might even be in no metabolite or chemical database at all (Kalisiak et al. 2009). In this case, two possibilities are available at the moment: (i) algorithms identifying metabolites in a reference database that are likely to be structurally very similar to the

[3] "Jaccard index—Wikipedia." https://en.wikipedia.org/wiki/Jaccard_index, Accessed 13 May 2017.

unknown metabolite. An example is iMet (Aguilar-Mogas et al. 2016), which uses MS/MS spectra and produces a sorted list of candidates, ranked by their similarity to the unknown metabolite. The algorithm also suggests the chemical transformation that is most likely to separate each of the candidates from the unknown metabolite; and (ii) *in silico* (or virtual) compound databases. This can start with an exhaustive enumeration of all structures for a given molecular formula with a software like MOLGEN (Benecke et al. 1997; Kerber et al. 2014). For most of the relevant metabolites, the purely combinatorial chemical space is vast, and some constraints need to be provided to limit the number of structural isomers to a reasonable number, and to accommodate further or prior knowledge. It might be known, e.g., that a metabolite has the backbone (or framework) of a particular compound class, or that one or more functional groups are known. Maybe additional UV data can give hints about aromatic ring systems, or some characteristic MS/MS fragment corresponds to a known substructure. This narrowing-down requires an iterative approach, but can be worthwhile for very promising and challenging unknowns.

Another alternative to generate novel metabolite structures is to incorporate biochemical rules and mechanisms, and expand an existing structure database by anticipated biochemical reactions. The MINE database ("MINE" n.d.) contains more than 500,000 compounds expanded from KEGG, EcoCyc or the Yeast Metabolome Database (YMDB). As the name suggests, LipidHome (Foster et al. 2013) is specialized in lipids, containing millions of theoretical structures.

Prediction of Retention Times and Retention Indices

The annotation of unknown compounds obtained by GC-MS and LC-MS is classically done by exploiting the information provided by the mass spectra and the retention time (RT) or retention index (RI, see Box 4.4) of the compound. In the LC-MS world, retention indices are not (yet) widely adopted, since the retention behavior on LC columns is less well understood, and hence the selection of universal index compounds covering all metabolites is difficult. The unknowns are either compared to a reference standard or to spectral libraries (for the mass spectra) and the literature (for the RI). However, since empirical data on RIs are scarce compared to the number of available mass spectra in spectral databases, many approaches have been developed to estimate the retention time or retention index.

A simple approach is the mapping of retention times. If the RTs for a set of compounds X are known on an LC system A, and for a set Y on another system B, and they have a non-empty intersection Z, then it is possible to create models based on Z to translate from system A to B and vice-versa. This approach is used in the PredRet database developed by Stanstrup et al. (2015). It is still required that the RT was measured for an unknown on some LC system.

Other approaches make use of quantitative structure–retention relationships (QSRRs), estimating the RI based on descriptors derived from the chemical structure of the compounds. Several types of descriptors can be used: physicochemical, quantumchemical, topological descriptors, etcetera (Héberger 2007). Many different QSRR models have been developed, mostly based on multilinear regression (MLR), partial least squares regression (PLS), support vector regression (SVR) or artificial neural networks (ANN). An overview of the different models is given by Héberger (2007) and Katritzki et al. (2010). While some QSRR models have been developed for very specific structurally similar groups of compounds, such as pyrazines, alkylbenzenes or aldehydes, other models aim to predict RIs for a broader range of structurally more diverse compounds (Katritzky et al. 2010; Mihaleva et al. 2009; Stein et al. 2007). QSSR models have also been applied for RI prediction for derivatized compounds, as well as for comprehensive GCxGC-MS and high resolution GC-MS (Dossin et al. 2016; Kumari et al. 2011).

Incorporating Network Information

Metabolite annotation can also be guided by knowledge on endogenous metabolic reactions leading to the synthesis/consumption of metabolites present in the data set. This information on reactions, available in Genome-Scale Metabolic Networks (see Chapter 8) and stored in databases such as KEGG (Kanehisa et al. 2014) or BioCyc (Caspi et al. 2014), provides clues on metabolites that should be present in the sample. Nevertheless, this predicted metabolome may contain thousands of candidates and thus requires to be filtered or ordered to guide more accurately the annotation. The overall strategy to take network information into account consists in detecting a cascade of reactions connecting metabolites already identified or annotated (Jourdan et al. 2010). Since these reactions are likely to be involved in the process under study, their substrates and products can be expected to be found in the sample. It thus allows creating a list of candidates which can be used to guide the annotation of the peak list. It is possible to go further by prioritizing these putative annotations. In fact, if a metabolite is close in the network to many identified ones, then it should have a higher probability to be found in the peak list (Rogers et al. 2009). The implementation of the approach, available as the R package ProbMetab (Silva et al. 2014), consists in spreading probabilities in the network starting from identified metabolites.

References

"AMDIS." n.d. Available at: http://chemdata.nist.gov/dokuwiki/doku.php?id=chemdata:amdis. [Accessed May 29, 2017]

"AnalyzerPro|SpectralWorks." n.d. Available at: http://www.spectralworks.com/products/analyzerpro/. [Accessed May 29, 2017].

"MINE." n.d. Available at: http://minedatabase.mcs.anl.gov. [Accessed October 26, 2017].

Aguilar-Mogas, A., M. Sales-Pardo, M. Navarro, R. Tautenhahn, R. Guimerà, and O. Yanes. 2016. "iMet: A Computational Tool for Structural Annotation of Unknown Metabolites from Tandem Mass Spectra." arXiv [q-bio. QM]. arXiv. Available at: http://arxiv.org/abs/1607.04122.

Allen, F., R. Greiner, and D. Wishart. 2015. "Competitive Fragmentation Modeling of ESI-MS/MS Spectra for Putative Metabolite Identification." *Metabolomics: Official Journal of the Metabolomic Society* 11 (1): 98–110.

Benecke, C., T. Grüner, A. Kerber, R. Laue, and T. Wieland. 1997. "MOLecular Structure GENeration with MOLGEN, New Features and Future Developments." *Fresenius' Journal of Analytical Chemistry* 359 (1): 23–32.

Biller, J. E., and K. Biemann. 1974. "Reconstructed Mass Spectra, A Novel Approach for the Utilization of Gas Chromatograph—Mass Spectrometer Data." *Analytical Letters* 7 (7): 515–528.

Bloemberg, T. G., J. Gerretzen, H. J. P. Wouters, J. Gloerich, M. van Dael, H. J. C. T. Wessels, L. P. van den Heuvel, P. H. C. Eilers, L. M. C. Buydens, and Ron Wehrens. 2010. "Improved Parametric Time Warping for Proteomics." *Chemometrics and Intelligent Laboratory Systems* 104 (1): 65–74.

Broeckling, C. D., F. A. Afsar, S. Neumann, A. Ben-Hur, and J. E. Prenni. 2014. "RAMClust: A Novel Feature Clustering Method Enables Spectral-Matching-Based Annotation for Metabolomics Data." *Analytical Chemistry* 86 (14): 6812–6817.

Busch, K. L. 2002. "Chemical Noise in Mass Spectrometry." *Spectroscopy* 17 (10): 32–37.

Caspi, R. et al. 2014. "The MetaCyc Database of Metabolic Pathways and Enzymes and the BioCyc Collection of Pathway/Genome Databases." *Nucleic Acids Research* 42 (Database issue): D459–D471.

Clément, F. et al. 2018. "Mind the Gap: Mapping Mass Spectral Databases in Genome-Scale Metabolic Networks Reveals Poorly Covered Areas." *Metabolites* 8 (3): 51.

Colby, B. N. 1992. "Spectral Deconvolution for Overlapping GC/MS Components." *Journal of the American Society for Mass Spectrometry* 3 (5): 558–562.

Di Marco, V. B., and G. G. Bombi. 2001. "Mathematical Functions for the Representation of Chromatographic Peaks." *Journal of Chromatography A* 931 (1–2): 1–30.

Domingo-Almenara, X. et al. 2016. "eRah: A Computational Tool Integrating Spectral Deconvolution and Alignment with Quantification and Identification of Metabolites in GC/MS-Based Metabolomics." *Analytical Chemistry* 88 (19): 9821–9829.

Dossin, E., E. Martin, P. Diana, A. Castellon, A. Monge, P. Pospisil, M. Bentley, and P. A. Guy. 2016. "Prediction Models of Retention Indices for Increased Confidence in Structural Elucidation during Complex Matrix Analysis: Application to Gas Chromatography Coupled with High-Resolution Mass Spectrometry." *Analytical Chemistry* 88 (15): 7539–7547.

Dromey, R. G., M. J. Stefik, T. C. Rindfleisch, and A. M. Duffield. 1976. "Extraction of Mass Spectra Free of Background and Neighboring Component Contributions from Gas Chromatography/Mass Spectrometry Data." *Analytical Chemistry* 48 (9): 1368–1375.

Du, X., and S. H. Zeisel. 2013. "Spectral Deconvolution for Gas Chromatography Mass Spectrometry-Based Metabolomics: Current Status and Future Perspectives." *Computational and Structural Biotechnology Journal* 4 (June): e201301013.

Dührkop, K., H. Shen, M. Meusel, J. Rousu, and S. Böcker. 2015. "Searching Molecular Structure Databases with Tandem Mass Spectra Using CSI:FingerID." *Proceedings of the National Academy of Sciences of the United States of America* 112 (41): 12580–12585.

Eilers, P. H. C. 2004. "Parametric Time Warping." *Analytical Chemistry* 76 (2): 404–411.

Eliasson, M., S. Rännar, R. Madsen, M. A. Donten, E. Marsden-Edwards, T. Moritz, J. P. Shockcor, E. Johansson, and J. Trygg. 2012. "Strategy for Optimizing LC-MS Data Processing in Metabolomics: A Design of Experiments Approach." *Analytical Chemistry* 84 (15): 6869–6876.

Foster, J. M., P. Moreno, A. Fabregat, H. Hermjakob, C. Steinbeck, R. Apweiler, M. J. O. Wakelam, and J. A. Vizcaíno. 2013. "LipidHome: A Database of Theoretical Lipids Optimized for High Throughput Mass Spectrometry Lipidomics." *PloS One* 8 (5): e61951.

Fuhrer, T., and N. Zamboni. 2015. "High-Throughput Discovery Metabolomics." *Current Opinion in Biotechnology* 31: 73–78.

Haug, K. et al. 2012. "MetaboLights—An Open-Access General-Purpose Repository for Metabolomics Studies and Associated Meta-Data." *Nucleic Acids Research*. doi:10.1093/nar/gks1004.

Héberger, K. 2007. "Quantitative Structure–(Chromatographic) Retention Relationships." *Journal of Chromatography. A* 1158 (1–2): 273–305.

Hiller, K., J. Hangebrauk, C. Jäger, J. Spura, K. Schreiber, and D. Schomburg. 2009. "MetaboliteDetector: Comprehensive Analysis Tool for Targeted and Nontargeted GC/MS Based Metabolome Analysis." *Analytical Chemistry* 81 (9): 3429–3439.

Hoffmann, N. et al. 2019. "mzTab-M: A Data Standard for Sharing Quantitative Results in Mass Spectrometry Metabolomics." *Analytical Chemistry* 91 (5): 3302–3310

Ipsen, A., E. J. Want, J. C. Lindon, and T. M. D. Ebbels. 2010. "A Statistically Rigorous Test for the Identification of Parent-Fragment Pairs in LC-MS Datasets." *Analytical Chemistry* 82 (5): 1766–1778.

Jellema, R. H., S. Krishnan, M. M. W. B. Hendriks, B. Muilwijk, and J. T. W. E. Vogels. 2010. "Deconvolution Using Signal Segmentation." *Chemometrics and Intelligent Laboratory Systems* 104 (1): 132–139.

Jourdan, F., L. Cottret, L. Huc, D. Wildridge, R. Scheltema, A. Hillenweck, M. P. Barrett, D. Zalko, D. G. Watson, and L. Debrauwer. 2010. "Use of Reconstituted Metabolic Networks to Assist in Metabolomic Data Visualization and Mining." *Metabolomics: Official Journal of the Metabolomic Society* 6 (2): 312–321.

Kalisiak, J., S. A. Trauger, E. Kalisiak, H. Morita, V. V. Fokin, M. W. W. Adams, K. Barry Sharpless, and G. Siuzdak. 2009. "Identification of a New Endogenous Metabolite and the Characterization of Its Protein Interactions through an Immobilization Approach." *Journal of the American Chemical Society* 131 (1): 378–386.

Kanehisa, M., S. Goto, Y. Sato, M. Kawashima, M. Furumichi, and M. Tanabe. 2014. "Data, Information, Knowledge and Principle: Back to Metabolism in KEGG." *Nucleic Acids Research* 42 (Database issue): D199–D205.

Katritzky, A. R., M. Kuanar, S. Slavov, C. Dennis Hall, M. Karelson, I. Kahn, and D. A. Dobchev. 2010. "Quantitative Correlation of Physical and Chemical Properties with Chemical Structure: Utility for Prediction." *Chemical Reviews* 110 (10): 5714–5789.

Kerber, A., R. Laue, M. Meringer, and C. Rücker. 2014. "Mathematical Chemistry and Chemoinformatics: Structure Generation, Elucidation and Quantitative Structure-Property Relationships." books.google.com. Available at: https://books.google.com/books?hl=en&lr=&id=hhHoBQAAQBAJ&oi=fnd&pg=PR5&dq=Mathematical+Chemistry+and+Chemoinformatics+Structure+Generation+Elucidation+and+Quantitative+Structure-Property+Relationships+Walter+de+Gruyter&ots=EK-4sWLXjI&sig=233CvY4K400ZHCa2dht9PclESxk.

Kim, S. et al. 2016. "PubChem Substance and Compound Databases." *Nucleic Acids Research* 44 (D1): D1202–D1213.

Kind, T., G. Wohlgemuth, D. Y. Lee, Y. Lu, M. Palazoglu, S. Shahbaz, and O. Fiehn. 2009. "FiehnLib: Mass Spectral and Retention Index Libraries for Metabolomics Based on Quadrupole and Time-of-Flight Gas Chromatography/Mass Spectrometry." *Analytical Chemistry* 81 (24): 10038–10048.

Kind, T., K.-H. Liu, D. Y. Lee, B. DeFelice, J. K. Meissen, and O. Fiehn. 2013. "LipidBlast in Silico Tandem Mass Spectrometry Database for Lipid Identification." *Nature Methods* 10 (8): 755–758.

Kirwan, J. A., R. J. M. Weber, D. I. Broadhurst, and M. R. Viant. 2014. "Direct Infusion Mass Spectrometry Metabolomics Dataset: A Benchmark for Data Processing and Quality Control." *Scientific Data* 1: 140012.

Kovats, E. 1958. "Gas-Chromatographische Charakterisierung Organischer Verbindungen. Teil 1: Retentionsindices Aliphatischer Halogenide, Alkohole, Aldehyde Und Ketone." *Helvetica Chimica Acta* 41 (7): 1915–1932.

Kuhl, C., R. Tautenhahn, C. Böttcher, T. R. Larson, and S. Neumann. 2012. "CAMERA: An Integrated Strategy for Compound Spectra Extraction and Annotation of Liquid Chromatography/Mass Spectrometry Data Sets." *Analytical Chemistry* 84 (1): 283–289.

Kumari, S., D. Stevens, T. Kind, C. Denkert, and O. Fiehn. 2011. "Applying in-Silico Retention Index and Mass Spectra Matching for Identification of Unknown Metabolites in Accurate Mass GC-TOF Mass Spectrometry." *Analytical Chemistry* 83 (15): 5895–5902.

Lange, E., K. Reinert, and C. Groepl. 2005. "OPENMS; a Generic Open Source Framework for Chromatography/MS-Based Proteomics." *Molecular & Cellular Proteomics* 4: S25.

Lange, E., R. Tautenhahn, S. Neumann, and C. Gröpl. 2008. "Critical Assessment of Alignment Procedures for LC-MS Proteomics and Metabolomics Measurements." *BMC Bioinformatics* 9: 375.

Libiseller, G. et al. 2015. "IPO: A Tool for Automated Optimization of XCMS Parameters." *BMC Bioinformatics* 16: 118.

Lu, H., Y. Liang, W. B. Dunn, H. Shen, and D. B. Kell. 2008/3. "Comparative Evaluation of Software for Deconvolution of Metabolomics Data Based on GC-TOF-MS." *Trends in Analytical Chemistry: TRAC* 27 (3): 215–227.

Luedemann, A., K. Strassburg, A. Erban, and J. Kopka. 2008. "TagFinder for the Quantitative Analysis of Gas Chromatography—Mass Spectrometry (GC-MS)-Based Metabolite Profiling Experiments." *Bioinformatics* 24 (5): 732–737.

Mihaleva, V. V., H. A. Verhoeven, R. C. H. de Vos, R. D. Hall, and R. C. H. J. van Ham. 2009. "Automated Procedure for Candidate Compound Selection in GC-MS Metabolomics Based on Prediction of Kovats Retention Index." *Bioinformatics* 25 (6): 787–794.

Mylonas, R., Y. Mauron, A. Masselot, P.-A. Binz, N. Budin, M. Fathi, V. Viette, D. F. Hochstrasser, and F. Lisacek. 2009. "X-Rank: A Robust Algorithm for Small Molecule Identification Using Tandem Mass Spectrometry." *Analytical Chemistry* 81 (18): 7604–7610.

Ni, Y., M. Su, Y. Qiu, W. Jia, and X. Du. 2016. "ADAP-GC 3.0: Improved Peak Detection and Deconvolution of Co-Eluting Metabolites from GC/TOF-MS Data for Metabolomics Studies." *Analytical Chemistry* 88 (17): 8802–8811.

O'Callaghan, S. et al. 2012. "PyMS: A Python Toolkit for Processing of Gas Chromatography-Mass Spectrometry (GC-MS) Data. Application and Comparative Study of Selected Tools." *BMC Bioinformatics* 13 (May): 115.

Pence, H. E., and A. Williams. 2010. "ChemSpider: An Online Chemical Information Resource." *Journal of Chemical Education* 87 (11): 1123–1124.

Ridder, L., J. J. J. van der Hooft, and S. Verhoeven. 2014. "Automatic Compound Annotation from Mass Spectrometry Data Using MAGMa." *Mass Spectrometry* 3 (Spec Iss 2): S0033.

Ridder, L., J. J. J. van der Hooft, S. Verhoeven, R. C. H. de Vos, R. van Schaik, and J. Vervoort. 2012. "Substructure-Based Annotation of High-Resolution Multistage MS(n) Spectral Trees." *Rapid Communications in Mass Spectrometry: RCM* 26 (20): 2461–2471.

Rogers, S., R. A. Scheltema, M. Girolami, and R. Breitling. 2009. "Probabilistic Assignment of Formulas to Mass Peaks in Metabolomics Experiments." *Bioinformatics* 25 (4): 512–518.

Ruttkies, C., E. L. Schymanski, S. Wolf, J. Hollender, and S. Neumann. 2016. "MetFrag Relaunched: Incorporating Strategies beyond in Silico Fragmentation." *Journal of Cheminformatics* 8: 3.

Salek, R. M. et al. 2013. "The MetaboLights Repository: Curation Challenges in Metabolomics." *Database: The Journal of Biological Databases and Curation* 2013.

Scheubert, K., F. Hufsky, D. Petras, M. Wang, L.-F. Nothias, K. Dührkop, N. Bandeira, P. C. Dorrestein, and S. Böcker. 2017. "Significance Estimation for Large Scale Metabolomics Annotations by Spectral Matching." *Nature Communications* 8 (1): 1494.

Silva, R. R., F. Jourdan, D. M. Salvanha, and F. Letisse. 2014. "ProbMetab: An R Package for Bayesian Probabilistic Annotation of LC–MS-Based Metabolomics." Available at: http://bioinformatics.oxfordjournals.org/content/30/9/1336.short.

Skogerson, K., G. Wohlgemuth, D. K. Barupal, and O. Fiehn. 2011. "The Volatile Compound BinBase Mass Spectral Database." *BMC Bioinformatics* 12: 321.

Smith, C. A., E. J. Want, G. O'Maille, R. Abagyan, and G. Siuzdak. 2006. "XCMS: Processing Mass Spectrometry Data for Metabolite Profiling Using Nonlinear Peak Alignment, Matching, and Identification." *Analytical Chemistry* 78 (3): 779–787.

Stacklies, W., H. Redestig, M. Scholz, D. Walther, and J. Selbig. 2007. "pcaMethods—A Bioconductor Package Providing PCA Methods for Incomplete Data." *Bioinformatics* 23 (9): 1164–1167.

Stanstrup, J., S. Neumann, and U. Vrhovšek. 2015. "PredRet: Prediction of Retention Time by Direct Mapping between Multiple Chromatographic Systems." *Analytical Chemistry* 87 (18): 9421–9428.

Stein, S. E. 1999/8. "An Integrated Method for Spectrum Extraction and Compound Identification from Gas Chromatography/Mass Spectrometry Data." *Journal of the American Society for Mass Spectrometry* 10 (8): 770–781.

Stein, S. E., V. I. Babushok, R. L. Brown, and P. J. Linstrom. 2007. "Estimation of Kovats Retention Indices Using Group Contributions." *Journal of Chemical Information and Modeling* 47 (3): 975–980.

Sud, M. et al. 2016. "Metabolomics Workbench: An International Repository for Metabolomics Data and Metadata, Metabolite Standards, Protocols, Tutorials and Training, and Analysis Tools." *Nucleic Acids Research* 44 (D1): D463–D470.

Sumner, L. W., Z. Lei, B. J. Nikolau, K. Saito, U. Roessner, and R. Trengove. 2014. "Proposed Quantitative and Alphanumeric Metabolite Identification Metrics." *Metabolomics: Official Journal of the Metabolomic Society* 10 (6): 1047.

Tikunov, Y. M., S. Laptenok, R. D. Hall, A. Bovy, and R. C. H. de Vos. 2012. "MSClust: A Tool for Unsupervised Mass Spectra Extraction of Chromatography-Mass Spectrometry Ion-Wise Aligned Data." *Metabolomics: Official Journal of the Metabolomic Society* 8 (4): 714–718.

Tsugawa, H., T. Cajka, T. Kind, Y. Ma, B. Higgins, K. Ikeda, M. Kanazawa, J. VanderGheynst, Oliver Fiehn, and Masanori Arita. 2015. "MS-DIAL: Data-Independent MS/MS Deconvolution for Comprehensive Metabolome Analysis." *Nature Methods* 12 (6): 523–526.

van den Dool, H., and P. Dec. Kratz. 1963. "A Generalization of the Retention Index System Including Linear Temperature Programmed Gas—Liquid Partition Chromatography." *Journal of Chromatography A* 11: 463–471.

Vinaixa, M., E. L. Schymanski, S. Neumann, M. Navarro, R. M. Salek, and O. Yanes. 2016/4. "Mass Spectral Databases for LC/MS- and GC/MS-Based Metabolomics: State of the Field and Future Prospects." *Trends in Analytical Chemistry: TRAC* 78: 23–35.

Wehrens, R., G. Weingart, and F. Mattivi. 2014. "metaMS: An Open-Source Pipeline for GC–MS-Based Untargeted Metabolomics." *Journal of Chromatography B* 966 (September): 109–116.

Wehrens, R., T. G. Bloemberg, and P. H. C. Eilers. 2015. "Fast Parametric Time Warping of Peak Lists." *Bioinformatics* 31 (18): 3063–3065.

Weisser, H. et al. 2013. "An Automated Pipeline for High-Throughput Label-Free Quantitative Proteomics." *Journal of Proteome Research* 12 (4): 1628–1644.

Wolf, S., S. Schmidt, M. Müller-Hannemann, and S. Neumann. 2010. "In Silico Fragmentation for Computer Assisted Identification of Metabolite Mass Spectra." *BMC Bioinformatics* 11: 148.

Yanes, O., R. Tautenhahn, G. J. Patti, and G. Siuzdak. 2011. "Expanding Coverage of the Metabolome for Global Metabolite Profiling." *Analytical Chemistry* 83 (6): 2152–2161.

5

Nuclear Magnetic Resonance Spectroscopy Data Processing

Maria Vinaixa, Naomi Rankin, Jeremy Everett and Reza Salek

CONTENTS

Introduction ... 101
NMR Data Analysis Pipeline ... 102
From Raw Data to Spectra ... 105
Pre-Processing in the Time Domain ... 105
Processing in the Frequency Domain ... 106
From Spectra to Data Matrices ... 109
Fingerprinting Approach ... 110
Regions of Interest (ROIs) Approach ... 111
Region Exclusion, Data Normalization and Scaling ... 112
Targeted Profiling Approach ... 113
Metabolite Assignment/Identification ... 114
Online Resource for Metabolite Assignments ... 116
NMR Metabolite Identification Using STOCSY ... 117
Combining NMR and MS for Metabolite Assignment ... 117
NMR File Formats and Standards ... 118
Other Applications of NMR-Based Metabolomics ... 119
NMR-Based Stable Isotope Resolved Metabolomics ... 119
NMR Structural Elucidation ... 120
High-Resolution Magic Angle Spinning (HR-MAS) NMR Spectroscopy ... 121
Advancement of NMR Spectroscopy ... 122
Conclusion ... 123
References ... 123

Introduction

NMR is a powerful tool for structure elucidation and has the ability to detect a wide variety of compounds with different physicochemical properties, with full quantitation, no need for reference standards, often with minimal or no batch effects, minimal sample preparation and a highly automated instrument optimization and operation. Major limitations of NMR spectroscopy for metabolomics are its lack of sensitivity and relatively low spectral dispersion. The field of metabolomics has witnessed significant technological improvements in NMR methodology during the last decade. Today, high-throughput NMR-based metabolomics has the ability to process of the order of 100 samples per day. However, data analysis remains a bottleneck. NMR spectra acquired from biological matrices such as cells, tissues or biofluids contain hundreds to thousands of resonances from hundreds of metabolites. It is of paramount importance to accurately process such spectra in order to obtain biologically relevant information. In this chapter, the focus is on different approaches or pipelines to deal with NMR data from a metabolomics experiment. It is not intended to review all possible applications in this field, but rather to concentrate on the essentials and prerequisites of its successful implementation as a metabolite profiling tool.

NMR Data Analysis Pipeline

NMR data analysis is based on detecting signals from individual metabolites in a mixture. The experimental Free Induction Decay (FID) decays to zero as the equilibrium in nuclear spin magnetization in the xy plane is itself reduced to zero by spin-spin relaxation processes (see Chapter 3 and lucid texts from Ziessow (1987), Kuchal (2002) and Keeler (2011). The FID encodes the resonance frequency for each nucleus in each spin system, which, in metabolomics, is a function of the structures of the metabolites. NMR-based, metabolomics data analysis extracts biologically meaningful information in several major stages summarized in Figure 5.1. The first stage in the analysis pipeline is the so-called spectrum pre-processing step, where the collection of FIDs is converted from the time domain to the frequency domain, leading to an interpretable set of NMR spectra: usually one 1D ^{1}H NMR spectrum for each of the samples in the particular study. Regions of the spectra that are either not required or not of use in the data analysis are then removed. For example, for urine spectral analysis, the regions of low frequency (up to 0.8 ppm), the regions of high frequency (over 10 ppm) and the water region (from 4.7 to 4.9 ppm) are all removed as they generally contain no useful information. In the next step, the spectra are reduced to data matrices (tables) containing the intensities from either identified metabolites, spectral regions (bins) (Smolinska et al. 2012; Vettukattil 2015) or even each data point from each peak (full data point resolution). Finally, the data matrices are used as input for statistical analysis. It should be stressed that NMR data processing steps are context-dependent (Craig et al. 2006), meaning that specific data analysis pipelines cannot be generalized to all types of experimental designs or conditions. Therefore, a number of variables and values need

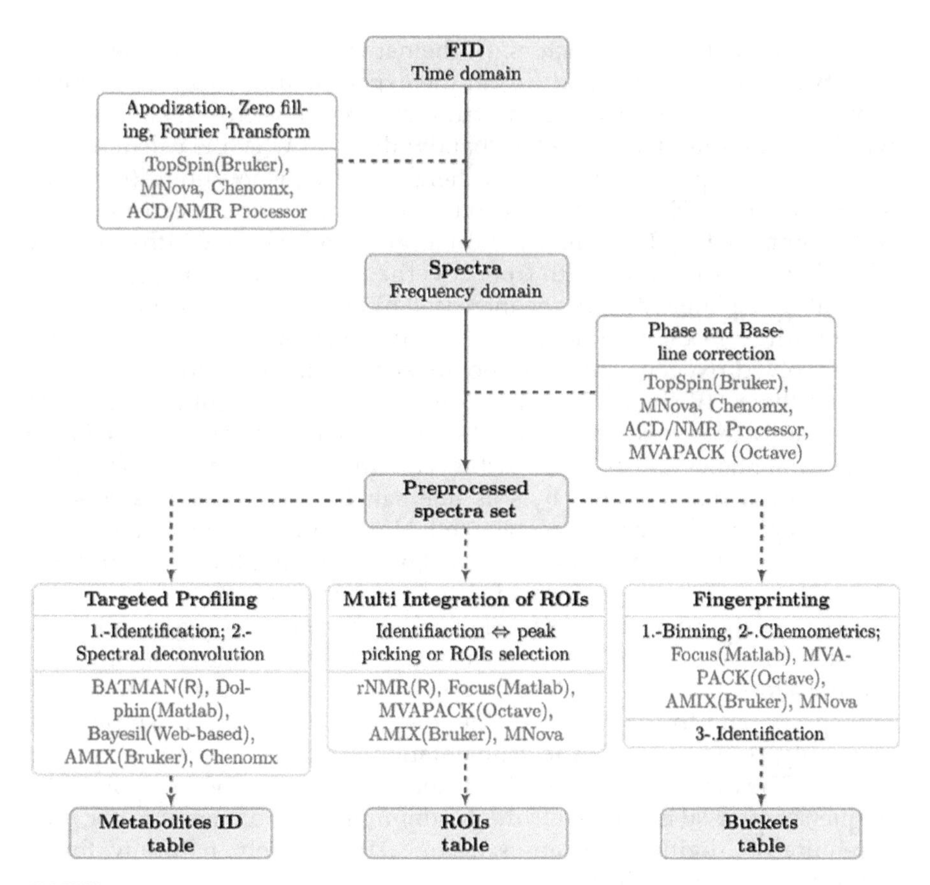

FIGURE 5.1
NMR data analysis pipeline schema. Green blocks represent different types of data generated across the entire NMR pipeline, while orange and yellows blocks represent pre-processing and processing steps, respectively. First, raw data, i.e., series of FIDs, are generated by NMR spectroscopy. Several different pre-processing steps are applied, usually one after another, on the raw NMR data converting the data set from the time domain to the frequency domain NMR spectrum. Then, three different approaches might be used, usually depending on the experimental design. This approach dictates the way in which metabolite identification is performed and the data output of the entire analysis. Blue text indicates open-source software or freely available solutions for either processing or pre-processing steps, in the entire workflow, while red text indicates those that are commercial, proprietary or third-party software solutions.

to be adjusted during the whole pipeline, which can affect the data analysis outcome. These variables are usually tuned according to the experts' criteria, e.g., by visual assessment of the spectra.

There are several spectral attributes commonly used to describe the quality of an NMR spectrum. Resolution (the minimum frequency difference between peaks that can be distinguished) is one of them.

It is determined by three factors: (i) the natural linewidth, (ii) the homogeneity of the magnetic field (B_0) in the experiment and (iii) the number of points used to digitize the spectrum and hence each signal. The linewidth is measured using the half bandwidth, $v_{1/2}$ which is calculated by measuring the peak width at half-height in frequency units (i.e., Hz). A TSP $v_{1/2}$ of <1 Hz represents good quality for a urine spectrum and a well shimmed 14.1 T magnet will often give TSP ^{1}H linewidths in urine of <0.5 Hz. Natural linewidth (i) is governed by relaxation processes, but the real signal linewidth will be increased by B_0 field inhomogeneities and by impurities in the sample, especially paramagnetic impurities, including dissolved oxygen. The more homogeneous the magnetic field B_0 the sharper the NMR signals and therefore the better the resolution, and the more closely the experimental NMR signal linewidth will approach that of the natural linewidth (ii). This is achieved with good shimming conditions i.e., ensuring that the field B_0 is as homogeneous as possible across the sample region in the probe receiver coil. Note, however, the signal resolution cannot be better than that governed by the natural linewidth and will always be worse in the real world with imperfect magnets and additional sources of relaxation in the sample such as dissolved oxygen and paramagnetic metal ions. Shortened relaxation times will cause line broadening and a loss of resolution. An expert eye is needed to examine the quality of a spectra and to spot artifacts due to sample preparation, sample type or quality, or as a result of instrumentation error or incorrect parameter settings. The digital resolution (iii) is merely the number of data points per frequency interval in the spectrum. A minimum of six data points per peak is required to digitize and represent a peak adequately: naturally, this is more demanding for sharp peaks.

Figure 5.1 shows a schematic diagram of several steps, often carried out on the original raw data to create data matrices amenable to statistical analysis. After converting the time domain FID data to spectral data by Fourier transformation, one can either look for specific patterns in the spectral data after identification and quantification of known metabolites (targeted profiling), define and integrate signals in regions of interest (ROIs), execute the statistical analysis across the full data points set in the spectra, or simply perform a binning (also known as bucketing) by dividing the full ppm axis into segments and integrating each segment, that will be used as variables later in the analysis. Commonly, NMR data handling pipelines consist of multiple software packages. An extensive range of NMR software packages are available for metabolomics data analysis (see Ellinger et al. 2013) for a detailed review). Only a few of them are open source and from those, not one covers the entire data analysis workflow (Spicer et al. 2017). In the next section, we will discuss different steps often carried out in NMR data processing, starting from raw data, to spectral interpretation and finally to metabolite identification.

From Raw Data to Spectra

Pre-Processing in the Time Domain

Several steps are usually carried out to convert raw time-domain FID signals to the frequency domain spectrum. The most commonly used data processing steps are: zero filling, apodization and Fourier transformation to generate the frequency domain NMR spectrum, as summarized in Figure 5.1. In the case of 1D-^{1}H-NMR for example, the FID consists of a complex sum of decaying sine waves with different frequencies representing ^{1}H nuclei in different chemical environments (different parts of different metabolites). Each one of these sine waves decays gradually as the protons relax and recover from the radio-frequency pulse (see Chapter 3). Interpreting data in the time domain or FID is very difficult and therefore the time domain data are converted to the frequency domain by means of Fourier Transformation (FT)—see Figure 5.2. The NMR spectra after FT displays peaks for protons in different environments. The frequencies of these spectral peaks are identical to those of the corresponding sine waves in the FID, their intensities correlated with the intensity of the sine wave and their linewidths dependent on the decay rate for each wave—think of a musical chord (three notes struck together) "ringing out" over time. FT of that chord can be used to express the frequency and amplitude (or loudness) of its constituent notes. This is possible because the three notes, each with their own sine wave, when added together produce a specific wave function, which can be deconstructed and reconstructed (Anon n.d.).

Prior to FT, steps are carried out in the time domain to improve the resolution of the resultant spectral signals. Like in most other forms of spectroscopy, to increase the resolution one would need to increase the acquisition time of the FID, thereby gaining more data points, but the drawback is an increase in overall data acquisition time and an increase in the noise. Luckily,

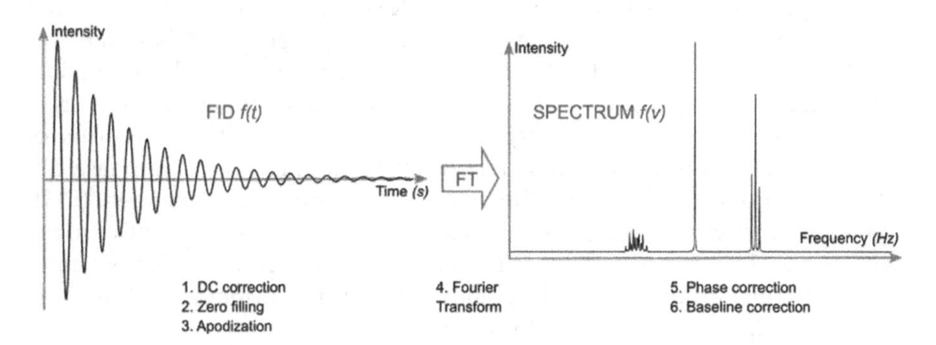

FIGURE 5.2
NMR raw FID file data in time domain (t1) are converted to frequency data (f1) by Fourier Transformation.

in the context of NMR, provided the FID has decayed to zero, a cheap alternative exists: adding zeros to the end of the FID resulting in enhanced spectral resolution. This step is known as zero-filling. Doubling the number of data points in the FID will improve the true resolution due to information loss in the Fourier Transform but further zero filling only improves the digital resolution of each peak.

NMR generally suffers from low signal-to-noise ratio (S/N). As S/N increases as the square root of the number of scans (e.g., 16 scans provide 4 times better S/N than 1 scan), S/N can be increased by acquiring more scans. However, this is time-consuming. An alternative approach is to apply apodization or a weighting function to the FID in order to improve the S/N. The simplest of these functions is an exponential decay [$W(t) = \exp(-\mathrm{lb}t)$], where lb is the line broadening factor in Hz, which indicates how much the half bandwidth of the peak will increase. Line broadening effectively reduces the instrumental noise, while at the same time increasing the linewidth in the spectrum. Apodization to increase S/N works by decreasing signal intensities at the end of the FID, where the S/N is worst as the signal has almost decayed to zero, but the noise is still present, thus eliminating some of the noise. Optimal S/N improvement occurs when the lb factor applied during exponential multiplication equals the resonances' natural line width. However, since in a metabolic profiling experiment, many metabolites with many signals with different linewidths are detected, this can be difficult to judge. There are many other functions used to enhance the signal for NMR spectra and for instance signal resolution can be enhanced (at the expense of lowering S/N) by transforming the natural Lorentzian NMR lineshape into a narrower Gaussian lineshape using Gaussian multiplication (Hoch and Stern 1996).

Processing in the Frequency Domain

Phasing and Baseline Correction

After applying the FT to generate the NMR spectrum, several further processing steps are carried out, most notably phasing and baseline correction. Phasing is carried out to achieve an absorptive signal (Lorentzian, see Figure 5.3) in order to achieve accurate signal intensities and integrals. There are two different types of phase correction. The zero-order or frequency-independent phase correction (PH0) applies when the phase difference is equal for all signals. The first-order or frequency-dependent (PH1) phase correction applies a phase change with a linear increase relative to the distance to the reference signal. Phase correction is usually carried out automatically but can be corrected manually if required (Montigny et al. 1990; Chen et al. 2002; Balacco and Cobas 2009).

Another common spectrum correction is baseline correction, often applied hand in hand with the phase correction. Distortion in an NMR baseline can result in inaccurate integrals, incomplete peak-picking or peaks becoming obscured. A flat baseline is important for correctly calculating the area

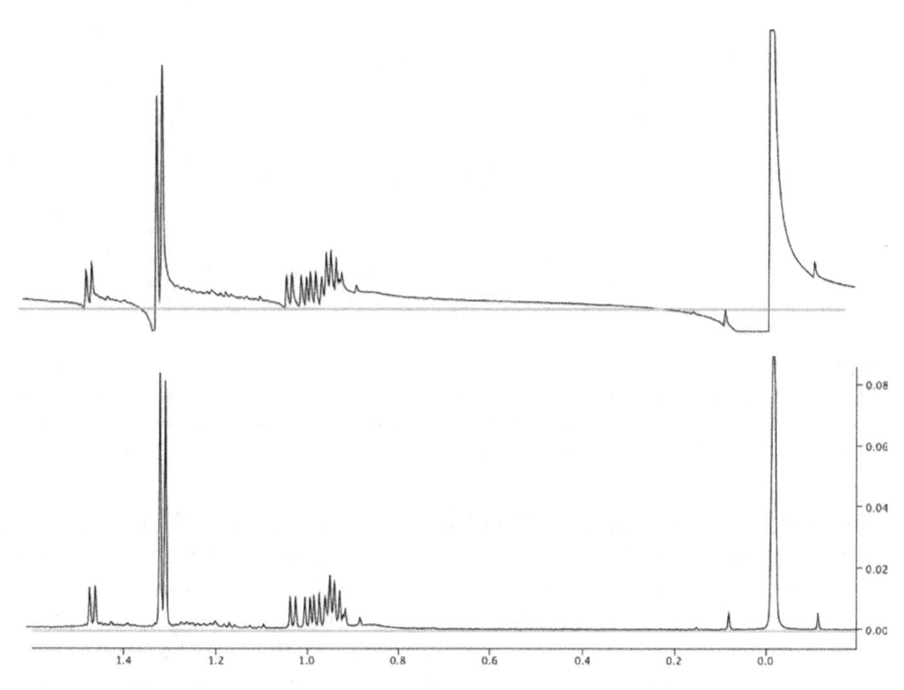

FIGURE 5.3
Phase correction in NMR. Top: spectra before phase correction. Bottom: spectra after PH0 phase correction. The interval from δ (−0.2 to 1.6 ppm) for 1D ^{1}H-NMR spectra acquired on human cells hydrophilic crude extract is displayed. Resonances correspond to CH_{3-} groups from TSP at δ (0 ppm), branched chain amino acids mostly (leucine, isoleucine and valine) at δ (0.9–1.1 ppm) lactate at δ (1.33 ppm) and alanine δ (1.46 ppm).

under the peak and to quantify metabolites (Figure 5.4). In the limit of a fully dispersive peak, the integral will be zero, so correct phasing is important. There are many reasons for distorted baselines, most commonly due to instrument issues, such as receiver overload. Receiver overload can happen if the "receiver gain" is set too high, saturating the analog-to-digital converter. As a result, the first points of the FID can get distorted. Many baseline problems can be manually or automatically corrected (Figure 5.4). Approaches used more often are linear or polynomial corrections. Linear prediction can correct distortions of the first few points in an FID, flattening the baseline. Many phase or baseline corrections rely on the visual feedback of the software to the user and best practice comes with experience. The objective is to achieve Lorentzian peak shapes with minimal distortion and a flat baseline. Best practice is to keep functions as simple as possible, e.g., using polynomial correction of first or second order only. More aggressive baseline correction methods can give flatter baselines, but there is a trade-off in that this often results in a loss of data (by subtracting "real" information from the spectrum). This will affect the final result and increase errors during data integration (Xi and Rocke 2008). There are a variety of baseline corrections, some

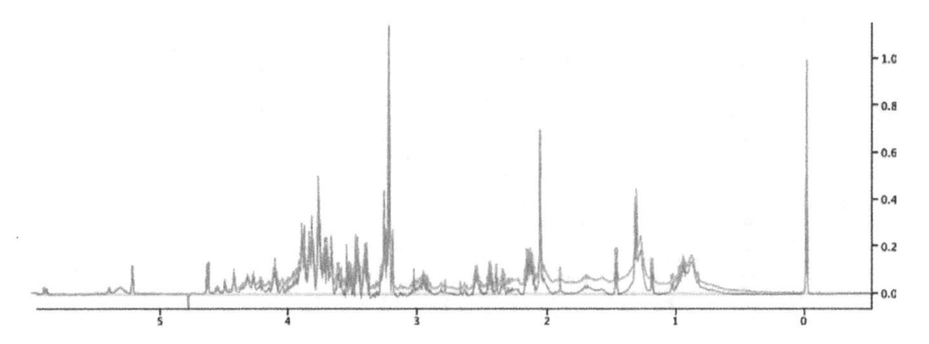

FIGURE 5.4
1D ^{1}H-NMR spectra acquired on rat liver hydrophilic crude extract prior to baseline correction/subtraction (grey) and after spline baseline correction/subtraction (blue). The global baseline correction is usually applied to the entire spectra.

are local, others are more global and automated (Bao et al. 2012), some require user parameter adjustment such as airPLS (adaptive iteratively reweighted penalized least squares) algorithm (Zhang et al. 2010), an automated baseline correction algorithm that requires parameter adjustment known as Lambda, which can vary within a very large range (from 10 up to 1.e + 06), therefore requiring caution and experience to use. There are many more different types of algorithms for baseline correction; for more examples, see Xi and Rocke (2008) and Jacob et al. (2017). For large scale metabolic phenotyping, automation of these processes is very important.

Chemical Shift Calibration

Chemical shifts across samples are calibrated using an internal standard for chemical shift referencing. These internal standards have multiple equivalent protons (and carbons) giving a single strong signal, are highly chemically stable and their signal frequencies are not significantly affected by concentration or pH changes. To this end, the most widely used internal standard is sodium 3-(trimethylsilyl)propionate-2,2,3,3-d_4 (TSP) with deuterated methylene groups to avoid interferences (additional chemical shift signals) in the ^{1}H-NMR spectrum. The nine equivalent protons of TSP give rise to a sharp, strong, singlet resonance at a chemical shift defined as δ(0 ppm), which therefore does not overlap with peaks from metabolites. However, TSP is not advised to be used when dealing with samples with high protein content such as plasma or serum (Beckonert et al. 2007). TSP binds to proteins resulting in broad bandwidths due to slowing of its motions in solution, thus increasing its linewidth and decreasing its signal intensity (Shimizu et al. 1994). In such cases, referencing to other peaks such as formate at δ(8.45 ppm) might be better suited. Other common peaks used as reference are the methyl doublet of Lactate at δ(1.33 ppm) or the anomeric glucose doublet at δ(5.23 ppm) (Verwaest et al. 2011) mainly used for plasma and serum biofluids (Pearce et al. 2008). However, the concentration of the metabolites cannot be calculated when referencing to internal metabolites since the concentration is not known. The common exception to this is glucose, whose concentration in plasma can often be accurately measured by standard biochemical tests.

Chemical Shift Misalignments Correction

Despite referencing or calibrating to TSP, chemical shift fluctuations for a particular metabolite may still occur across different samples due to the sensitivity of some NMR signals to small changes in environmental conditions such as pH or ionic strength, or the concentrations of various metal ions, particularly for hydrogens in the vicinity of ionizable groups such as carboxylate groups (such as formate, acetate, lactate and citrate). Proper alignment of spectral peaks is a crucial pre-processing step prior to downstream statistical analysis, in order to avoid potential confounding statistical analysis. However, there are some controversies surrounding the appropriateness of alignment methodologies (Veselkov et al. 2009; Spraul et al. 2015) and there is no gold standard method for alignment. Various methods for handling spectral alignments, especially in urine samples where variation in pH and ionic strength routinely effect the spectra, have been proposed. It is even possible to use the information encoded in the chemical shifts of a variety of metabolites to extract information about the concentrations of metal ions and other metabolites in a biofluid that are not even visible by NMR. Deconvoluting interrelationships between concentrations and chemical shifts in urine provides a powerful analysis tool (Takis et al. 2017). For a comprehensive review see Vu and Laukens (2013). It should be noted that the fluctuations used for chemical shift in NMR spectroscopy are very different in nature than the retention time shifts in chromatography, and methods that work well in one context may not be very suitable in another (Giskeødegård et al. 2010). To address this, several workflow packages have developed or implemented a set of alignment methods. The nmrProcflow has two methods, one based on a Least Squares algorithm, and the other based on a Parametric Time Warping to deal with spectra alignment (Bloemberg et al. 2010; Vu et al. 2011; Jacob et al. 2017). Focus, another NMR workflow tool has developed a RUNAS (Recursive Unreferenced Alignment of Spectra) algorithm, which efficiently aligns NMR peaks, while avoiding the use of a reference spectrum (Alonso et al. 2014).

From Spectra to Data Matrices

Similar to the situation in MS-based metabolomics, a crucial step in the analysis pipeline consists of converting the data into features (or peak list in NMR terminology). Currently, there are three different approaches to generating a peak list from pre-processed spectra sets, which are often complementary (Wishart 2008/3): *(i) fingerprinting (bin/bucket table), (ii) multi integration of regions of interest (ROI table) and (iii) targeted profiling (metabolite ID table).* Differences among these three approaches rely basically on the order in which compound identification is performed in a workflow (Figure 5.1). For example, in the *fingerprinting* approach, spectral patterns are statistically compared based on the measured intensities. Relevant features (generally those that are discriminatory) are selected and subsequently annotated. In the other two approaches all detectable metabolites are identified (and quantified) prior to the statistical analysis. Each approach has its own advantages and disadvantages, and it is important to be aware of them in order to critically evaluate the results.

Fingerprinting Approach

In a fingerprinting approach, signal intensities are integrated in a number of intervals, known as bins or buckets. This leads to a data matrix, where each feature corresponds to a part of the ppm axis (see Figure 5.5). Depending upon the width of the bins, several of the bins could belong to the same metabolite or alternatively, the same bin could include several peaks from different metabolites. This approach is quite agnostic to peak relationship to a metabolite, and simply slices the peaks regions. Typical bin widths are between 0.01 and 0.04 ppm or in some instances lower to 0.001 ppm, i.e., at full spectral resolution. The advantage of binning is that the number of variables can be reduced quite drastically: In a typical 1D ^{1}H NMR experiment used for metabolomics, the number of data points are usually between 16 and 64 k, whereas after binning this will typically be an order of magnitude smaller. However, modern high-speed computers enable full data point processing without technical issues. The major disadvantage is therefore a statistical one, in that the user is sampling a much higher dimensional data space and so false discovery of biomarkers from the data must be more carefully avoided.

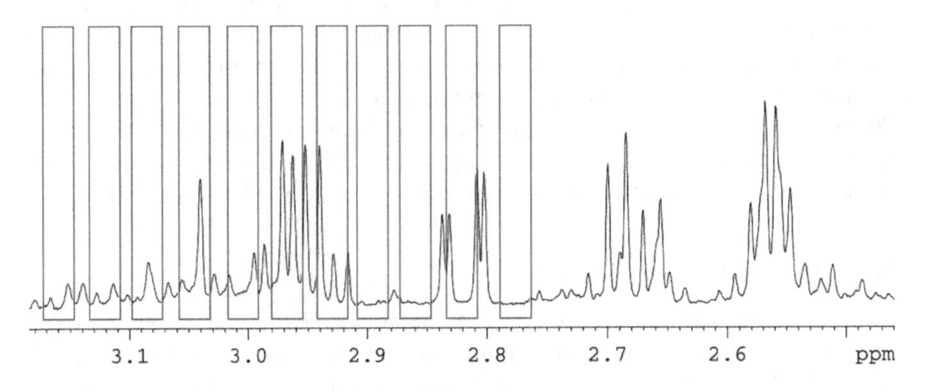

FIGURE 5.5
Bucketing or binning consists of segmenting a spectrum into small areas called bins or buckets. The bins or buckets are often fixed size usually between 0.01 and 0.04 ppm or in some instances 0.001 ppm or smaller (Uniform bucketing). In this example, the bin sizes are 0.03 ppm and there are gaps between the bins; this is for clarity only.

Fingerprinting is ideal for identifying global trends in spectral patterns or to compare the overall metabolic composition of the samples (e.g., cases versus controls) under investigation. The fingerprinting approach is the less demanding in terms of identification requirements since spectral resonances are not immediately annotated. Instead, samples are compared based on their pre-processed spectra and only those regions showing statistically significant differences between groups are retained for identification. Often chemometric approaches (including supervised and unsupervised methods—see Chapter 6) are used for statistical comparison of samples.

One additional advantage of binning is that it prevents small NMR shift variations from affecting the statistical processing, provided such shifts are confined within a single bucketed region. Conventional bucketing procedures use fixed-width bins, which can cause resonances from the same metabolite to be split into two or more bins, hampering interpretation of results. To overcome this drawback, some commercial software such as ACD/Labs™ (Toronto, Canada) have implemented a solution termed intelligent bucketing. This method allows bin boundaries to be chosen based on local minima.[1] Other strategies proposed involve calculating the bin boundaries by optimizing an objective function using a dynamic programming strategy (Davis et al. 2007; Anderson et al. 2010; Sousa et al. 2013). All of them have demonstrated improved performance respective to traditional uniform binning. Overall, fingerprinting is the simplest and least demanding approach for analysis of NMR spectra data sets. It enables a straightforward overview of the NMR data with minimal user intervention. Using a fingerprinting approach, the main patterns and trends in the data set can often be highlighted quite easily. This allows focusing only on variables retaining important variance for the scientific question at hand.

Regions of Interest (ROIs) Approach

Another analysis method would be to define regions of interest (ROIs) or interesting resonances in a supervised manner (which could be considered a special case of intelligent binning, although in this case knowledge of (most of) the metabolite identities is important). Hence, positions and size of the bins are defined by user (Figure 5.6) and they can be inferred from existing literature or online NMR reference databases (for example the Biological Magnetic Resonance Bank). ROIs may correspond to annotated resonances (known compounds) or even compounds that are unknown at the time of analysis. The area under such ROIs can be considered as a surrogate of the relative abundance of these metabolites in a biological sample. Provided that the pulse sequence in use to acquire the NMR spectra

[1] "ACD/Labs.com :: ACD/NMR Processor Academic Edition." 2010. 14 Dec. 2015 http://www.acdlabs.com/resources/freeware/nmr_proc/

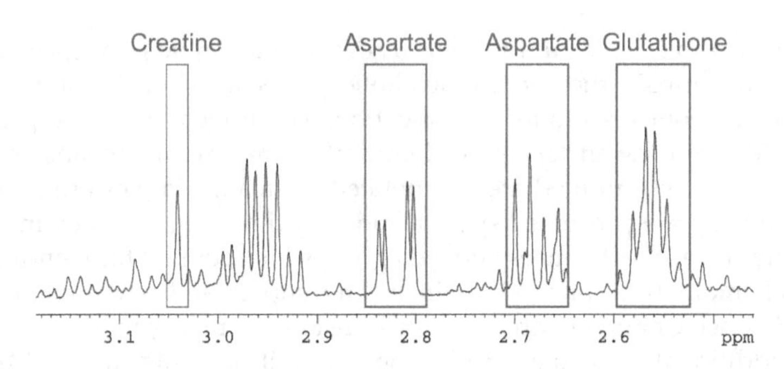

FIGURE 5.6
NMR annotation of the peaks using multi-integration of ROIs based on prior knowledge. This is mainly a manual process, requiring user skill in identifying regions of interest. The ROI are potentially metabolites of interest. However, in this approach equally a region can be selected without being an identifiable metabolite (Unknown metabolite). It can be considered a special case of intelligent binning where the user decides the size of each bin based on prior knowledge. Areas of the spectra which are not of interest are not binned.

allows quantitative acquisition (in terms of phasing, baseline, receiver gain, solvent suppression, relaxation delay and acquisition time), this area is proportional to the concentration of metabolites attributed to each specific resonance (see Chapter 3) (Barding et al. 2012). Therefore, relative concentrations of metabolites can be obtained by simple integration of the NMR peaks. The key issue here is the correct assignment of ROIs resonances to a metabolite structure, which is not a trivial task. NMR structure assignment in complex matrices, such as in cell cultures or biological fluids, currently requires a skilled NMR specialist, and the assignments have to be confirmed by 2D NMR or reference libraries of pure compounds (Dona et al. 2016).

Region Exclusion, Data Normalization and Scaling

Prior to binning, any areas of the spectra containing random noise variance are excluded. Among these regions, water suppression areas around 4.8 ppm, as well as regions of spectra containing reference compounds such as TSP δ(0 ppm) or regions containing non-relevant metabolic information, are typically excluded. The bucketed spectra are then normalized, for instance, to a constant sum (e.g., total sum of integral/bin intensities), probabilistic quotient normalization, dry weight tissue or protein content, to minimize the biological and technical variation resulting from differences in the amount of biological specimens (e.g., for dissolved tissue samples); or differences in dilution (e.g., for urine samples where variable dilution depends on hydration status of the subject) or differences in detector response. Probabilistic Quotient Normalisation (PQN) normalization (Dieterle et al. 2006) has recently gained ground as the standard normalization method because of its better capacity to minimize dilution differences, particularly for urine samples.

To prepare samples for chemometric analysis by Principal Component Analysis (PCA) or similar multivariate methods (see Chapter 6), the data must be mean centered and scaled (column-wise). The aim is to ensure that large variations in the intensities of small peaks, which could be of biological significance, are not overlooked because they are swamped by small variations in very large signals. There are several different scaling methods such as Pareto-scaling or standardization to zero mean and unit variance, also called auto-scaling (see Chapter 6). Each scaling method affects the multivariate modeling result in a particular and usually case-dependent way (Craig et al. 2006), although Pareto scaling seems to be often used for NMR. Other scaling methods (e.g., log scaling) can also be used to transform data in order to make it better approximate a normal distribution and minimize the effect of large peaks and outliers on the discriminatory power (Craig et al. 2006). The choice of scaling method depends on the scientific question in hand, experimental design, properties of the data set or the data analysis method selected (Craig et al. 2006; Gromski et al. 2014).

Targeted Profiling Approach

The final approach, targeted profiling, aims to identify and quantify (in relation to an internal standard) specific compounds in a sample set. Peaks or known resonances are modeled (for example by peak-fitting) based on spectral reference libraries of pure compounds. Deconvolution of peaks from an ^{1}H-NMR spectrum is performed in order to obtain concentration estimates for all identified compounds. This allows a more exact and elegant quantification of overlapping resonances (see Figure 5.7). In both fingerprinting and ROI-based approaches, it is currently impossible to assign *all* resonances arising from complex biological matrices, in particular when

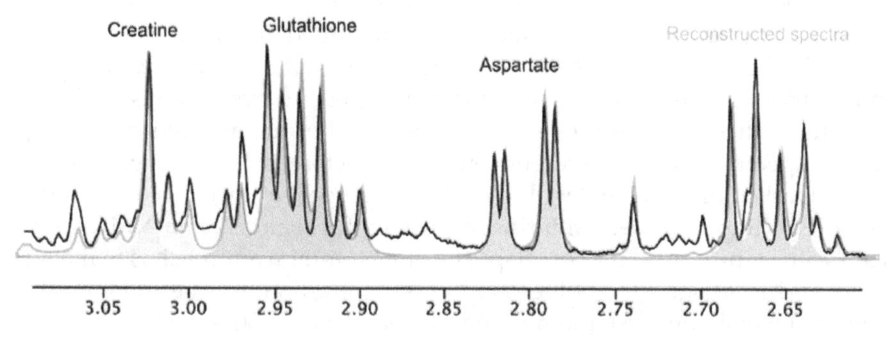

FIGURE 5.7
NMR data analysis and assignment of the peaks using a targeted profiling approach. The profiles of known compounds are fitted manually allowing deconvolution of the NMR spectra in biofluid samples (or biological mixtures) into their individual components. The experimentally obtained NMR spectrum is displayed in black and the reconstructed spectrum is shown in blue.

peaks are overlapping, since the complexity of the spectra are quasi-fractal. The researcher can therefore be left with many unknown resonances, making interpretation difficult. Targeted profiling, on the other hand, focuses only on the known peaks of interest, i.e., the ones in the reference library.

Several tools use a targeted profile approach; for example, commercial solutions are provided by Bruker AMIX package (Anon n.d.) or Chenomx (Anon n.d.) that incorporate libraries of pure standard compounds. The reference libraries can be fully searchable and contain hundreds of different compounds for different proton resonance frequencies, and pH values.

There are open-source solutions that can perform targeted analysis. BAYESIL (http://bayesil.ca/) can automatically identify and quantify metabolites from a 1D ^{1}H NMR spectra of ultra-filtered plasma, serum or cerebrospinal fluid (Ravanbakhsh et al. 2015). BATMAN (http://batman.r-forge.r-project.org/) uses a Bayesian model to automatically quantify metabolites (Hao et al. 2014). This Bayesian model needs some template information such as chemical shifts, J-couplings, multiplicity and intensity ratios derived from spectral database. BATMAN has a chemical-shift sorting method to assist with assignments. DOLPHIN uses both 1D and 2D NMR data for targeted profiling (Gómez et al. 2014). 2D J-resolved NMR spectra are often used, in combination with 1D spectra to assist with metabolite identification, especially in crowded regions of the spectra (Dona et al. 2016).

Metabolite Assignment/Identification

In NMR-based metabolite annotation, the most common approaches are to use either experimental or simulated reference library compounds to match or to fit the existing biological spectra or alternatively to use chemical shifts and multiplicity matching for metabolite identified in literature (Lindon et al. 1999; Beckonert et al. 2007). Such steps are usually an iterative process and require several matching and correction steps to overcome issues that are usually due to spectral overlap. The approaches are assisted by having specific spectral data for different conditions such as pH and concentration to gradually annotate metabolites. Other approaches use deconvolution methods to improve spectral fitting, especially for regions of an NMR spectrum with significant signal overlap. The most common approach is to match peaks and to assign metabolites step by step, eliminating the more easily identified metabolite first and moving to more difficult ones next.

1D NMR data often is not sufficient for a confident assignment of the metabolite peaks (Beckonert et al. 2007; Ludwig and Viant 2010; Pudakalakatti et al. 2014), and therefore 2D spectral information from complementary multidimensional NMR spectroscopy is required to confirm the assignment. The most commonly used 2D NMR methods involve homonuclear (JRES, COSY, TOCSY) and heteronuclear (HSQC, HMBC) 2D NMR experiments. 2D NMR is powerful as it provides significant new information not present in a 1D ^{1}H NMR spectrum and also reduces signal congestion often seen in 1D NMR by spreading resonances across a second dimension. Dona et al. (2016) provides a good guide for metabolite identification. Similar to 1D reference libraries, several 2D NMR reference libraries exist for aiding resonance assignment. These complementary 2D experiments can have long-acquisition times and therefore are not routinely acquired for every sample; just selected representative samples of each spectral type. An exception to this is the powerful 2D J-resolved ^{1}H NMR experiment that separates homonuclear ^{1}H–^{1}H couplings into the second dimension whereas chemical shifts remain in the first dimension. The experiment also separate homonuclear coupling from heteronuclear coupling, mitigates the effects of magnetic field inhomogeneity through the 1800 refocusing pulse and enables the derivation of proton broadband-decoupled proton NR spectra via projection of the 2D surface onto the chemical shift axis. Once the peaks/resonances have been accurately annotated, these identifications can be applied to other spectra of the same type (provided experimental conditions remain constant) (Figure 5.8).

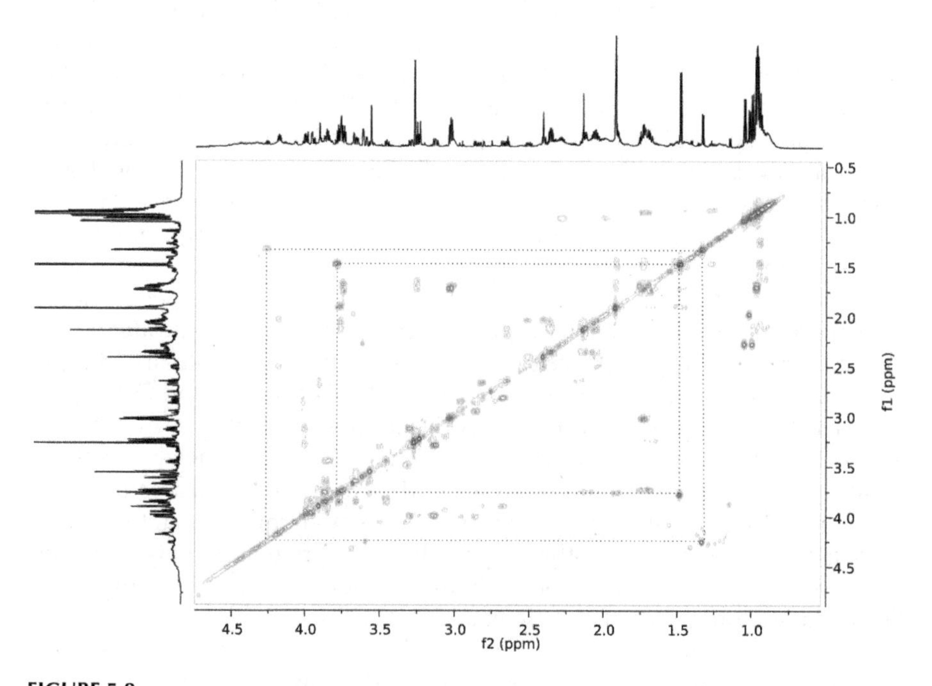

FIGURE 5.8
An example of a 2D COSY spectrum (sample from *E. coli* extracellular medium), with ^{1}H-^{1}H coupling for threonine and alanine connected with the dotted lines. The corresponding 1D ^{1}H NMR, matching the 2D correlation cross peaks (in orange) are at top and left hand-side of the 2D COSY spectrum.

Online Resource for Metabolite Assignments

Several resources are now available that can assist with metabolite identification by storing NMR data submitted by researchers and making such information accessible to the community. Overall, such resources could be categorized as a database that stores: (i) peak tables detailing the chemical shift and multiplicity of different reference chemical compounds, often classified by sample type or (ii) experimentally obtained spectra from reference chemical compounds as raw and processed files. Such reference compounds (including metabolites) are either available commercially, or are synthesized, and used to make standard solutions that can be analyzed using 1D or 2D NMR spectroscopy. Often these are analyzed at different magnetic field strengths, concentrations, temperatures, with different pH or salt contents, and with different experimental NMR parameters to allow better matching with metabolites in NMR spectra obtained in different laboratories, in different sample types, with different experimental protocols. Peak table resources often do not have any reference or raw files but can be used to search for specific chemical shifts to enable metabolite identification. Spectral databases allow visual comparison, and in some cases data can be downloaded and compared with the experimentally obtained NMR spectrum under analysis. Examples include the Human Metabolome Database (HMDB, http://www.hmdb.ca/), NMR Metabolomics Database of Linkoping, Sweden (MDL, http://www.liu.se/hu/mdl/main/) and the Spectral Database for Organic Compounds (SDBS, http://sdbs.db.aist.go.jp/sdbs/). Some spectral databases are specific for the identification of metabolites from 2D experiments. For example, TOCSY Customized Carbon Trace Archive (TOCCATA) is specific for TOCSY and natural abundance $^{1}H(^{13}C)$-HSQC-TOCSY spectra using an isomer-specific database (http://spin.ccic.ohio-state.edu/index.php/toccata/index) (Bingol et al. 2014). The same authors also proposed a method based on unified and isomer-specific database interrogation that provides improved HSQC based metabolite identification (COLMAR) (http://spin.ccic.ohio-state.edu/index.php/colmar) (Bingol et al. 2015). The isomer-specific features of $^{1}H(^{13}C)$- TOCCATA and COLMAR are useful as they overcome the potential failure to identify a metabolite due to low abundance isomer peaks being absent and the metabolite then failing to be identified. In addition, COLMAR contains useful confidence terms including a peak matching ratio and a peak uniqueness statistic that decrease false positives. Another approach is collection and analysis of the 2D J-resolved spectra of metabolite standards, for example, the Birmingham Metabolite Library (BML) (Ludwig et al. 2011).

Tools that make use of freely available public repositories are MetaboMiner (Xia et al. 2008), SpinAssign (Chikayama et al. 2010) and MetaboHunter (Tulpan et al. 2011); MetaboMiner (http://wishart.biology.ualberta.ca/metabominer/) is a web service interface performing semi-automated metabolite identification of 2D TOCSY and HSQC spectra.

SpinAssign (http://prime.psc.riken.jp/), developed by RIKEN Plant Science Centre, provides batch-annotations against user NMR peaks based on a database of over 1700 ^{13}C-HSQC peaks, corresponding to 270 metabolites. MetaboHunter (http://www.nrcbioinformatics.ca/metabohunter/) is a web service providing automatic metabolite identification based on spectra or peak lists using stratified (pH, solvent, NMR frequency) data from publicly available reference libraries such as the Human Metabolome Database (HMDB) and the Madison Metabolomics Consortium Database (MMCD).

NMR Metabolite Identification Using STOCSY

A well-established method assisting with metabolite identification and, at the same time, highlighting significant inter-molecular correlations in metabolic profiling data is Statistical Total Correlation Spectroscopy (STOCSY) (Cloarec et al. 2005). The STOCSY method takes advantage of multi-collinearity of the resonances in ^{1}H NMR spectra. In other words, if two peaks in an NMR spectrum come from the same metabolite, the abundance of those peaks from sample to sample will be correlated. STOCSY generates a pseudo 2D NMR spectrum that displays the correlation among the NMR peaks across the whole sample set. Assuming the correlation is due to intramolecular correlations or covariance (same metabolite), then the information that these peaks both come from the same metabolite allows reduction of the number of potential metabolites that these peaks be attributed to. However, correlation between peaks could also indicate that the peaks belong to different metabolites in the same biochemical pathway (intermolecular or biochemical covariance) (Holmes et al. 2006). The combination of STOCSY with a supervised chemometrics analysis method, such as the orthogonal projection on latent structure-discriminant analysis (O-PLS-DA, Chapter 6), has shown to be a powerful tool in identifying metabolites playing a statistically significant role in metabolic profiling (Cloarec et al. 2005).

Combining NMR and MS for Metabolite Assignment

Metabolite identification can be improved by combining NMR and MS spectroscopy. This is self-evidently a good idea as the two methods are entirely orthogonal in the information that they provide. Indeed, the current metabolite identification recommendations from the Metabolomics Standards Initiative requires: "A minimum of two independent and orthogonal data (such as retention time and mass spectrum) compared directly relative to an authentic reference standard" are obtained in order to declare a metabolite to be confidently identified (Sumner et al. 2007a). Bingol and Brüschweiler developed a method that confirms metabolite identifications by combining high-resolution MS with NMR for the identification of unknown components in complex mixtures (Bingol et al. 2015). These authors used accurate mass MS data to produce chemical formulas of the mixture components and generated a list of all feasible structures. Next, they predicted (*in silico*) possible NMR spectra for each structure and compared it with experimental data, matching the information obtained from both the MS and NMR techniques (Bingol et al. 2015). Later the authors carried out an inverted approach by first identifying metabolite candidates from 1D or 2D NMR spectra using NMR database querying, e.g., the COLMAR HSQC database, and then predicting the mass to charge ratios (m/z) of their possible ions, adducts, fragments and characteristic isotope distributions from these candidate structures that would be expected to be observed in mass spectrometry (Bingol and Brüschweiler 2015; Bingol et al. 2015). In the next step, the expected MS peaks were compared to the experimentally obtained MS (1) spectrum for the direct assignment of the signals belonging to the same molecule from both the MS and NMR spectra. This approach, called NMR/MS Translator, facilitates the assignment of MS data with high confidence, and at the same time validates the NMR results. The NMR/MS Translator is fully automated and takes only a few seconds to accomplish on a computer workstation (Bingol and Brüschweiler 2015; Bingol et al. 2015). In general, it is important to confirm NMR identifications by an orthogonal method such as high-resolution LC-MS when identifying metabolites that are critical for a biological interpretation of the experiment, or metabolites for which library databases contain no spectral data. Confidence in the identification of known metabolites can also be problematic and needs approaching with care (Sumner et al. 2007b; Everett 2015). The identification of truly novel metabolites requires careful work to the exacting standards of natural product discovery.

NMR File Formats and Standards

Altogether, both spectral repository matching and software tools can be used to interrogate NMR data, in order to discover new information from samples under analysis. The scope and availability of reference compounds are limited, and many laboratories have created their own in-house libraries with additional reference spectra to achieve a wider identification. The larger the reference compound libraries are, the better the chance of matching the metabolite to the reference compound for identification. However, various tools used for NMR data analysis create their own specific output formats, often not compatible with other software. Unless shared publicly, and in an open, accessible format, the use of in-house libraries and tools will not benefit the wider research community. We need NMR spectral repositories to make depositions of data for metabolomics more collective and accessible. There have been great strides in the standardization of NMR-based metabolomics particularly in capturing, reporting and exchange of data in metabolomics (Rubtsov et al. 2007; Emwas et al. 2015; Rocca-Serra et al. 2016). One such effort is EMBL-EBI MetaboLights (Haug et al. 2012), a data repository for metabolomics experiments and reference compounds. In addition to a data deposition platform, we also need an open access NMR file format to achieve better exchange of data in NMR-based metabolomics.

Raw data formats are not yet standardized, though some initiatives such as an open markup language for NMR or nmrML (Anon n.d.) are working toward this. This lack of an NMR raw data format has prevented further software development. Currently, the most widely used open source export format is the Joint Committee on Atomic and Molecular Physical Data (JCAMP-DX version 6.0) (Davies and Lampen 1993), but it does not possess a flexible hierarchical structure and is not easily extendable to capture supplementary information. Such extensibility (ability to handle future growth), as shown in XML formats such as AnIML (http://animl.source forge.net/), is a highly desired feature of a standard format in the fast-paced and dynamic world of metabolomics data-capture. Among the more persistent efforts, with wider community back-up, is the minimum information reporting guidelines by the MSI workgroups (www.metabolomics-msi.org/); see also Chapter 10. The Metabolomics Standards Initiative (MSI) has provided detailed suggestions about the metadata that should be captured and reported for NMR experiments. In particular, the MSI has put forth recommendations to report instrument descriptions and configurations, instrument-specific sample preparation and data acquisition parameters (Sansone et al. 2007; Rubtsov et al. 2007; Sumner et al. 2007b). These efforts resulted in the first round of NMR XML data standard development, focusing on raw and processed one- and two-dimensional NMR experiments and associated metadata. The BML-NMR repository of metabolite spectra has already implemented these suggestions with relatively minimal (and fully documented) modifications (Ludwig et al. 2011).

Other Applications of NMR-Based Metabolomics

NMR-Based Stable Isotope Resolved Metabolomics

Due to the tight homeostatic regulation of intracellular metabolite concentrations and high levels of interconnectivity of metabolic networks, the profiling of metabolites and their steady-state concentrations are impractical to resolve by metabolomics using metabolic pathways. To provide detailed maps of biochemical pathways and to map metabolic flux through them, tracer studies are needed. These studies exploit an isotopically labeled enriched precursor, generally glucose or glycine universally labeled with the stable isotope ^{13}C. This is introduced to the biological system (living cells) of interest, which takes up the labeled precursor and uses it to produce endogenous metabolites which themselves become labeled. Metabolomics analysis can be used to profile (identify) the labeled metabolites. Over time, this allows investigation of the pathways and kinetics of specific metabolites. Due to its unique capabilities in determining atomic positions of isotope labels (positional isotopomers), NMR (specifically ^{13}C NMR) has been used in stable isotope tracer studies for decades (Cohen et al. 1979; 1981). These types of studies allow dynamic characterization of metabolic pathways with known stoichiometry and composition (previously delineated). In fact, some of the major advances in understanding how biochemical pathways become rewired in cancer have been facilitated by this type of study (Mullen et al. 2012; Benjamin et al. 2012). DeBerardinis et al. (2007) for example, used glioblastoma cell lines and real-time ^{13}C-NMR to demonstrate that glutamine drives tricarboxylic acid (TCA) cycle anaplerosis primarily to fulfill the cell proliferation demands for both reducing equivalents (NADH) and building blocks.

To date, most of these studies have been limited to the targeted profiling of known or anticipated metabolic pathways. However, building on comprehensive metabolic profiling technologies (Fan 1996) has nowadays broadened the coverage of stable isotope tracer studies. In this regard, so-called Stable Isotope-Resolved Metabolomics (SIRM) has emerged as a powerful non-targeted approach to stable isotope tracing (Fan and Lane 2011). Non-hypothesis-driven SIRM makes use of fully isotope-labeled substrates since they require extensive metabolite labeling (Chokkathukalam et al. 2014). Either uniformly labeled glucose [U-^{13}C]-Glc or uniformly labeled glutamine [U-^{13}C]-Gln are used to trace back the incorporation of their labeled atoms to a wide range of central metabolite intermediates. The cellular fate of these labeled precursors is monitored using comprehensive profiling techniques allowing an unbiased mapping of fluxes through multiple metabolic pathways (Creek et al. 2012). SIRM has enabled the dynamic study of cancer cell metabolism (Fan et al. 2009) including important observations such as MYC regulation of proline metabolism (Liu et al. 2012) and MYC-dependent hypoxic metabolism of glutamine in the absence of glucose (Le et al. 2012).

NMR Structural Elucidation

Identification of known metabolites using databases and software was discussed above (see **Metabolite Assignment/Identification**). If a metabolite cannot be identified using these methods (for example if it is a novel metabolite), full structural elucidation is required if the metabolite is to be identified. Since the peaks in an ^{1}H-NMR spectrum represent resonances/signals from each different type of proton (based on the molecular environment) in the sample and splitting patterns (multiplets) give information about the neighboring protons (see Chapter 3), an NMR spectrum provides structural information. In addition to the chemical shift (δ, in ppm) providing information as to the type of proton generating the peak (alkane, alkene, benzene, aldehyde, etc.), the integrals of particular resonances also represent relative numbers of each type of proton in the molecule (with caveats, see Chapter 3). Not only does the spin-spin splitting give the number of hydrogens on adjacent carbons, the coupling constant *J* (usually in frequency units, Hz) is a measure of the interaction between a pair of protons and gives information about the arrangement of the atoms involved (how close they are in space and their relative orientations). Even more structural information, including through-bond and through-space connectivities, to allow metabolite identification can be obtained from ^{13}C and 2D NMR. NMR spectroscopy is therefore a powerful method for structural elucidation. However, structural elucidation of a novel metabolite is extremely challenging in a complex sample matrix and therefore several time-consuming extraction and purification steps are often required in order to isolate and concentrate the metabolite of interest.

The information obtained from the range of 1D and 2D NMR experiments performed can then be combined in order to identify key substructures of the metabolite by matching to databases. Computer Aided Structural Elucidation (CASE) software programs are available to make this process more efficient and less subject to user error or bias. The total amount of NMR-based information accrued can be analyzed relative to the size of the metabolite in order to calculate the confidence of metabolite identification according to the MICE (metabolite identification of carbon efficiency) methodology (Everett, 2015). For metabolomics "natural product-likeness scores" are useful for identifying structures most likely to occur in a biologic sample. NMR spectroscopy and high mass accuracy MS can also be combined for further structural information as knowledge of the molecular formula is crucial for structural elucidation.

The use of NMR spectroscopy for structural elucidation is a topic and book chapter on its own; therefore, we will not expand on this topic further and refer the reader to the following references for more information on this topics: Kim et al. 2011; Robinette et al. 2012; van der Hooft et al. 2013; Dona et al. 2016.

High-Resolution Magic Angle Spinning (HR-MAS) NMR Spectroscopy

So far, we have discussed only "solution based" NMR spectroscopy. However, solid state NMR has also been widely used in metabolomics, generally to acquire NMR spectra from tumor biopsies or intact tissues. High-resolution magic angle spinning (HR-MAS) NMR spectroscopy (Beckonert et al. 2010) has been used on tissue biopsies to differentiate between benign and malignant tissue with high sensitivity and specificity (Chan et al. 2009). Sample preparation and analysis are relatively simple; depending on the size of the insert, an average of 10–30 mg of tissue is packed into disposable 30–50 μL Teflon inserts. Depending on the tissues either D_2O saline solution or D_2O phosphate buffer are added to the insert and introduced into ZrO_2 rotors. The rotor is placed in the probe, spinning the sample (usually about

5–10 kHz for biological samples) at the magic angle θm (ca. 54.74°, where $\cos^2 \theta m = 1/3$) with respect to the direction of the magnetic field. Spinning samples at this angle results in normally broad lines becoming narrower and increasing the resolution due to dipole-dipole interaction averaging to zero (Tycko and Dabbagh 1990). However, care must be taken to ensure that the sample is not damaged as a consequence of the high g forces generated by rapid spinning of the sample.

Advancement of NMR Spectroscopy

NMR spectroscopy is less sensitive in comparison to MS and can only detect metabolites in μM or mM concentration range. Therefore, numerous efforts have been focused on enhancing the sensitivity and resolution of NMR, in order to improve metabolite identification and quantification. One such approach is adding certain chemical compounds to the NMR tube in order to assist with identification of certain analytes with specific physicochemical properties known as "blind source separation (BSS)," which also can also help with decomposing the NMR signal into simpler elements (Bingol et al. 2016). Another strategy is using charged nanoparticle-based chemicals to selectively suppress metabolites with specific charge resulting in simplified NMR spectra (Zhang et al. 2015). The same group showed that addition of nanoparticles for removal of proteins, from serum for example, prior to ultrafiltration can result in excellent metabolite recovery and cleaner NMR spectra (Zhang et al. 2016). Methods to overcome the low signal-to-noise ratio (S/N) problem in ^{13}C NMR have also been proposed. For example, signal enhancement by spectral integration (SENSI) facilitates integration of a large number of spectra (Misawa et al. 2016). The authors demonstrated that combining 181 different spectra to create a synthesized single spectrum, achieved a 10-fold increase in S/N. This allowed the detection of 100 additional peaks (not visible in the individual spectra) without any need to increase data acquisition time for individual samples (Misawa et al. 2016). This method could be useful in metabolomics studies, where researchers are already collecting spectra from large sample sets. Another promising area is in nuclear polarization methods that can achieve significant sensitivity enhancement, while at the same time enabling real-time *in vivo* monitoring of metabolism. Two most common approaches are parahydrogen induced polarization (PHIP) and dynamic nuclear polarization (DNP) which also offer hyperpolarization of nuclei such as ^{13}C or ^{15}N and detection of downstream reactions providing great promise for metabolomics applications (Reile et al. 2016). There are many more advancements in the NMR-based metabolomics; for a more comprehensive review see (Nagana Gowda and Raftery 2017; Markley et al. 2017).

Conclusion

NMR is a highly reproducible and inherently quantitative technique with powerful molecular structure elucidation powers making it an extremely useful tool for metabolomics studies. Careful attention to pre-processing and processing is required in order to obtain good spectra required for robust analysis. There are three approaches to the analysis of spectra: fingerprinting, multi-integration of ROIs and targeted profiling. In fingerprinting the spectra are split into bins/buckets and statistical analysis is performed to identify which bins are able to discriminate between sample types. Only identification of the metabolites accounting for differences between those bins is required, making this the simplest method. In the ROI approach, knowledge of the spectra is used to split the spectra and again discriminatory regions are identified. In targeted profiling, known metabolites are identified and quantified using various software tools. The metabolite concentration is then compared between groups to identify significant changes. In all these methods, metabolite identification (whether at the beginning or the end or the pipeline) is crucial. NMR spectroscopy (^{1}H, ^{13}C and 2D) provides structural information that can be used to identify known metabolites through the use of various databases and tools. It can also be used in combination with MS for structural elucidation of novel metabolites with tremendous power. With the advent of Biobanks and Phenome Centers around the globe, the requirement for very large sample set metabolic profiling is growing rapidly. The stability and quantitative nature of NMR spectroscopy means it is ideally placed to meet this growing need. The future for NMR-based metabolomics is therefore very bright.

References

Alonso, A. et al. 2014. "Focus: A Robust Workflow for One-Dimensional NMR Spectral Analysis." *Analytical Chemistry* 86 (2): 1160–1169.

Anderson, P. E. et al. 2010. "Dynamic Adaptive Binning: An Improved Quantification Technique for NMR Spectroscopic Data." *Metabolomics: Official Journal of the Metabolomic Society* 7 (2): 179–190.

Balacco, G., and C. Cobas. 2009. "Automatic Phase Correction of 2D NMR Spectra by a Whitening Method." *Magnetic Resonance in Chemistry: MRC* 47 (4): 322–327.

Bao, Q. et al. 2012. "A New Automatic Baseline Correction Method Based on Iterative Method." *Journal of Magnetic Resonance* 218: 35–43.

Barding, G. A., Jr, R. Salditos, and C. K. Larive. 2012. "Quantitative NMR for Bioanalysis and Metabolomics." *Analytical and Bioanalytical Chemistry* 404 (4): 1165–1179.

Beckonert, O. et al. 2007. "Metabolic Profiling, Metabolomic and Metabonomic Procedures for NMR Spectroscopy of Urine, Plasma, Serum and Tissue Extracts." *Nature Protocols* 2 (11): 2692–2703.

Beckonert, O. et al. 2010. "High-Resolution Magic-Angle-Spinning NMR Spectroscopy for Metabolic Profiling of Intact Tissues." *Nature Protocols* 5 (6): 1019–1032.

Benjamin, D. I., B. F. Cravatt, and D. K. Nomura. 2012. "Global Profiling Strategies for Mapping Dysregulated Metabolic Pathways in Cancer." *Cell Metabolism* 16 (5): 565–577.

Bingol, K. et al. 2014. "Customized Metabolomics Database for the Analysis of NMR 1–1 H TOCSY and 13 C—1 H HSQC-TOCSY Spectra of Complex Mixtures." *Analytical Chemistry* 86 (11): 5494–5501.

Bingol, K. et al. 2015. "Unified and Isomer-Specific NMR Metabolomics Database for the Accurate Analysis of (13)C-(1)H HSQC Spectra." *ACS Chemical Biology* 10 (2): 452–459.

Bingol, K. et al. 2016. "Emerging New Strategies for Successful Metabolite Identification in Metabolomics." *Bioanalysis* 8 (6): 557–573.

Bingol, K., et al. 2015. "Metabolomics Beyond Spectroscopic Databases: A Combined MS/NMR Strategy for the Rapid Identification of New Metabolites in Complex Mixtures." *Analytical Chemistry* 87 (7): 3864–3870.

Bingol, K., and R. Brüschweiler. 2015. "NMR/MS Translator for the Enhanced Simultaneous Analysis of Metabolomics Mixtures by NMR Spectroscopy and Mass Spectrometry: Application to Human Urine." *Journal of Proteome Research* 14 (6): 2642–2648.

Bloemberg, T.G. et al. 2010. "Improved Parametric Time Warping for Proteomics." *Chemometrics and Intelligent Laboratory Systems* 104 (1): 65–74.

Bruker. Overview AMIX—Exploring of Spectroscopic Data: Unsurpassed Integration of Spectroscopy and Metabolomics. *Bruker.com*. Available at: https://www.bruker.com/products/mr/nmr/nmr-software/software/amix/overview.html [Accessed October 7, 2016].

Chan, E.C.Y. et al. 2009. "Metabolic Profiling of Human Colorectal Cancer Using High-Resolution Magic Angle Spinning Nuclear Magnetic Resonance (HR-MAS NMR) Spectroscopy and Gas Chromatography Mass Spectrometry (GC/MS)." *Journal of Proteome Research* 8 (1): 352–361.

Chen, L. et al. 2002. "An Efficient Algorithm for Automatic Phase Correction of NMR Spectra Based on Entropy Minimization." *Journal of Magnetic Resonance* 158 (1–2): 164–168.

Chenomx Inc. Metabolite Discovery and Measurement: Home. Available at: http://www.chenomx.com/ [Accessed October 7, 2016].

Chikayama, E. et al. 2010. "Statistical Indices for Simultaneous Large-Scale Metabolite Detections for a Single NMR Spectrum." *Analytical Chemistry* 82 (5): 1653–1658.

Chokkathukalam, A. et al. 2014. "Stable Isotope-Labeling Studies in Metabolomics: New Insights into Structure and Dynamics of Metabolic Networks." *Bioanalysis* 6 (4): 511–524.

Cloarec, O. et al. 2005. "Statistical Total Correlation Spectroscopy: An Exploratory Approach for Latent Biomarker Identification from Metabolic 1H NMR Data Sets." *Analytical Chemistry* 77 (5): 1282–1289.

Cohen, S. M., P. Glynn, and R. G. Shulman. 1981. "13C NMR Study of Gluconeogenesis from Labeled Alanine in Hepatocytes from Euthyroid and Hyperthyroid Rats." *Proceedings of the National Academy of Sciences of the United States of America* 78 (1): 60–64.

Cohen, S.M., S. Ogawa, and R. G. Shulman. 1979. "13C NMR Studies of Gluconeogenesis in Rat Liver Cells: Utilization of Labeled Glycerol by Cells from Euthyroid and Hyperthyroid Rats." *Proceedings of the National Academy of Sciences of the United States of America* 76 (4): 1603–1609.

Craig, A. et al. 2006. "Scaling and Normalization Effects in NMR Spectroscopic Metabonomic Data Sets." *Analytical Chemistry* 78 (7): 2262–2267.

Creek, D.J. et al. 2012. "Stable Isotope-Assisted Metabolomics for Network-Wide Metabolic Pathway Elucidation." *Analytical Chemistry* 84 (20): 8442–8447.

Davies, A.N. & Lampen, P. 1993. "JCAMP-DX for NMR." *Applied Spectroscopy* 47 (8): 1093–1099.

Davis, R.A. et al. 2007. "Adaptive Binning: An Improved Binning Method for Metabolomics Data Using the Undecimated Wavelet Transform." *Chemometrics and Intelligent Laboratory Systems* 85 (1): 144–154.

DeBerardinis, R.J. et al. 2007. "Beyond Aerobic Glycolysis: Transformed Cells Can Engage in Glutamine Metabolism that Exceeds the Requirement for Protein and Nucleotide synthesis." *Proceedings of the National Academy of Sciences of the United States of America* 104 (49): 19345–19350.

Dieterle, F. et al. 2006. "Probabilistic Quotient Normalization as Robust Method to Account for Dilution of Complex Biological Mixtures. Application in 1H NMR Metabonomics." *Analytical Chemistry* 78 (13): 4281–4290.

Dona, A.C. et al. 2016. "A Guide to the Identification of Metabolites in NMR-Based Metabonomics/Metabolomics Experiments." *Computational and Structural Biotechnology Journal* 14: 135–153.

Ellinger, J.J. et al. 2013. "Databases and Software for NMR-Based Metabolomics." *Current Metabolomics* 1 (1): 28–40.

Emwas, A.-H. et al. 2015. "Standardizing the Experimental Conditions for Using Urine in NMR-Based Metabolomic Studies with a Particular Focus on Diagnostic Studies: A Review." *Metabolomics* 11 (4): 872–894.

Everett, J.R. 2015. "A New Paradigm for Known Metabolite Identification in Metabonomics/Metabolomics: Metabolite Identification Efficiency." *Computational and Structural Biotechnology Journal* 13 (Supplement C): 131–144.

Fan, T. 1996. "Metabolite Profiling by One-and Two-Dimensional NMR Analysis of Complex Mixtures." *Progress in Nuclear Magnetic Resonance Spectroscopy* 28 (2): 161–219.

Fan, T.W.M. & Lane, A.N. 2011. "NMR-Based Stable Isotope Resolved Metabolomics in Systems Biochemistry." *Journal of Biomolecular NMR* 49 (3–4): 267–280.

Fan, T.W.M. et al. 2009. "Altered Regulation of Metabolic Pathways in Human Lung Cancer Discerned by 13 C Stable Isotope-Resolved Metabolomics (SIRM)." *Molecular Cancer* 8 (1): 1–19.

Giskeødegård, G.F. et al. 2010. "Alignment of High Resolution Magic Angle Spinning Magnetic Resonance Spectra Using Warping Methods." *Analytica Chimica Acta* 683 (1): 1–11.

Gómez, J. et al. 2014. "Dolphin: A Tool for Automatic Targeted Metabolite Profiling Using 1D and 2D (1)H-NMR Data." *Analytical and Bioanalytical Chemistry* 406 (30): 7967–7976.

Gromski, P.S. et al. 2014. "The Influence of Scaling Metabolomics Data on Model Classification Accuracy." *Metabolomics: Official Journal of the Metabolomic Society* 11 (3): 684–695.

Hao, J. et al. 2014. "Bayesian Deconvolution and Quantification of Metabolites in Complex 1D NMR Spectra Using BATMAN." *Nature Protocols* 9 (6): 1416–1427.

Haug, K. et al. 2012. MetaboLights: An Open-Access General-Purpose Repository for Metabolomics Studies and Associated Meta-Data. *Nucleic Acis Research* 41: D781–D786. doi:10.1093/nar/gks1004.

Hoch, J. C., and A. S. Stern. 1996. NMR Data Processing. Available at: http://eu.wiley.com/cda/product/00471039004,00.html.

Holmes, E., O. Cloarec, and J. K. Nicholson. 2006. "Probing Latent Biomarker Signatures and in Vivo Pathway Activity in Experimental Disease States Via Statistical Total Correlation Spectroscopy (STOCSY) of Biofluids: Application to $HgCl_2$ Toxicity." *Journal of Proteome Research* 5 (6): 1313–1320.

http://nautil.us/blog/the-math-trick-behind-mp3s-jpegs-and-homer-simpsons-face.

Jacob, D. et al. 2017. "NMRProcFlow: A Graphical and Interactive Tool Dedicated to 1D Spectra Processing for NMR-Based Metabolomics." *Metabolomics: Official Journal of the Metabolomic Society* 13 (4): 36.

Keeler, J. 2011. *Understanding NMR Spectroscopy*, Chichester, UK: John Wiley & Sons.

Kim, H. K., Y. H. Choi, and R. Verpoorte. 2011. "NMR-Based Plant Metabolomics: Where Do We Stand, Where Do We Go?" *Trends in Biotechnology* 29 (6): 267–275.

Kuchal, P. W. 2002. "Spin Dynamics: Basics of Nuclear Magnetic Resonance, by Malcolm H. Levitt." *NMR in Biomedicine* 15 (4): 301–304.

Le, A. et al. 2012. "Glucose-Independent Glutamine Metabolism via TCA Cycling for Proliferation and Survival in B Cells." *Cell Metabolism* 15 (1): 110–121.

Lindon, J.C., Nicholson, J.K., and Everett, J.R. 1999. "NMR Spectroscopy of Biofluids." In *Annual Reports on NMR Spectroscopy*, edited by G.A. Webb. San Diego, CA: Academic Press, pp. 1–88.

Liu, W. et al. 2012. "Reprogramming of Proline and Glutamine Metabolism Contributes to the Proliferative and Metabolic Responses Regulated by Oncogenic Transcription Factor c-MYC." *Proceedings of the National Academy of Sciences of the United States of America* 109 (23): 8983–8988.

Ludwig, C. et al. 2011. "Birmingham Metabolite Library: A Publicly Accessible Database of 1-D 1H and 2-D 1H J-Resolved NMR Spectra of Authentic Metabolite Standards (BML-NMR)." *Metabolomics: Official Journal of the Metabolomic Society* 8 (1): 8–18.

Ludwig, C., and M.R. Viant. 2010. Two-Dimensional J-Resolved NMR Spectroscopy: Review of a Key Methodology in the Metabolomics Toolbox. *Phytochemical Analysis: PCA* 21 (1): 22–32. doi:10.1002/pca.1186/full.

Markley, J.L. et al. 2017. "The Future of NMR-Based Metabolomics." *Current Opinion in Biotechnology* 43: 34–40.

Misawa, T. et al. 2016. "SENSI: Signal Enhancement by Spectral Integration for the Analysis of Metabolic Mixtures." *Chemical Communications* 52 (14): 2964–2967.

Montigny, F. et al. 1990. "Automatic Phase Correction of Fourier-Transform NMR Data and Estimation of Peak Area by Fitting to a Lorentzian Shape." *Analytical Chemistry* 62 (8): 864–867.

Mullen, A.R. et al. 2012. "Reductive Carboxylation Supports Growth in Tumour Cells with Defective Mitochondria." *Nature* 481 (7381): 385–388.

Nagana Gowda, G.A., and D. Raftery. 2017. "Recent Advances in NMR-Based Metabolomics." *Analytical Chemistry* 89 (1): 490–510.

nmrML—home. Available at: http://nmrml.org/ [Accessed October 9, 2016].

Pearce, J.T.M. et al. 2008. "Robust Algorithms for Automated Chemical Shift Calibration of 1D 1H NMR Spectra of Blood Serum." *Analytical Chemistry* 80 (18): 7158–7162.

Pudakalakatti, S.M. et al. 2014. "A Fast NMR Method for Resonance Assignments: Application to Metabolomics." *Journal of Biomolecular NMR* 58 (3): 165–173.

Ravanbakhsh, S. et al. 2015. "Accurate, Fully-Automated NMR Spectral Profiling for Metabolomics." *PLoS One* 10 (5): e0124219.

Reile, I. et al. 2016. "NMR Detection in Biofluid Extracts at Sub-μM Concentrations via para-H2 Induced Hyperpolarization." *The Analyst* 141 (13): 4001–4005.

Robinette, S.L. et al. 2012. "NMR in Metabolomics and Natural Products Research: Two Sides of the Same Coin." *Accounts of Chemical Research* 45 (2): 288–297.

Rocca-Serra, P. et al. 2016. Data Standards can Boost Metabolomics Research, and if There is a Will, There is a Way. Metabolomics 12: 14. doi:10.1007/s11306-015-0879-3.

Rubtsov, D.V. et al. 2007. "Proposed Reporting Requirements for the Description of NMR-Based Metabolomics Experiments." *Metabolomics: Official Journal of the Metabolomic Society* 3 (3): 223–229.

Sansone, S.A. et al. 2007. "The Metabolomics Standards Initiative." *Nature Biotechnology* 25 (8): 846–848.

Shimizu, A., M. Ikeguchi, and S. Sugai. 1994. "Appropriateness of DSS and TSP as Internal References for 1H NMR Studies of Molten Globule Proteins in Aqueous Media." *Journal of Biomolecular NMR* 4 (6): 859–862.

Smolinska, A. et al. 2012. "NMR and Pattern Recognition Methods in Metabolomics: From Data Acquisition to Biomarker Discovery: A Review." *Analytica Chimica Acta* 750: 82–97.

Sousa, S. A. A., A. Magalhães, and M. M. C. Ferreira, 2013. "Optimized Bucketing for NMR Spectra: Three Case Studies." *Chemometrics and Intelligent Laboratory Systems* 122: 93–102.

Spicer, R. et al. 2017. "Navigating Freely-Available Software Tools for Metabolomics Analysis." *Metabolomics: Official Journal of the Metabolomic Society* 13 (9): 106.

Spraul, M. et al. 2015. Wine Analysis to Check Quality and Authenticity by Fully-Automated 1H-NMR. *BIO Web of Conferences*, 5, pp. 02022. Available at: http://oiv.edpsciences.org/articles/bioconf/abs/2015/02/bioconf_oiv2015_02022/bioconf_oiv2015_02022.html [Accessed October 9, 2018].

Sumner, L.W. et al. 2007a. "Proposed Minimum Reporting Standards for Chemical Analysis." *Metabolomics: Official Journal of the Metabolomic Society* 3 (3): 211–221.

Sumner, L.W. et al. 2007b. "Proposed Minimum Reporting Standards for Chemical Analysis Chemical Analysis Working Group (CAWG) Metabolomics Standards Initiative (MSI)." *Metabolomics: Official Journal of the Metabolomic Society* 3 (3): 211–221.

Takis, P. G. et al. 2017. "Deconvoluting Interrelationships Between Concentrations and Chemical Shifts in Urine Provides a Powerful Analysis Tool." *Nature Communications* 8 (1): 1662.

The math trick behind MP3s, JPEGs, and Homer Simpson's Face—Facts so Romantic—nautilus. *Nautilus*. Available at: http://nautil.us/blog/the-math-trick-behind-mp3s-jpegs-and-homer-simpsons-face [Accessed October 7, 2016].

Tulpan, D. et al. 2011. "MetaboHunter: An Automatic Approach for Identification of Metabolites from 1H-NMR Spectra of Complex Mixtures." *BMC Bioinformatics*, 12: 400.

Tycko, R., and G. Dabbagh. 1990. "Measurement of Nuclear Magnetic Dipole—Dipole Couplings in Magic Angle Spinning NMR." *Chemical Physics Letters,* 173(5): 461–465.

van der Hooft, J.J.J. et al. 2013. "Structural Elucidation of Low Abundant Metabolites in Complex Sample Matrices." *Metabolomics: Official Journal of the Metabolomic Society* 9 (5): 1009–1018.

Verwaest, K. A. et al. 2011. 1H NMR-Based Metabolomics of CSF and Blood Serum: A Metabolic Profile for a Transgenic Rat Model of Huntington Disease. *Biochimica et Biophysica Acta (BBA)—Molecular Basis of Disease* 1812 (11): 1371–1379.

Veselkov, K. A. et al. 2009. "Recursive Segment-Wise Peak Alignment of Biological 1 H NMR Spectra for Improved Metabolic Biomarker Recovery." *Analytical Chemistry* 81 (1): 56–66.

Vettukattil, R. 2015. Preprocessing of Raw Metabonomic Data. *Methods in Molecular Biology* 1277: 123–136.

Vu, T.N. et al. 2011. "An Integrated Workflow for Robust Alignment and Simplified Quantitative Analysis of NMR Spectrometry Data." *BMC Bioinformatics* 12: 405.

Vu, T.N., and K. Laukens. 2013. "Getting Your Peaks in Line: A Review of Alignment Methods for NMR Spectral Data." *Metabolites* 3 (2): 259–276.

Wishart, D. S. 2008/2003. "Quantitative Metabolomics Using NMR." *Trends in Analytical Chemistry: TRAC* 27 (3): 228–237.

Xi, Y., and D.M. Rocke. 2008. "Baseline Correction for NMR Spectroscopic Metabolomics Data Analysis." *BMC Bioinformatics* 9: 324.

Xia, J. et al. 2008. "MetaboMiner—Semi-automated Identification of Metabolites from 2D NMR Spectra of Complex Biofluids." *BMC Bioinformatics* 9: 507.

Zhang, B. et al. 2015. "Use of Charged Nanoparticles in NMR-based Metabolomics for Spectral Simplification and Improved Metabolite Identification." *Analytical Chemistry* 87 (14): 7211–7217.

Zhang, B. et al. 2016. "Nanoparticle-assisted Removal of Protein in Human Serum for Metabolomics Studies." *Analytical Chemistry* 88 (1): 1003–1007.

Zhang, Z.-M., S. Chen, and Y.-Z. Liang. 2010. "Baseline Correction Using Adaptive Iteratively Reweighted Penalized Least Squares." *The Analyst* 135 (5): 1138–1146.

Ziessow, D. 1987. Book Review: Principles of Nuclear Magnetic Resonance in One and Two Dimensions (International Series of Monographs on Chemistry 14). By R. E. Ernst, G. Bodenhausen, and A. Wokaun. *Angewandte Chemie* 26 (11): 1192–1195.

6

Statistics: The Essentials

Ron Wehrens and Pietro Franceschi

CONTENTS

Introduction 129
Distributions 130
Inference 133
Confidence Intervals 133
Statistical Testing 134
Multiple Testing 136
Linear Regression 140
Multivariate Statistics 142
Principal Component Analysis 145
Partial Least Squares Regression 148
Discussion 153
References 155

Introduction

Once you have done your experiments and have the data preprocessed and nicely organized, you will want to extract the information that should be in there. This sometimes can be a complicated procedure. If an experiment is well designed, the analysis strategies are usually already discussed in advance, but still the data may need to be cleaned or preprocessed. In other cases, especially common in untargeted experiments, it is not clear *a priori* what questions to ask and what the data analysis is supposed to lead to. In such a case, the idea is usually to start with summarizing and visualizing the data, and then take things from there. Visualization, in particular, is of extreme importance. It will show you whether there are unexpected results, is useful as an informal—but powerful!—means of quality control and can provide pointers to new questions and even suggestions for new experiments. Such an exploratory analysis forms one of the two main approaches to statistics.

The other approach is focused more on inference, on formulating deductions and conclusions on the basis of the data. These conclusions usually are about a property of a population, whereas we are looking at only a part, a subset, of the population. In statistics, such a subset is referred to as a sample.

Be careful not to confuse this with the other use of the word, the sample that is analyzed in the lab: a statistical sample will usually consist of several physical lab samples. A key property of the sample is that it should be representative of the population, for obvious reasons. We have already discussed sampling issues in Chapter 2—here, we will concentrate on the essentials of the statistical techniques often applied in metabolomics research, in order to draw conclusions about the biological system at hand. We focus on concepts and ideas rather than on formulas. More details can be found in textbooks and in the references.

We start with a section on distributions, with the main aim to explain why in statistics one often assumes a particular distribution and why it is important to do so. The next section is on inference: we can use the fitted distributions to derive properties and answer questions by statistical hypothesis testing. Another possibility is to build statistical models that allow predictions. The chapter concludes with a discussion of the most common forms of multivariate analysis, Principal Component Analysis (PCA) and Partial Least Squares (PLS) regression. For these, distributional assumptions are less important (simply because they cannot be made in a sensible way) and much more stress is placed on model validation procedures.

Distributions

Very often in the description of a statistical analysis, one sees a phrase like "we assume a normal distribution." Such an assumption allows one to summarize the data points, possibly quite a lot of data points, in a few key numbers. The normal distribution, for example, is completely defined by two parameters: the mean μ and the variance σ^2 (or the square root of the variance, the standard deviation); also other distributions are determined by a very limited number of parameters. We can never know the "true" values of these parameters, but we can try to estimate them. The estimates for mean and variance usually are indicated with $\bar{x}$ and s^2, respectively, and are given by the well-known formulae

$$\bar{x} = \Sigma_n x_n / N$$

and

$$s^2 = \Sigma_n \left(x_n - \bar{x}\right)^2 / (N-1)$$

where N is the total number of observations, and n runs from 1 to N. Note that these estimates are independent of any distributional assumptions: they are just formulae.

What distribution to use depends on the (usually unknown) mechanism generating the data. For real-valued data, often a normal distribution works quite well—the so-called central-limit theorem shows that the result of a large number of independent disturbances of the measurement usually leads to a normal distribution, whatever the form and distributions of the individual disturbances. This can be visualized by the results of throwing dice in Figure 6.1—throwing one dice leads to a flat, uniform distribution; throwing two dice and adding the numbers already leads to a triangular distribution with a clear peak in the center, and adding the results of throwing ten dice already is very close to a normal distribution. Indeed, the agreement with the normal distribution with empirical mean and standard deviation, shown in the right plot of Figure 6.1, is excellent. The fact that the overall influence of many unrelated effects on the eventual measurement often leads to normally distributed data is a major reason why the normal distribution plays such a central role in many statistical procedures; another is that its mathematical form leads to convenient equations.

Assessing the agreement of an empirical distribution (the bars in Figure 6.1) with a theoretical one (such as the curves in Figure 6.1) is actually more conveniently done using quantile-quantile (QQ) plots, where quantiles of both the empirical and theoretical distributions are plotted against each other. For a good agreement, the points are close to a straight line. For example, the data from the rightmost plot in Figure 6.1 are shown in a QQ plot in Figure 6.2. Indeed, we see that the agreement with the normal distribution is excellent. Note that such plots can be made for any type of distribution, not just for the normal distribution.

Once we have a distribution that fits the data well, we can use this to calculate many things that cannot easily be calculated from the data themselves; confidence intervals, for one (see below). There are situations where other distributions are better suited than the normal distribution. For small

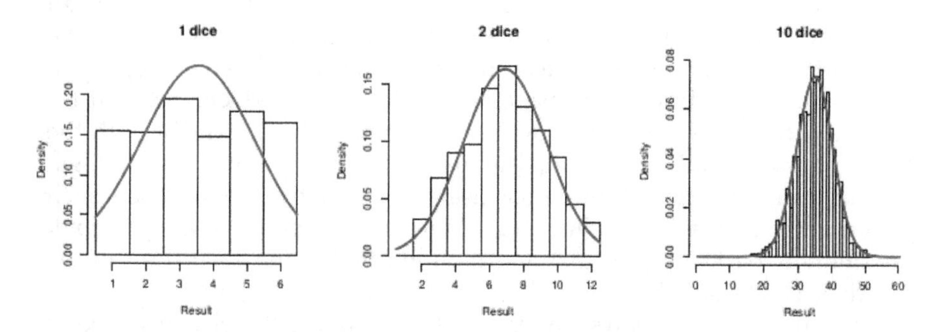

FIGURE 6.1
Illustration of the central limit theorem: adding several independent sources of variation leads to a distribution that is close to normal. Left plot: simulated outcome of 100 throws with one dice; middle plot: 100 throws of two dice (the sum is shown); right plot: 100 throws of ten dice.

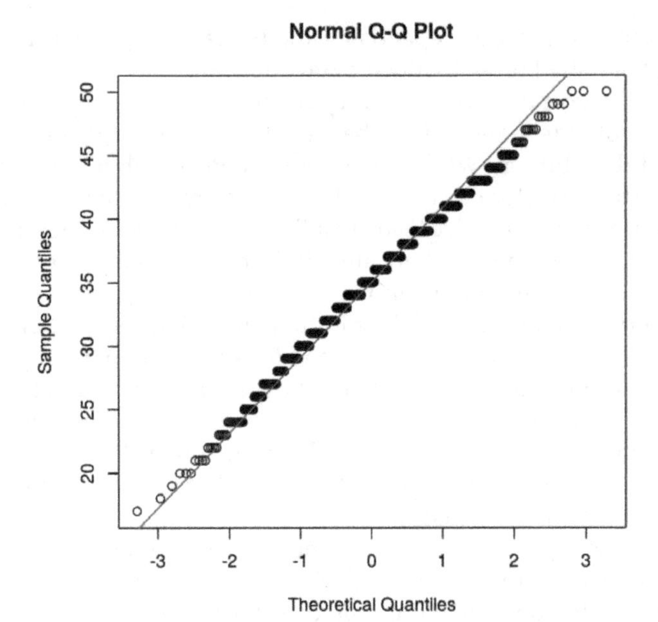

FIGURE 6.2
QQ plot for simulated dice data (100 throws of 10 dice).

numbers of measurements, the t-distribution is usually preferred to the normal distribution because it is wider, representing more uncertainty about the location of the unknown mean. For larger numbers of measurements, the t-distribution gets more narrow, as one would expect; in the limit (an infinite number of measurements), the t-distribution is equal to the normal distribution. Counts are often well described by a Poisson distribution; when the counts are really large (as typically is the case in metabolomics experiments where the measured signal often is a count statistic) we can also treat them as real numbers on a continuous scale. However, since the measurement error in these cases usually is proportional to the intensity of the signal, a logarithmic transformation is often employed. On the logarithmic scale, the measurement error then is constant, in agreement with the assumptions behind many statistical analyses. Also in other cases a logarithmic transformation of the data is useful. For positive real numbers such as concentrations, where values below zero are not possible, a log-normal distribution, basically a normal distribution fitted to the logarithms of the raw data, can be suitable.

A careful statistical analysis often involves studying the physical process generating the data, or choosing the most appropriate description of the data on the basis of visual inspection, e.g., using the QQ plots shown above. The chosen distribution can then be used for inferential purposes.

Inference

With the term "inference" we mean the process of deriving logical conclusions from the data. Examples are the derivation of confidence intervals for a sample mean or for a prediction, and statistical testing.

Confidence Intervals

In the natural sciences it is quite common to report both an average and a standard deviation as a summary of a series of repeated experiments. An alternative is to provide an average and a confidence interval, often covering 95% of the probability of the values of a variable (derived from the observations). For a normally distributed variable *X*, a 95% confidence interval is given by

$$CI = \bar{x} \pm z_{0.05} sd(x) = \bar{x} \pm z_{0.05} s$$

where the z value, taken from a table or a computer program, determines the desired coverage. The z value is usually given as the complement of the coverage of the confidence interval, so an interval of 95% corresponds to a z value at level 0.05. The corresponding interval for a t-distribution will be very similar:

$$CI = \bar{x} \pm t_{df,0.05} s$$

where the t value that replaces z is determined by an additional parameter, the number of degrees of freedom. Here, we use one degree of freedom in estimating the mean (the standard deviation is calculated with respect to the mean), so the number of degrees of freedom here is $n-1$. The larger this number, the more similar t becomes to z; with $df = \infty$ the two are equal. These confidence intervals are intervals for individual data points; that is, they represent the intervals within which we expect new data to be.

One can also calculate confidence intervals for the mean. These are more narrow and get narrower still when more points are used to calculate the mean. This is in line with intuition: when we have more data points, we are more certain about the location of the mean. The formulas are almost similar to the ones seen before—the one using the t statistic is

$$CI_{\mu} = \bar{x} \pm t_{df,0.05} s/\sqrt{N}$$

where the subscript μ indicates that this is a confidence interval for the mean. From the formula one can see that the size of the confidence interval halves when we have four times more data points.

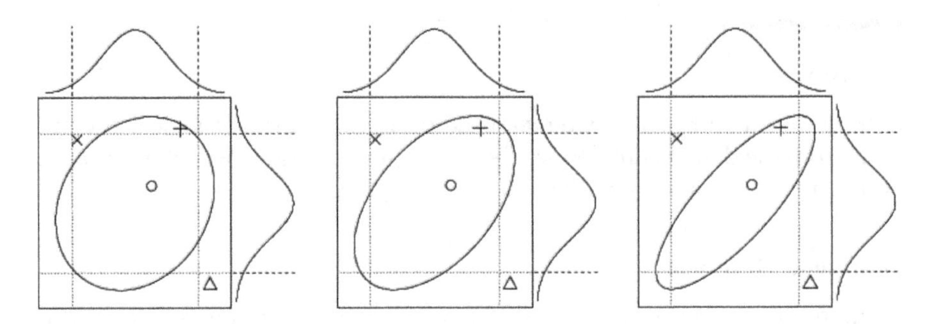

FIGURE 6.3
Univariate (dashed and dotted lines) and bivariate (ellipses) 95% confidence intervals for data with low, medium and high correlations (from left to right, respectively).

Also confidence intervals for two or more variables simultaneously can be calculated. The covariance (or correlation) between the variables then needs to be taken into account. Examples for a couple of bivariate normal distributions with different covariances are given in Figure 6.3.

When two variables do not show a very high correlation, the bivariate confidence ellipse (for individual data points) is close to a circle, but with increasing correlation the ellipse becomes more and more narrow. This makes sense: given information on one variable, we can also hazard a guess on what the other variable may look like. The symbols in the plot show four different situations that can occur. The circle corresponds to a point that is within the 95% confidence intervals for both the univariate and bivariate cases; the triangle gives a point that is outside all intervals. The interesting points are the plus and the cross signs—they show that it is possible to be outside (at least one of) the univariate confidence intervals and yet be within the bivariate interval, and the opposite. The higher the correlation, the easier it is to achieve such a situation. If you are thinking: "how bad can it be?" consider that in metabolomics very often one metabolite is represented in the data by several features (mass peaks in MS-based methods, ppm values in NMR data) that by definition will be very highly correlated ($r>0.95$ is not uncommon!)—one very good reason to try and aggregate the features into relative metabolite abundances and do the statistical analyses on these.

Statistical Testing

The ultimate objective of scientific research is to derive general rules from the results of a limited number of experiments. The common setting for this process is that one wants to use the data to support a specific hypothesis generated in the context of a broad scientific question. We could, for example, want to check the hypothesis that there is a difference on the average concentration of a specific pigment between white and red roses. To verify this hypothesis we design an experiment, collect several flower samples and analyze the concentration of that pigment in the samples. Our investigation ends up with the mean concentration of the pigment in the two classes. As expected, the two means are different. The question is whether this difference supports our initial hypothesis or it can be attributed to chance alone. Since we have analyzed only a limited number of samples and not the complete population of roses, the question has to be asked in a statistical way. This means that it will never be possible to be completely sure of the answer; the best that can be done is to quantify this unavoidable uncertainty. This step is where statistical hypothesis testing comes into play.

First of all, it is necessary to define in a precise way what we want to test. The usual approach is to assess if the difference between the two experimental means is likely to occur by chance, assuming the samples arise from the same population of roses. In other words, we assume that the distribution of the pigment is the same for all the roses. This "one population" hypothesis states that chance alone is responsible for the observed results and it is called null hypothesis (H0). If the null hypothesis is unlikely it will be reasonable to conclude the opposite, namely that the two groups have actually different means. In that case one can reject H0 and accept the alternative hypothesis (H1). This way of looking at the problem has the advantage that to test H0 we do not need to make any assumption about the means of the two populations under the alternative hypothesis. In a sense, there is only one H0 for an infinite number of H1 hypotheses. The actual testing of H0 can be performed in many ways and the interested reader is referred to more specific statistical resources (Armitage et al., 2008). The testing process will return the probability of obtaining a difference at least as large as the observed one under the null hypothesis, the famous (or infamous) p value. Such a p value will never be exactly zero: there is always the possibility that the two samples appear different only by chance and that they are in reality coming from the same population. It is therefore necessary to decide a cut-off value, α: a p-value smaller than α means one rejects H0 at the $1-\alpha$ confidence level. For historical—and formerly practical—reasons, cut-off values of 0.05, 0.01 or 0.001 are most common. It is important to point out that these limits should not be used as guillotines: the chosen α does not represent a distinction between absolute truth and a respectable opinion. The chosen confidence level does enable others to replicate the reasoning during the analysis, and therefore must be explicitly indicated when presenting analysis results.

Differences between two mean values $\bar{x}_1$ and $\bar{x}_2$ are usually assessed using a so-called t-test, where the t statistic is given by

$$t = \frac{|\bar{x}_1 - \bar{x}_2|}{sd\left(\bar{x}_1 - \bar{x}_2\right)}$$

that is, the (absolute) difference expressed in terms of standard deviation. Depending on the number of samples available and whether or not the standard deviations in the two groups can be assumed to be equal, there are different ways of calculating the denominator in this equation—modern software packages like R will take care of that for you. The calculated t value is then compared to what you would expect to find under the null hypothesis, which will lead to the p values mentioned earlier. Again, leave this step to the software. For other operations than comparing two means, other tests are available; one example is the F-test for comparing two variances. In all cases, a test statistic is calculated and compared to the expected values under the null hypothesis.

In the statistical tests discussed so far, we have focused on so-called parametric tests, in which the distribution of the test statistic (a difference, a ratio of variances, etc.) is described quite precisely. If such a description is incorrect, obviously the outcome of the test cannot be trusted. That is why testing for normality (the null hypothesis says that the data are normally distributed) is such a popular pastime. Its value is rather overstated in many cases: with few samples it is very hard to detect a deviation from normality, and conversely, with very many samples almost any irregularity will lead to a rejection of the null hypothesis and will point to a significant departure from the normal distribution. Non-parametric tests circumvent the problem by relying on much less stringent assumptions. The Mann-Whitney U test, for example, is the non-parametric counterpart of the t-test. It does not test whether the mean values are significantly different, but assesses whether the two groups come from the same population. It does this by considering the order of the data points (the ranks) rather than the measurements themselves, and as a result the influence of outliers is greatly diminished. For most statistical tests, non-parametric versions are available, for which the assumptions are much less important.

The catch is (and of course there is a catch) that these tests are less powerful than their parametric equivalents, especially for low numbers of cases—this means that it is more difficult to find a difference that is there. This is normally referred to as the "power" of a test, defined by $1-\beta$ where β is the fraction of false negatives (FNs, also referred to as Type-II errors), i.e., cases where the null hypothesis is false but is not rejected. Conversely, cases where the null hypothesis is incorrectly rejected lead to false positives (FPs, Type-I errors). The meaning of true positives and true negatives is analogous. Note that defining α for assessing significance automatically, for a given difference, defines β as well: we can minimize or even eliminate all FPs (by choosing α very small, basically accepting the null hypothesis in all cases) but that by definition leads to a large number of FNs and a low power.

Multiple Testing

The scenario just described to introduce statistical testing is difficult to apply directly in an untargeted metabolomics context. If one is interested in finding what is differentiating red and white roses, it is common to measure up to several thousands of variables (features or metabolites) in a number of roses of both colors and try to identify which variables are different in the two groups, using thousands of tests. These variables are then potential biomarkers. This shift of perspective brings the statistical framework from "one hypothesis testing" to "multiple hypothesis testing," an apparently small change that has a profound impact on the statistical aspects of the problem. The natural extension of the reasoning just described passes through the definition of a new H0, which now assumes that the observed differences in the means of any of the measured variables are the result of chance alone. To test this hypothesis it is possible to combine the p-values of the single variable tests, but it is clear that the possibility of finding at least one variable which is significantly different "by chance" grows with the number of variables. To better clarify this point, consider a hypothetical experiment where our red and white roses are compared for two independent metabolites (variables), each tested at the 0.05 level. For the first variable there is a 5% chance of wrongly concluding that the two populations are different, and the very same reasoning holds for the second variable. Because of that, and since the two tests are independent, the probability of concluding that the two types of roses are different in at least one variable grows to 9.75% ($1 - 0.95 \times 0.95$). It is easy to imagine the impact of this problem in an untargeted metabolomics experiment dealing with thousands of different experimental variables.

To cope with this multiple testing problem, it is possible to follow two approaches. The first one is based on the idea that if the two types of roses are not different, none of the variables should show a statistically significant difference. To control this family wise error rate (FWER) at a specific level, the more straightforward way is to reduce α proportionally to the number of variables: α/N. In the previous example, to control the FWER at 5% level, single variables should be tested at the 2.5% level.

This leads to the well-known Bonferroni correction, which is appealing for its simplicity, but in experiments with thousands of variables it is extremely strict (to control the FWER at 5% level with 1000 variables, each test should be performed at a $0.05/1000 = 5 \times 10^{-5}$ level). A more liberal way to deal with multiple testing issues is to relax the previous criterion and accept the presence of some false positives (also named false discoveries) in the list of variables. How to practically achieve the control of the false discovery rate (FDR) has been the subject of extensive statistical research and its treatment goes beyond the scope of this book. For a more detailed coverage of the topic in the context of mass spectrometry based-omics technologies the reader can refer to reference Franceschi et al. (2013).

It should be clear that there is always a chance of incorrectly rejecting the null hypothesis for some experimental variables, which are then wrongly labeled as biomarkers. The concentration of those variables in the data set at hand will appear to be clearly different; the point is that it is not possible to ascertain whether the same variable will be confirmed as "biomarker" in a new and independent validation experiment. In other words, false positives show apparent differences at the "sample" level where there are no differences at the "population" level.

To show the statistical testing machinery at work, consider the positive ionization data from MTBLS18. The full data set contains 202 samples and 5930 variables. For our first example let us focus on two specific variables: variables M_2696 (m/z = 487.151, rt = 160 sec) and M_1329 (m/z = 319.24, rt = 547 sec). We want to test if the concentration of these two "metabolites" is different in the two treatment types ("water" and "phytophthora") regardless the genotype. To make the distributions of the intensities of the samples more close to normality, we log transform the data. In both cases, the null hypothesis is that the effect of the two treatments is the same, so under H0 the mean value of the two variables in the "water" and "phytophthora" groups should be the same.

To test this hypothesis we perform a t-test. As discussed before, we decide a cut-off confidence level, and to be quite strict, we set it at 0.01. The output of the t-test tells us that the difference in the mean intensities of variable M_2969 is statistically significant at the chosen level, while it is not for variable M_1329. In other words, the chance of finding such a difference for M_2969 is smaller than 1%, under the null hypothesis of no difference, whereas for M_1329 it is more likely. The actual concentrations of the two metabolites in the data set are shown in the stripchart[1] plot presented in Figure 6.4. The outcomes of the plots are exactly in line to what one could expect: the means of variable M_2969 looks indeed different in the two treatment groups, while in the other case the two groups look really similar.

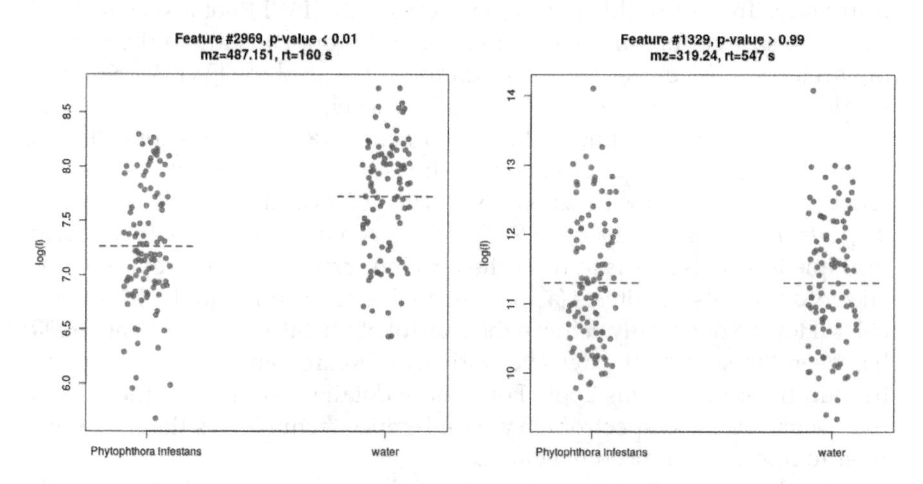

FIGURE 6.4
Stripchart plot of the log transformed intensities of the variables M_2969 (left) and M_1329 (right) of MTBLS18. The horizontal lines highlight the mean intensity values. The mean intensity of M_2926 in the two groups is statistically different at the 0.01 level.

[1] A stripchart randomly spreads the individual data points on the X-axis to better show how values are distributed.

Let's use the same data set to illustrate the "multiple testing issue." First of all, we randomly allocate each sample to either the "water" or "phytophthora" treatment group in order create an artificial situation where there should be no difference in the two groups. Suppose now that this new data set is the result of an untargeted metabolomics experiment designed to investigate whether the two treatment groups are different or not.

The question, now, is radically different from the previous case: there we were focusing on the values of variables M_2969 and M_1329; here we want to perform the testing on at least one of the 5930 variables. The more natural way to do that is to perform 5930 t-tests and look to their p-values, setting an α level of 0.05.

Considering that the two treatment types ("water" and "phytophthora") are now not different due to the random labeling, one would be tempted to expect that no variable should show a statistically significant difference. However, if we actually perform analysis we find that out of the 5930 variables, 268 show a p-value lower than α, so we have 268 potential biomarkers. The histogram showing the distribution of the p-values is displayed in Figure 6.5.

As already discussed, the reason for that goes back to the essentials of statistics. A p-value lower than α does not mean that we are sure that this variable is different at the population level, but only that there is a low probability of finding a difference of this size (or bigger) under the null hypothesis. It is clear that the possibility of finding at least one variable which is significantly different "by chance" grows if we do 5930 trials. The Bonferroni correction we have just described is able to reduce the number of false biomarkers. The problem is that with such a big number of variables, to control the FWER at the 0.01 level, the "single test" threshold goes down to $0.05/5930 = 1 \times 10^{e-6}$. This very low level could make it difficult to preserve

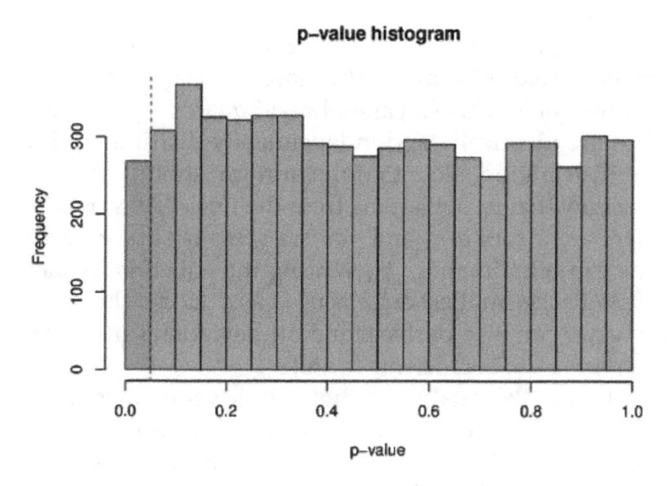

FIGURE 6.5
Histogram showing the distribution of the p-values in the random-labeled data set. The vertical red line and the blue bar show the fraction of false positives.

also any true biomarkers: indeed, trying harder to avoid false positives will always lead to more false negatives (we could avoid false positives altogether by always accepting the null hypothesis, but there would be no true positives as a result, only false negatives). As discussed in the section dedicated to theory, there are less conservative ways to deal with the multiple testing issue, which are based on the control of the false discovery rate.

Linear Regression

When we have two variables, we are often interested in the relation between them. We could, for instance, use the values of one to predict the other. A classical example from analytical chemistry is to try to calculate the concentration of a compound (y) in one sample from the measurement of a measured spectroscopic quantity such as absorption (x). To do that it is necessary to define the general relation between the two quantities. In this specific case, the relation can be expressed in a linear form:

$$y = b_0 + xb_1 + \varepsilon$$

where y is the quantity to be predicted (the concentration), x is the observed quantity (the absorption), and b_0 and b_1 are the regression coefficients, corresponding to intercept and slope, respectively. Finally, ε is the residual, the difference between the true, measured value and the value according to the linear model. Given a series of measurements of the absorption at specific concentrations (like the one which could be obtained in a series of dilutions of a chemical standard) (x_i, y_i) one can fit the line and obtain estimates for b_0 and b_1. This fitting is done by minimizing the differences between the model values and the measured values: to be more exact, by minimizing the sum of the squared residuals. This so-called Least-Squares regression assumes that the residuals are identically and independently distributed (i.i.d) according to a normal distribution. Note that the minimization is done on the residuals in y direction: that is, deviations from the line are assumed to occur only because there are errors in y, and not in x. This is almost never true, but in practice one can often manage by writing the equation in such a way that the variable with the smallest error is used as x. Under the above-mentioned assumptions, we can also derive standard deviations and confidence intervals for the regression coefficients b_0 and b_1.

We can estimate the quality of the fit in several ways. Often, a p-value for the regression is given, which is obtained by comparing the variance of the residuals with the variance explained by the line, using an F-test. In univariate regression, this is equivalent to testing whether b_1 differs from zero: a regression line with a slope of zero clearly indicates that there is no relation between x and y, and therefore is not significant.

One does not have to limit oneself to just one independent variable: often, a better result is obtained using more information. The taste of a tomato, for example, may be adequately modeled by the concentrations of a number of metabolites:

$$y = b_0 + b_1 x_1 + b_2 x_2 + \ldots + b_M x_M$$

As long as M, the number of variables, is smaller than N, the number of samples, this equation can be solved in exactly the same way as with univariate regression. If $M > N$ (but also if the predictors are correlated) we need special tricks—these will be discussed in the next section.

A final remark about regression: both the independent and dependent variables need not be numerical in nature. If y, the dependent variable, is a class label (e.g., treated or untreated, sample or control), the corresponding modeling technique is often called Discriminant Analysis, or more generally, classification. Using the above equation in the analysis of independent variables that are categorical rather than numerical leads to Analysis of Variance (ANOVA) models; for data containing both continuous and discrete independent variables the technique is called Analysis of Covariance (ANCOVA). Such data often occur in the analysis of experimental designs. In the MTBLS18 data, for example, one may want to model the influence of the Treatment and Genotype experimental factors, and their interaction, on a specific feature, in this case a mass peak. The equation (in so-called Wilkinson–Rogers notation) is:

$$\text{feature} \sim \text{Genotype} + \text{Treatment} + \text{Genotype} \times \text{Treatment}$$

When fitting such a model, we get coefficients for each of the experimental levels and associated p-values. This can be done for all individual mass peaks in the data set. As an example, the summary of one of these (*viz.*, the one with the largest F statistic) is given in Table 6.1.

The first column, labeled "Estimate," gives the size of the coefficient. The next columns give the standard error, t-value and p-value, respectively. Of the six genotypes, only five are visible: that is because the software (in this case R) has chosen to use the first genotype as the reference level, and to include that in the intercept. The same is true for the treatment factor: only the Phytophthora treatment is visible, while the Control level is included in the intercept. We see that the p-values for two genotypes, erp2 and pen2, are smaller than 0.05, which means they are significantly different from zero. Likewise, we see that the coefficient for Phytophthora is significantly different from zero. This means that the intensity of this feature depends on genotype and treatment, and since the last interaction coefficient is also significant, it means that the effect of the treatment is not the same across all genotypes. Note that you can choose a different genotype as a reference as well, or, alternatively, perform all pairwise combinations to see which genotypes are similar and which are different (using, again, multiple testing corrections!). Very often such statistical rigor is not really needed for a useful interpretation of the data.

TABLE 6.1

Regression Coefficients for the Linear Model Describing the Phytophthora Data

	Estimate	Std. Error	t value	Pr(>|t|)
(Intercept)	8330.7985	732.5024	11.37	**0.0000**
D	1201.7847	1035.9149	1.16	0.2492
erp1	-1648.0566	1035.9149	-1.59	0.1153
erp2	-2524.5253	1107.4396	-2.28	**0.0251**
gl1	1588.9920	1107.4396	1.43	0.1549
pen2	3707.2581	1107.4396	3.35	**0.0012**
phytophthora	-2784.8944	1035.9149	-2.69	**0.0086**
D:Phytophthora	-2126.0686	1465.0049	-1.45	0.1503
erp1:Phytophthora	-100.9596	1465.0049	-0.07	0.9452
erp2:Phytophthora	587.4456	1566.1561	0.38	0.7085
gl1:Phytophthora	-2643.1873	1538.3797	-1.72	0.0893
pen2:Phytophthora	-4174.8513	1516.4241	-2.75	**0.0072**

p-values are shown in the last column; The first five lines after the "Intercept" line are the genotypes; "Phytophthora" is the treatment (see text).

Building linear models for large numbers of features is maybe not the most intelligent strategy, and one has to take care that the assumptions of linear modeling (such as the dynamic range of the measurement equipment, errors being independently and identically distributed according to a normal distribution) are met. This means looking at diagnostic plots and summary statistics, something that is very difficult to automate. As a result, these (and other, much more sophisticated) linear models are most useful in later stages of the analysis, when one has limited the set of interesting features to a much lower number. Initially, one typically tackles large data matrices with multivariate methods, the topic of the next section.

Multivariate Statistics

In modern life sciences, we typically analyze few samples very extensively: it is not uncommon to measure many more variables than samples. On the one hand this is good: we learn a lot by measuring a lot. Variables are usually correlated to some extent, so that, e.g., a measurement error in one variable can be compensated for by other variables—random errors tend to cancel out. On the other hand it also creates problems. The very same correlation that can be helpful in some cases is a nuisance in other cases; in particular, it becomes virtually impossible to interpret model coefficients when correlations are present and no measures are taken to "regularize" things.

Obviously it is a good idea to assess the levels of correlations between variables. For small numbers of variables, interesting visualization approaches

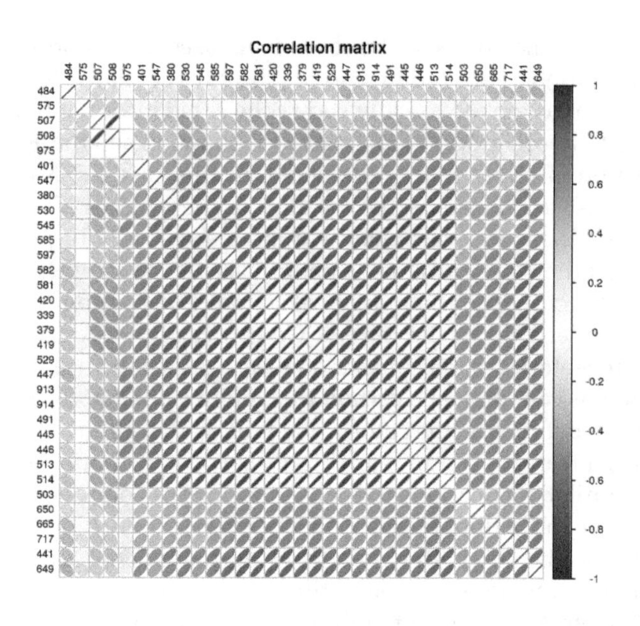

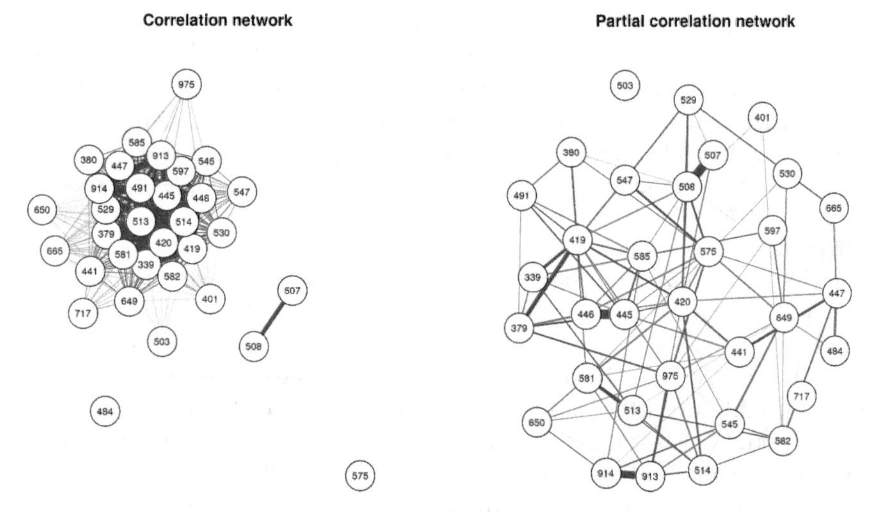

FIGURE 6.6
Visualization of correlations between the 33 features (mass peaks) eluting between 550 and 552.5 seconds (negative mode). Data have been log-scaled before calculating correlations. Left plot: simple visualization of the correlation matrix. Middle plot: correlation network. Right plot: partial correlation network. The labels are the (rounded) masses of the mass peaks.

are available. In Figure 6.6, for example, correlations between a set of mass peaks, all eluting at approximately the same time, are shown. The left plot shows a classical visualization of the correlation matrix, where large positive correlations are shown by narrow ellipsoids with positive slopes, and negative correlations by ellipsoids with negative slopes. Small correlations lead to almost circular shapes. The coloring focuses the attention to the large correlations. Variables are ordered so that highly correlated features cluster together. The middle plot basically shows the same information, but now in the form of a network graph: connections are shown when the absolute value of a correlation is larger than a threshold, in this case 0.5. Many different algorithms exist to determine the location of the vertices. From these two plots one can clearly see that many variables show very high correlations, indicating that they follow the same pattern across the samples. This is what one would expect if, e.g., the mass peaks would be coming from the same metabolite—indeed, from the masses we can easily recognize a couple of cases that are isotopes. Such correlation measures can be used to determine whether there is coelution at a certain time point (see Chapter 3), although it is very hard to define specific cutoffs that will work well in all cases.

Obviously, if features A and B are highly correlated, a third feature C will show very similar correlations to both A and B. This does not really tell us much about the individual relationships between C and A on the one hand, and C and B on the other. To obtain this information, one can calculate so-called partial correlations, compensating for the correlations between other variables (Steuer 2006). The network of partial correlations is shown on the right, here with a threshold of 0.125. Again we see a couple of isotopes being connected by strong positive correlations. Although these visualization methods have their usages, for large sets of variables (say, more than fifty) they quickly lose appeal. Note that these correlation networks should not be confused with metabolomic networks (see Chapter 8). In general it is impossible to infer metabolic networks from correlation networks—estimates of the concentration levels of many metabolites at one point in time do not contain enough information to do this.

Another problem in multivariate statistics is in statistical testing: as we have seen, the danger of false positive results increases rapidly with an increasing number of tests. In a way this is logical: the number of independent pieces of information equals the number of samples (in the optimal situation), and that is it. Without further assumptions, we cannot hope to increase the information content. Multivariate statistics typically relies less on statistical testing and more on approximate methods and visualization.

In contrast to the situation in the previous section, where we used quite well-defined descriptions and assumptions of the data in order to take decisions or make predictions, here we usually are in the situation where we know nothing, or only very little: typically in metabolomics, the number of variables far exceeds the number of independent samples. The first problem is already the visualization of the data. If we measure hundreds or even thousands of information bits about one sample, how can we show relationships between samples? Are some samples more alike than others? Are there groups in the data? If so, is the grouping visible in all variables, or only in some? Can we find which variables contribute to the grouping? These questions will be dealt with in the section on PCA, the dominant technique for the visualization of high-dimensional data (and not just in the area of metabolomics). For making predictive models, techniques like Partial Least Squares (PLS) are very popular. These explicitly use the information in the dependent variables (e.g., information on whether a sample is a treatment or a control) to define the model coefficients. Careful validation of such models is a must, since they are prone to overfitting and can easily lead to false conclusions. These methods have been popularized in particular in the field of chemometrics (Kjeldahl and Bro 2010), where multivariate data have been common since the seventies.

Principal Component Analysis

Visualization of large data sets can be difficult, especially if the size of the data is dominated by the number of variables. This is typically the case for data sets in the omics sciences in general and metabolomics in particular. What you would typically be looking for are similarities and dissimilarities between samples, possibly groupings of samples, and also relationships between the variables. The technique of Principal Component Analysis (PCA) allows one to compress the information in a data matrix into a much smaller number of variables, hopefully without losing too much—the underlying assumption is that a big part of the variation in the data is just noise, measurement error, and that by focusing on the signal, i.e., the part that does contain information, one can get clearer insights in the data.

The way that PCA achieves this dimension reduction is by combining variables into so-called Principal Components (PCs) (Jackson 2005; Jolliffe 1986). Of course, there is an infinite number of ways in which this combination can be done. The particular choice behind PCA is that the variables are combined in such a way that the first PC captures as much variation as possible. That is, this first axis in the new PCA coordinate system shows the individual samples as far apart as possible. If more than one PC is needed (and usually that is the case), subsequent PCs are defined in such a way that they are at right angles (orthogonal) to any previous PCs, and that they capture as much as possible of the remaining variation, after the variation captured by earlier PCs has been removed. One can also describe it as a rotation that takes a viewpoint showing the measurement points spread out as far as possible.

This procedure has several strong points. Firstly, it is unique: apart from the sign (the first axis can point to the left, or to the right; the samples are equally spread out in both cases), the compression is completely defined by the data and can be done automatically. Secondly, it is quite fast; even large data sets can be handled easily. Thirdly, the underlying mechanics are available in all statistical packages, and even in less specialized software like spreadsheets. Finally, because a very clear criterion has been optimized, the interpretation is quite straightforward.

In more mathematical terms, PCA can be written as follows:

$$X = UDV^T = TP^T$$

The first part describes the decomposition of data matrix $\boldsymbol{X}$, with N rows corresponding to samples, and M columns corresponding to variables, into a product of three separate matrices. These are the left and right singular vectors $\boldsymbol{U}$ and $\boldsymbol{V}$, respectively, and the diagonal matrix $\boldsymbol{D}$ containing singular values. This process is called the singular value decomposition (SVD). In PCA (the second part of the equation), one simply combines $\boldsymbol{U}$ and $\boldsymbol{D}$ into a new matrix $\boldsymbol{T}$, called the score matrix. This matrix describes the samples (rows of $\boldsymbol{X}$) in a new coordinate system, given by the PCs, and is often visualized with a score plot (see below). The variances of the columns in $\boldsymbol{T}$, the PCs, are equal to the squared diagonal elements in matrix $\boldsymbol{D}$, and they are decreasing so that the PCs are ordered in the direction of decreasing importance. The loading matrix $\boldsymbol{P}$ is equal to $\boldsymbol{V}$ and describes the contributions of the original variables (columns in $\boldsymbol{X}$) to the PCs. The maximal number of PCs is equal to the smallest dimension of $\boldsymbol{X}$: if there are fewer samples than variables, common in metabolomics, this means that at most N PCs can be defined. In that situation, the equalities in Eq. 6.1 are exact: no information is lost by writing $\boldsymbol{X}$ as these products of two or three matrices.

Nothing is lost, but nothing is gained either: the goal of PCA is to decrease the dimensionality of the data. In virtually all cases, one achieves this by discarding later PCs containing less information, and one is no longer describing $\boldsymbol{X}$ exactly:

$$\boldsymbol{X} \approx \tilde{\boldsymbol{T}}\tilde{\boldsymbol{P}}^T$$

where the tilde indicates that the number of PCs is smaller than the maximal number, given by min(N, M).[2] The assumption is that the relevant information, the signal, is present in the first PCs, and that later PCs contain mostly noise. Even for high-dimensional data sets with hundreds of variables, PCA can reveal interesting patterns in plots showing only two dimensions.

As an example, consider the positive ionization data from MTBLS18, containing 202 rows (samples) and 5930 columns (features). To avoid features with high intensities to dominate completely (remember, PCA is about variance!) we take logarithms (after adding one, to get rid of some zero values) and then mean-center the data before doing PCA. The resulting score and loading plots are shown in Figure 6.7.

The score plot on the left shows the location of the samples in the two dimensional space formed by the first two PCs. One can use different colors and/or symbols to indicate groups of samples—here, we indicated whether or not the samples were inoculated with Phytophthora. No obvious pattern is visible. The two-group structure may be related to other influences, or can be the result of a coincidence. The numbers on the axes indicate how much

[2] Note that if $N < M$ and the data are mean-centered, the Nth component has a singular value of zero—sometimes people therefore state that the maximal number of components that can be extracted is min(N–1, M): even though you can extract the Nth component, it is meaningless.

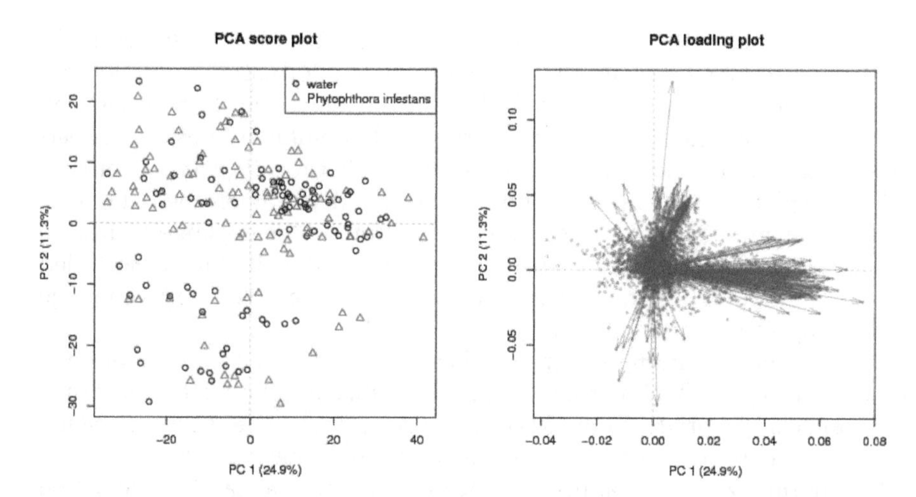

FIGURE 6.7
A PCA score plot (left) and loading plot (right) for the positive ionization data of MTBLS18, after log-scaling and mean-centering.

variance of the original data is explained: these two PCs together cover over a quarter of the total variation in the data, rather impressive considering that the original data matrix has almost 6,000 columns, and here we are only looking at two PCs.

The loading plot, right in the figure, shows the contributions of the original variables, the features from the MS data, to the two PCs. Larger loadings are shown as arrows from the origin, smaller ones by circles (to avoid an even more cluttered picture). Arrows pointing in more or less the same direction correspond to variables that are highly correlated in this subspace, and so the loading plot is an alternative to the visualization methods shown in Figure 6.4. Note that an apparently high correlation in a loading plot does not automatically mean that in the full data set there is a high correlation, too: in Figure 6.7, e.g., the two axes cover less than 40% of the variation, so even perfectly aligned arrows could still belong to variables that are not highly correlated in the full data. Furthermore, the length of the loading vector for a specific feature is a measure of how much of the variation in that feature is explained by the PCA model. Note that the 40% is an average over all features.

There is a direct relation between the position of the samples in the score plot and the positions of the variables in the loading plot. We can, e.g., consider that there are two distinct groups in the score plot, one above the other, roughly separated by a horizontal line at –5. The samples in the top group will have relatively high values for the feature that is pointing straight up in the loading plot, and low values for the features pointing downward. Loading plots and score plots can also be shown directly on top of each other—this is called a biplot, and especially when the number of variables (or samples!) is not too large, it can be very illuminating.

It is clear that these plots are powerful means of visualizing high-dimensional data. One can ask whether there is information in higher PCs as well. There usually is: simply looking at the plots, with appropriate colors, will in many cases show patterns that could point to interesting properties. Other possibilities are to consider all PCs that together lead to at least 80% of explained variance—the number of 80% depends on the data, or to choose the number of PCs after which a big jump in the explained variance is visible. The latter criterion is usually implemented by looking at so-called scree plots. In practice, they often are inconclusive. Examples are given in the online supplementary material.

An extremely important point is what scaling to use. Mean-centering is obligatory in PCA (Wehrens 2011), but many other, additional, forms of scaling can be appropriate. In the MTBLS18 example we have used log-scaling and subsequent mean-centering. Instead, we could have used autoscaling, which normalizes each feature to a mean of zero and a standard deviation of one. All variables are equally important in this case. Another alternative is Pareto scaling, in which high-variance features remain more important than low-variance features, something that in the interpretation sometimes is an advantage. We could also consider autoscaling or Pareto scaling after log-scaling. In each case, a different PCA result is to be expected, sometimes drastically so. Generally, log-scaling is appropriate in cases where the variation depends on the intensity (in mass spectrometry, where peak intensities basically are counts, this is often the case, as indicated by the popularity of the use of relative standard deviations to indicate variation). If zeros or even negative numbers show up in the data, a generalized log-transform could be an alternative—basically that adds a constant to the data so that the smallest value is larger than zero. Be aware that this is probably fine if the range of your data is in the thousands or even millions (again, something one sees in MS quite often) but not if the largest value is in the same order of magnitude as the constant added to the data.

One should realize that in PCA plots only those patterns will show up that are associated with large variances. This includes global effects, consistently present in many or even all variables, or more isolated but very large effects. In a sense, one has to be lucky to see something in PCA—more subtle differences are often not shown by the linear combinations formed in PCA. This is where the methods of the next section come in handy (Box 6.1).

Partial Least Squares Regression

The dimension reduction in PCA serves to concentrate relevant information into a relatively low number of PCs for visualization purposes, and to a lesser extent to separate the real signal from noise. Often, however, we know something about the cause of variation: if we are looking at two different groups of samples, for example, part of the variation may be associated with this grouping, and we may be interested to see what variation is associated

BOX 6.1 PCA DO'S AND DONT'S

- Never do PCA on data that have not been mean-centered (other scaling operations may be done, too, but at least column means should be zero)—otherwise the orthogonality of the PCs, so important in the interpretation, is lost.
- In score and loading plots, always show PC numbers, and percentage of variance explained for that PC.
- Never do a direct comparison of two score plots or two loading plots from different PCAs—they more likely than not will represent different subspaces. If you do want to see the location of data from set A in a PC space from set B, use projections.
- Many software packages plot ellipses in score plots. These are based on the assumption of multivariate normality, something that often is not the case[3] (e.g., when your data have a clear group structure).
- Be aware that conclusions from score and loading plots are based on a subspace: if you see, e.g., very high correlation between two variables in a loading plot, this does not mean that these are correlated in the real data. Especially when the corresponding PCs cover little variance, you may be up for some surprises.
- Resist the temptation to keep on trying different variants (e.g., different forms of scaling) until you find the plot that you like. Decide *a priori* what is the most appropriate form of scaling, and stick to that.

[3] To put it mildly...

with it. Sometimes, when we are lucky, such an effect is also visible in a PCA plot, and we can clearly see the two groups of samples separated in a score plot. We could even use the scores of $\boldsymbol{X}$, $\boldsymbol{T}$, in a regression equation to predict something about a dependent variable Y:

$$y = \boldsymbol{T}\boldsymbol{B} + \boldsymbol{E}$$

Here, $\boldsymbol{B}$ is the vector of regression coefficients, which can be estimated with least-squares regression easily: the columns of $\boldsymbol{T}$ are orthogonal, and the number of columns is usually quite a lot smaller than the number of rows. This kind of model is called Principal Component Regression (PCR).

More often, however, the variation associated with the groups is only a small part of the total variation, and will show up only in later PCs, or even not at all. That means that a relatively large number of PCs needs to be taken into account. Partial Least Squares (PLS) is a multivariate regression method using the same trick as PCA: a combination of the original variables into a smaller number of new ones, in this context called latent variables. Again, the regression model has the form $y = TB + E$.

Rather than focusing on capturing the variance in X, however, PLS dimension reduction tries to capture as much as possible of the covariance between X and y, where y contains the information on the grouping (in this example); alternatively, y can be a numerical variable, and then we are talking about a regression problem rather than a classification problem. The latent variables defined by PLS again can be used for score plots and loading plots. The result of taking y into account in the dimension reduction phase is that we are more or less guaranteed that the relevant information to predict y ends up in the first couple of latent variables.

PLS is an extremely powerful method, and unless used wisely is able to "learn" the data almost too well. That is, it is remembering the data, rather than fitting a model with predictive power. The more components you allow in your model, the better the fit, to the extent that even random data can be fitted extremely well. This is often called overfitting. What we would like to achieve is a balance: a good fit (but not too good) as well as predictive properties. This is usually achieved in a procedure called cross-validation: a part of the data is left out, and the model is built on the remainder, the training set. In the case of overfitting, the prediction for the left-out data will be very poor. Repeatedly building models in such a way that all samples have been part of the test set exactly once and recording the successes for different choices of the model parameter, the number of latent variables to include, leads to estimates of the cross-validation error. Very often one takes the number of latent variables for which this error is minimal, or, more conservatively, the smallest number of latent variables for which the error is close to the minimal value. Note that the cross-validation error associated with this choice is not a good estimate of the prediction error, i.e., the error we expect when making predictions for new samples. The reason is that the cross-validation error is biased downward: it is too optimistic, since it has specifically been chosen for being small. If one is interested in an unbiased estimate (and one usually is) one should apply the "optimized" PLS model to a new (independent) data set (often called a test set) and see how well it performs there.

There are many PLS variants: the original one by Wold (Martens and Naes 1989; Wold et al., 1983) is based on an iterative procedure, and is commonly referred to as "orthogonal PLS"; later, faster procedures were proposed (Dayal and MacGregor 1997; de Jong 1993; De Jong and Ter Braak 1994). Sometimes there are slight differences between the model coefficients, but these are "unlikely to be of any practical difference" (de Jong 1993). In metabolomics, particularly popular variants are OPLS (Trygg and Wold 2002) and OPLSDA (Bylesjö et al., 2006)—the latter is used for Y variables corresponding to discrete class variables, rather than continuous variables. These constitute methods in which the loadings (and therefore also the scores) of a PLS model are rotated to facilitate interpretation. One should note, however, that the fitted model in OPLS (and OPLSDA) is exactly the same as in PLS (or PLSDA) (Tapp and Kemsley 2009), and that claims that one performs better than the other are based on a misunderstanding. In particular, the importance of finding the right number of latent variables is as crucial as in "normal" PLS. A further sign that these methods are poorly understood is that often in the discrimination between two classes OPLSDA plots are shown containing two factors, whereas the whole principle of the method is that for this case the relevant information is present in the first latent variable only. Further relationships between OPLSDA and other forms of multivariate discriminant methods exist (Brereton and Lloyd 2014).

Rather than going into the mathematics (see the literature for that), we show how to use PLS and interpret the results, focusing on the MTBLS18 example. Suppose we are primarily interested in those features that show a big response to the Phytophthora treatment. We are again in familiar territory, now predicting the treatment status of a sample from the metabolite levels. Since the treatment is a discrete factor, we would usually call the technique PLSDA (Partial Least Squares Discriminant Analysis) (Barker and Rayens 2003) rather than PLS, but basically we are doing exactly the same thing. If the Treatment level is coded as a "1" and the Control as "0," we would assign a sample to the Treatment class if the predicted value is larger than 0.5, and to the Control class otherwise.

The PLS scores and loadings for the MTBLS18 data, again after log-scaling and mean-centering, are shown in Figure 6.8. The score plot, on the left, shows that the treated samples are predominantly present in the top right of the figure. Loadings pointing in this or the opposite direction refer to features that

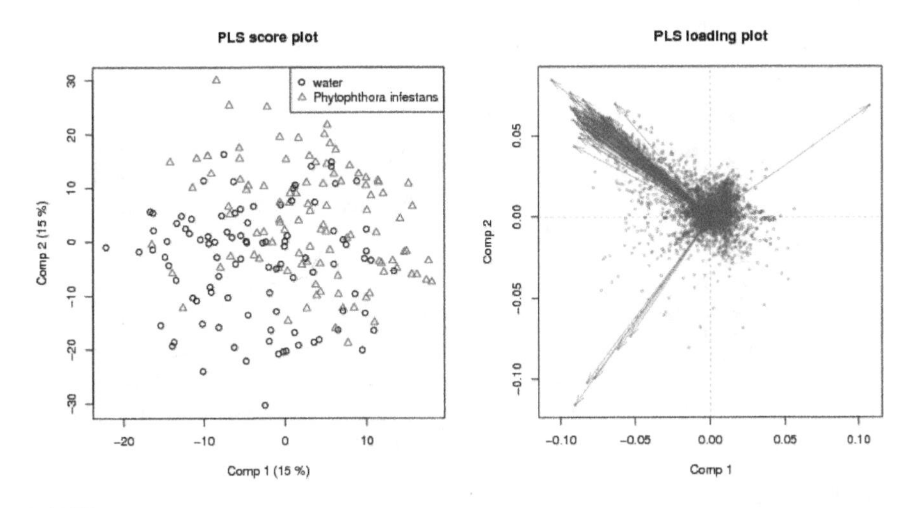

FIGURE 6.8
Results of applying PLS on the MTBLS18 data, using factor "Treatment" as response variable. The left plot shows the scores for *X*; the right plot the loadings.

may be influenced by the treatment. Interestingly, most of the largest loadings point in a direction that is not associated to the treatment difference at all. Often, it is easier to ignore the loading plot (especially if one has more than a few latent variables) and focus on the regression coefficients in $\boldsymbol{B}$ instead. Just like in the linear regression case, large coefficients are more influential. Unfortunately, in the case of PLS there is no analytical way to determine confidence intervals of the coefficients, so statistical significance cannot be deduced without further calculations, and then only approximately.

Figure 6.8 only focuses on the first two latent variables. Although there clearly is some separation between the Treatment levels, in two dimensions it is not perfect. Perhaps more components would improve the result. To determine how many components are optimal, a cross-validation is performed, and the result is shown in Figure 6.9. Clearly, it is very hard to take a decision as to how many components are necessary here: the cross-validation error keeps going down when more components enter the model. Note that the differences between the different models are very small: the range of the y-axis runs from 0.44 to 0.50 (on a scale of 0–1!). Basically, none of the models is able to make good predictions. We could go to even more latent variables but that again leads to an increased danger of overfitting.

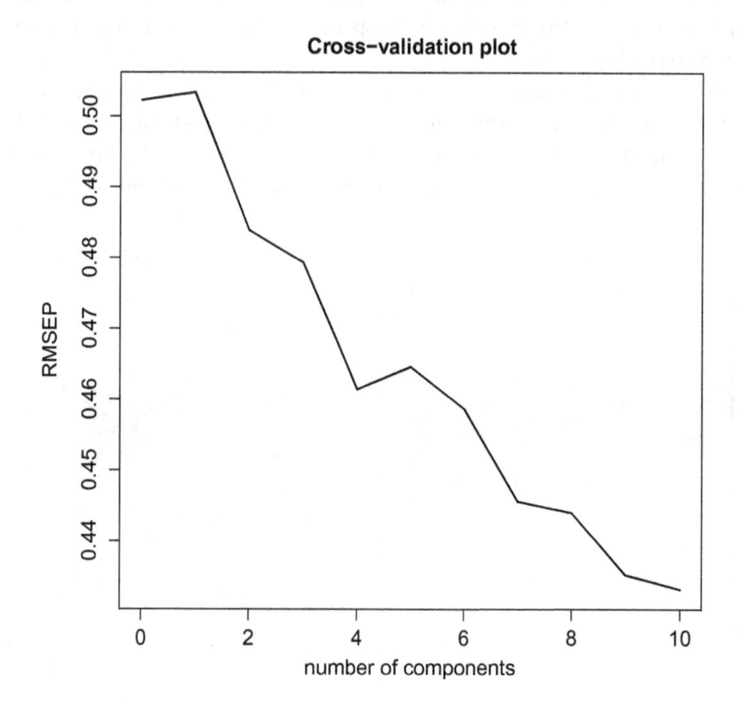

FIGURE 6.9
Cross-validation curve for the MTBLS18 data, analyzed with PLS. The y-axis shows the RMSEP value (which stands for root-mean-square error of prediction). No clear minimum is visible in the first ten latent variables.

Discussion

With increasing data set sizes, the expectation, usually not explicitly stated, is that now it is becoming easier to find answers to common questions. After all, we have measured so many numbers—surely you will be able to "find something." Unfortunately, reality is different. One reason is that data sets tend to grow in width rather than height—we know more about a few samples. This also means that the number of independent bits of information is not increasing: we assume that the few samples we have are representative of the complete population, and that summary statistics like averages are a good guess of what the population mean is. Fundamentally, nothing has changed since we had only one variable. Obviously, we have gained in one respect: being able to measure large numbers of variables cheaply and quickly allows us to consider many possible models, using different combinations of predictors. This in turn can help to gain biological understanding, not only because some variables are more highly correlated to the phenomenon of interest, but also because of the correlation patterns we see between the variables. Even when taking into account that it is always correlation we are looking at and not causation (yes, the number of births has dropped in the Western world, as has the number of storks, but that does not mean that there is a causal relation between the two), such patterns can be very informative and can lead to new, perhaps more specific, hypotheses that can be tested in follow-up experiments.

The price we pay for this enormous increase in data (not necessarily information) is that the character of statistical modeling has changed drastically. Whereas in the data-scarce era a lot of thought used to go into establishing the right model, assessing the model assumptions, and in the end the fitting, nowadays people prefer model-free methods that are easily accessible in open-source computer packages. It is important to note that both approaches have their place, and that one is not necessarily superior to the other. In the initial data analysis phase, one may be interested in the global view: which samples are similar? Which metabolites or features are correlated, which are anticorrelated? Do I see patterns? Which metabolites are associated with a change in the response variable, i.e., which have predictive power? These are questions that can conveniently be tackled by multivariate chemometric methods such as PCA and PLS (and a host of others). Typically, one quickly focuses on a much smaller subset of features or metabolites, associated with the behavior of interest.

Variable selection, just to take one question as an example, is often done on the basis of PLS regression coefficients or loadings, or measures like the Variable Importance of Projection (VIP) (Svante Wold et al., 2001). Unfortunately, for none of these a standard cut-off value can be calculated to distinguish between interesting and not-so-interesting variables, so data-dependent thresholds need to be defined (Wehrens and Franceschi 2012). One should realize that as soon as the number of variables exceeds the number of samples, there is an infinite number of combinations that leads to essentially the same predictions, much like there are an infinite number of lines through one point. The solutions obtained by all different modeling techniques are the result of particular choices and assumptions of the methods; no single method will ever be better than all other methods. In the same way, the variables selected by the different methods will differ, too, but it is hard to say which set is better. Alternatively, global optimization methods like Simulated Annealing or Genetic Algorithms can be employed (Wehrens 2011), although also here one is likely to find just one of many equivalent sets of variables.

An important way to decrease the number of variables without losing information is to analyze the data at the metabolite level rather than the mass peak level. While this is more or less standard in GC/MS, for LC/MS the much larger overlap problems make the assignment of mass peaks to (possibly unknown) metabolites a difficult process. As a result, many will simply analyze tables of mass peaks, as we have done here as well. By now it should be clear what the disadvantages of such an approach are: the much larger number of variables will increase the number of false positives; the loss of power by applying multiple testing corrections will even be bigger; features from the same metabolite will show high correlation, leading to difficulties for statistical testing; random variation that to some extent will cancel out when going from the feature level to the metabolite level is still present; and finally, the interpretation of the results is much more difficult. It is clear that no one is accepting this extra load of difficulties with pleasure, indicating the importance of grouping features into sets corresponding to metabolites.

By now, the modeling toolkit for high-dimensional data has expanded to contain many different methods and it can be quite a challenge, especially when confronted with this richness for the first time, to select the most suitable tools. A few rules of thumb can be given, though. For the common $M > N$ situation in metabolomics (more variables than samples), one should consider powerful non-linear methods only as a last resort (to avoid the phrase "not at all"). The reason is that this class of methods, like neural networks, support vector machines, and others (Wehrens 2011) have many tuning parameters and are very prone to overfitting. An added disadvantage is that they are essentially black boxes, not allowing any meaningful model inspection that can aid in interpretation. One exception here is the random forest (Breiman 2001), which with its built-in validation procedures is remarkably resistant to overfitting and currently is one of the most powerful off-the-shelf (one rarely has to modify the standard settings) modeling paradigms available.

A second class of models that can be considered is based on dimension reduction and includes the PLS and PCA-based methods introduced in this chapter. Ignoring for the moment the small variations on the general PLS theme, these methods are well-defined, usually perform well (provided they are used correctly) and through the model coefficients give insight in the underlying mechanics of the fitted model. There usually is only one explicit tuning parameter, the number of components, although the usual additional choices need to be made on the scaling and pretreatment of the data.

Finally, one can simply do the analysis on a variable-by-variable basis: typically, those metabolites for which a significant relation with the trait of interest is found are singled out for follow-up studies. One advantage of such an approach is that much more sophisticated models can be employed such as generalized linear models (McCullagh and Nelder 1989) or mixed models (West et al., 2014), incorporating information that would be hard to use for less structured methods like PLS. Approaches like the lasso and the elastic net (Hastie et al., 2013) take middle ground: by putting penalties on the sizes of the model coefficients, these so-called sparse methods are able to fit models with only few non-zero coefficients, thereby performing an implicit variable selection. Because of very efficient estimation procedures, they are much faster than step-wise selection methods.

In conclusion: classical statistics have the advantage of seeming to give clear-cut answers—there is a difference between groups A and B! (But read the fine print: at a confidence level of $\alpha = 0.05$.) In the regression context, predictions automatically come with prediction intervals, and statistical significance can be tested. The results are extremely good, as long as your assumptions are correct (check them!)—for this it may be necessary to spend quite some time and attention to the quality of the data: choosing the right transformations, eliminating outliers, determining how to handle non-detects and missing values (see Chapter 2) will be key to a successful data analysis. In the multivariate case, you get very intuitive overviews, and you do not have to worry about your assumptions (there are none) but answers depend on settings and are rarely as clear-cut as in the univariate case. They are, however, extremely useful as hypothesis-generating devices.

References

Armitage, P., G. Berry, and J. N. S. Matthews. 2008. *Statistical Methods in Medical Research*. Hoboken, NJ: John Wiley & Sons.

Barker, M., and W. Rayens. 2003. "Partial Least Squares for Discrimination." *Journal of Chemometrics* 17 (3): 166–173.

Breiman, L. 2001. "Random Forests." *Machine Learning* 45 (1): 5–32. doi:10.1023/A:1010933404324.

Brereton, R. G., and G. R. Lloyd. 2014. "Partial Least Squares Discriminant Analysis: Taking the Magic Away." *Journal of Chemometrics* 28 (4): 213–225.

Bylesjö, M., M. Rantalainen, O. Cloarec, J. K. Nicholson, E. Holmes, and J. Trygg. 2006. "OPLS Discriminant Analysis: Combining the Strengths of PLS-DA and SIMCA Classification." *Journal of Chemometrics* 20 (8–10): 341–351.

Dayal, B. S., and J. F. MacGregor. 1997. "Improved PLS Algorithms." *Journal of Chemometrics* 11 (1): 73–85.

Franceschi, P., M. Giordan, and R. Wehrens. 2013. "Multiple Comparisons in Mass-Spectrometry-Based-Omics Technologies." *Trends in Analytical Chemistry: TRAC* 50: 11–21.

Hastie, T., R. Tibshirani, and J. Friedman. 2013. *The Elements of Statistical Learning: Data Mining, Inference, and Prediction*. Berlin, Germany: Springer Science & Business Media.

Jackson, J. E. 2005. *A User's Guide to Principal Components*. Hoboken, NJ: John Wiley & Sons.

Jolliffe, I. T. 1986. "*Principal Component Analysis*." New York: Springer.

de Jong, S. 1993. "SIMPLS: An Alternative Approach to Partial Least Squares Regression." *Chemometrics and Intelligent Laboratory Systems* 18 (3): 251–263.

de Jong, S., and C. J. F. Ter Braak. 1994. "Comments on the PLS Kernel Algorithm." *Journal of Chemometrics* 8 (2): 169–174.

Kjeldahl, K., and R. Bro. 2010. "Some Common Misunderstandings in Chemometrics." *Journal of Chemometrics*. 24 (7–8): 558–564. doi: 10.1002/cem.1346/full.

Martens, H., and T. Naes. 1989. *Multivariate Calibration*. Chichester, UK: John Wiley & Sons.

McCullagh, P., and J. A. Nelder. 1989. "An Outline of Generalized Linear Models." In *Generalized Linear Models*, edited by P. McCullagh and J. A. Nelder, pp. 21–47. Boston, MA: Springer US.

Steuer, R. 2006. "Review: On the Analysis and Interpretation of Correlations in Metabolomic Data." *Briefings in Bioinformatics* 7(2): 151–158.

Tapp, H. S., and E. K. Kemsley. 2009. "Notes on the Practical Utility of OPLS." *Trends in Analytical Chemistry: TRAC* 28 (11): 1322–1327.

Trygg, J., and S. Wold. 2002. "Orthogonal Projections to Latent Structures (O-PLS)." *Journal of Chemometrics* 16 (3): 119–128.

Wehrens, R. 2011. *Chemometrics With R: Multivariate Data Analysis in the Natural Sciences and Life Sciences*. Berlin, Germany: Springer Science & Business Media.

Wehrens, R., and P. Franceschi. 2012. "Thresholding for Biomarker Selection in Multivariate Data Using Higher Criticism." *Molecular BioSystems* 8 (9): 2339–2346.

West, B. T., K. B. Welch, and A. T. Galecki. 2014. *Linear Mixed Models: A Practical Guide Using Statistical Software*, 2nd ed., Boca Raton, FL: CRC Press.

Wold, S., H. Martens, and H. Wold. 1983. "The Multivariate Calibration Problem in Chemistry Solved by the PLS Method." In *Matrix Pencils*, edited by Bo Kågström and Axel Rube, 286–293. Berlin, Germany: Springer.

Wold, S., M. Sjöström, and L. Eriksson. 2001. "PLS-Regression: A Basic Tool of Chemometrics." *Chemometrics and Intelligent Laboratory Systems* 58 (2): 109–130.

7

Data Fusion in Metabolomics

Johan A. Westerhuis, Frans van der Kloet and Age K. Smilde

CONTENTS

Introduction 157
Definition of Data Fusion 157
Goals of Data Fusion 159
Methods 160
Correlation-Based Methods 160
Interpretation of Correlation-Type Methods 161
SCA Type of Methods 162
Interpretation of SCA Type of Methods 163
Methods for Common and Distinct Information 164
DISCO-SCA 165
JIVE 166
OnPLS 167
Interpretation of Common and Distinctive Methods 167
Orthogonality Aspects 168
Results 168
Discussion 172
General Issues in SCA Type of Data Fusion 172
Sample Alignment 173
Block Scaling 173
Model Complexity 173
Conclusion 174
Experimental 174
Data Pre-Processing 174
References 175

Introduction

Definition of Data Fusion

In bioinformatics the term *data integration* refers to the situation where multiple sources of data are available of the same (living) system and where these sources are combined, together with the corresponding databases, to give a biological background to the measured features (Gomez-Cabrero et al. 2014, I1). The main goal of data integration is an improved understanding of the interactions that exist between the different components in the system. For this, many new approaches are currently being developed but no unified definition of data integration or a taxonomy for data integration methodologies yet exists.

The main topic of this chapter is data fusion, one specific subset of methods of data integration. In metabolomics research one often has the situation where multiple platforms have been used to measure various components of the same set of samples e.g., NMR, GC-MS, LC-MS, or negative mode and positive mode of the same LC-MS system. It can be that besides metabolomics measurements, transcriptomics or proteomics measurements of the same samples have been obtained as well. Combining these data sets should improve the interpretation about the differences between the samples and should also provide information on which features are most responsible for those differences, and how these features relate to each other.

The combination of data sets is sometimes referred to as data integration, or data fusion. The difference between these approaches is not well defined. Data integration methods are qualitative approaches that collect measurements of different origins and puts them on a well-defined scaffold, e.g., a metabolic network. Data integration thus considers the identity of the variables in the different data sets. For example, gene expression data, proteomics and metabolomics data can be integrated by placing their intensity values at the corresponding position in the metabolic network. The identity of the different variables of different sources is taken into account when collecting the data. When a metabolite is measured on two platforms, both variables are linked to the same position on the network. Thus for data integration, network information and ontologies on the meaning of the variables are necessary.

Data fusion methods are quantitative methods that combine multiple data sets that are measured on the same set of samples. The models that are developed are based mainly on correlations between the variables. In data fusion methods, the identity of the variables of the data sets is not directly used in the analysis. Only after the models have been developed can the identity of the variables be used when interpreting the data analysis results. In this chapter the focus is fully on data fusion methods.

The data fusion methods discussed in this chapter are very general and do not put restrictions on the interactions that could exist. For example, genes known to catalyze specific metabolic reactions are not restricted to interact with the corresponding metabolites only. Various types of applications for data fusion can be envisioned (Richards et al. 2010).

In *inter-platform* applications, the same samples are measured on two or more metabolomics platforms such that a more complete set of measured metabolites is used to explore the differences between the samples. This is a common situation since in many studies a combination of metabolomics platforms is used.

In *inter-sample* applications, samples from different compartments (e.g., urine, blood or tissue) are used, but they come from the same individual. The conclusions of such applications will focus on the differences between the individuals and how these are represented in the different compartments. An example of such an application is the analysis of plasma and cerebrospinal fluid (CSF) by NMR (Smolinska et al. 2012).

Inter-omics applications form a combination of the two previous cases: different omics platforms are used to analyze samples, either measured on the same samples or coming from different compartments of the same biological system. Again, the goal of data fusion is to obtain information on the differences between the biological systems. Here the conclusions will be based on interactions between features of the different omics levels.

As mentioned earlier, the connections between the features of the different omics levels are sometimes defined in a (Genome-Scale) Metabolic Network, but the data fusion methods discussed here ignore these networks: they are based on the correlation between the features and do not directly use the biological relationship between a specific gene and the metabolic reaction it catalyzes.

Goals of Data Fusion

Data fusion can have many goals that, however, are not always clearly defined at the beginning of data analysis. Making these goals explicit will help in performing the analysis and making it successful. Here are some typical cases.

- *Exploratory analysis:* Exploration means looking for patterns in the data such as clusters of objects or variables, outliers, influential data points, etc., mostly using visualization methods. In the context of data fusion, the aim is to explore a set of objects described by multiple sets of measurements, enabling feature exploration *within* and *between* blocks.
- *Comprehensive biomarkers:* Combining correlated features from several data sets may be useful in finding predictive information, e.g., for disease status. These features can even come from different compartments within the same objects. In data consisting of several feature sets, these sets can interact differently. Interactions can exist between variables of the same set or between variables from different sets. It can also be that the summaries of the different sets show interaction. Interpretation of such interactions in metabolomics experiments often requires biological information. Usually, data fusion does not take this into account, but the methods can be extended to incorporate biological relationships, e.g., by using biological network visualization tools (Pavlopoulos et al. 2008) and enrichment analysis (Xia and Wishart 2010a, 2010b; Kamburov et al. 2011; Xia and Wishart 2011).
- *Common versus distinctive variation:* Besides looking for correlations within different data sets (common variation), it can also be important

knowing what systematic information they do *not* share (distinctive information). The separation of variation (ANOVA-style) in the data of the common part and the distinctive part can greatly simplify interpretation of the results.

In this chapter three groups of methods will be introduced: correlation-based methods, simultaneous component analysis (SCA) based methods and methods that are able to distinguish common and distinctive information. The properties of these methods and their interpretation will be discussed, as well as general aspects of data fusion that could influence the results. We will perform data fusion on the MTBLS18 data mentioned earlier. Here we will not only consider the metabolomics data, but also the transcriptomics data, obtained using Affymetrix microarrays. Feature selection was applied to both positive mode LC-MS and transcriptomics data in such a way that differences between the genotypes were highlighted. Besides the selected metabolite features, 15 identified features were added although they were not able to discriminate between the genotypes. Note that negative mode LC-MS features were not used in this chapter. Multiple data fusion methods will be applied on the data and we will discuss the differences.

Methods

For simplicity we will consider only two data sets with, e.g., $\mathbf{X}_1(N \times M_1)$ representing LC-MS data obtained from N objects with M_1 ($m_1 = 1...M_1$) features (either metabolite levels or mass peaks) and $\mathbf{X}_2(N \times M_2)$ representing gene activities for the same samples. In both cases, the features are represented as columns of the respective data matrices. Furthermore, we will use the term object to indicate the sample or individual that represents the data on the same row of the data sets. Most of the methods can easily be extended to more than two blocks. We assume that all blocks have been standardized (autoscaled), so that the mean of each column equals 0 and the standard deviation of each column equals 1. For methods that aim to describe variances as described in the section on SCA methods, the variance of each variable plays a major role in the final model. When the variances are made equal for each variable, these methods focus on correlation between variables, which is easier to interpret and usually more biologically relevant.

Correlation-Based Methods

The first set of methods makes use of the correlation between the features within each block and between the blocks to improve interpretation and identification. Originally presented as correlation spectroscopy (Noda 1993; Noda et al. 2000), this method was introduced in metabolomics as Statistical Total Correlation Spectroscopy (STOCSY) (Cloarec et al. 2005). The method can be used to visualize highly correlated features in a single data set. If matrix $\mathbf{X}_1$ is standardized, the correlation matrix of features in the same data set are estimated as follows:

$$C_1 = \frac{\mathbf{X}_1^T \mathbf{X}_1}{N-1} \tag{7.1}$$

C_1 is now a correlation matrix of size $M_1 \times M_1$ with ones on the diagonal. Element $\mathbf{C}_{1ij}$ gives the correlation between variables $\mathbf{X}_{1i}$ and $\mathbf{X}_{1j}$. The correlation matrix can be visualized such that the correlation pattern in the data is easily presented. However, the interpretation of correlations of metabolomics data is not straightforward (Camacho et al. 2005; Steuer 2006).

The STOCSY approach can be extended to two platforms of the same type or of different types. Statistical HeterospectroscopY (SHY) is the variant that combines NMR and MS data (Crockford et al. 2008). SHY calculates the correlation of features across the two platforms. The crosscorrelation matrix $\mathbf{CC}_{1,2}$ of size $(M_1 \times M_2)$ can be estimated from two data sets $\mathbf{X}_1(N \times M_1)$ and $\mathbf{X}_2(N \times M_2)$ that have the same samples.

$$\mathbf{CC}_{1,2} = \frac{\mathbf{X}_1^T \mathbf{X}_2}{N-1} \tag{7.2}$$

The correlation between feature m_1 of $\mathbf{X}_1$ and m_2 of $\mathbf{X}_2$ can be found at position (m_1, m_2) of the $\mathbf{CC}_{1,2}$ cross-correlation matrix. Again, high cross-correlations can indicate that the features of the two platforms describe the same metabolite, while lower (but non-zero) correlations can indicate biologically regulated effects. An example of SHY where correlations between peaks of different platforms are indicated is presented in Figure 7.1. Here, the correlation maps of some identified metabolite LC-MS peaks and selected genes are provided together with their cross correlation plot. Malony-Hydroxycamalexin-Hexoside levels clearly correlate negatively with Cumaroyl and Feruloyl levels in the different mutants. The selected group of genes can be separated into three groups. In the cross-correlation plot (upper right) there is not a clear distinction in correlation between the identified metabolites and the selected genes. Some high correlations between Feruloyl levels and few genes can be observed.

Interpretation of Correlation-Type Methods

The correlation type of data fusion methods mainly focuses on the features and not on the objects. Therefore, these methods are tools for feature identification, visualization and to find chemically related or biologically related features. Subsequent feature selection can be performed to study the variation within the objects given the selected features. In the STOCSY paper the

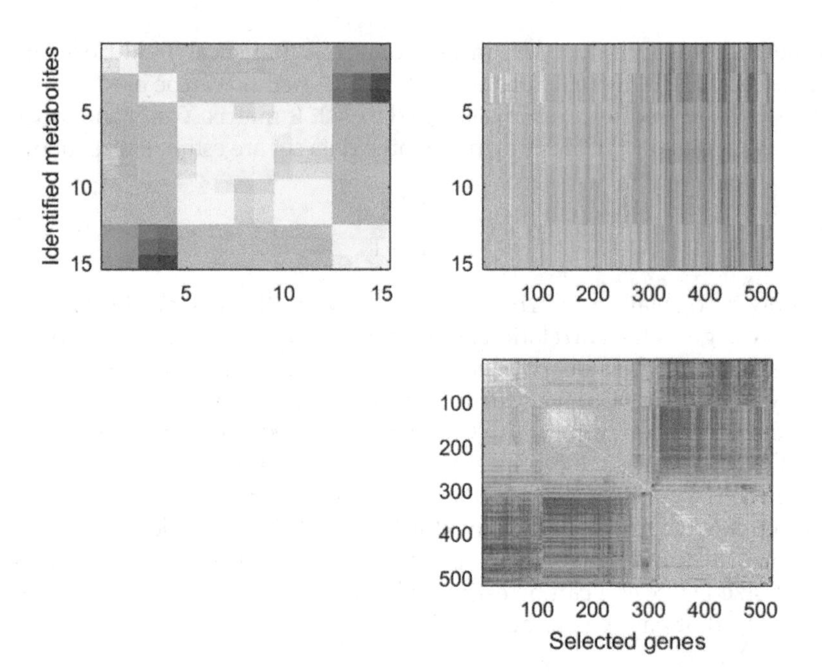

FIGURE 7.1
Correlation map of some identified metabolite peaks (Cumaroyl-Agmatin peaks 1-2,Feruloyl agmatin peaks 3-4, Camalexic peaks 5-12, and Malony-Hydroxycamalexin-Hexoside peaks 13-15) of LC-MS positive mode (top left), a correlation map of selected genes important to distinguish between the different mutant groups (bottom right), while the top right plot shows cross correlations between the identified metabolic peaks and the selected genes. Yellow indicates positive correlation and blue indicates negative correlation.

correlation visualization was followed by a classification method to indicate the importance of the selected features.

SCA Type of Methods

Another group of methods, also based on two data sets that have the objects in common or are measured on the same individual, aims to find the principal components underlying both data sets. They form the so-called simultaneous component analysis (SCA) type of methods. The "common" PCA components aim to describe as much as possible of the variation between the samples—they provide information on which features are most important to explore the differences between the samples. Many different methods are available that have a similar goal, e.g., multiblock PCA, hierarchical PCA, consensus PCA, SUMPCA, etc. All these methods are variants with slightly different properties, due to the underlying algorithms. Many overviews of such methods are already available in the literature (Westerhuis et al. 1998; Smilde et al. 2003; Van Deun et al. 2009). The general idea of the SCA type of methods is that the score values that are calculated represent the samples or individuals in both sets:

$$\begin{aligned} X_1 &= TP_1^T + E_1 \\ X_2 &= TP_2^T + E_2 \end{aligned} \tag{7.3}$$

Here X_1 and X_2 again have the same objects or are related to the same subject but they have different features. The scores representing the objects (T of size $N \times R$) are the same for both data sets. The various methods differ in the way these scores are calculated and their mathematical or statistical properties. The loadings $P_1(M_1 \times R)$ and $P_2(M_2 \times R)$ can directly be compared for interpretation purposes because they are based on the same set of scores. Another way of writing this type of data fusion is:

$$[X_1 \mid X_2] = T\left[P_1^T \mid P_2^T\right] + [E_1 \mid E_2] \tag{7.4}$$

Here the concatenation of the blocks is visualized. It is clear that the number of objects should be the same, but also the objects on each row of X_1 and X_2 have to be directly related.

Interpretation of SCA Type of Methods

One of the nice properties of data fusion is that features of different blocks can be directly related to each other. The scores define a relationship between the objects which forms the same basis for all data blocks. Thus, the scores can be interpreted similar to a score plot of a normal PCA model. Because the scores form the same basis for all data sets, the relationship between features in different blocks can be compared directly. The loadings of all blocks can be shown simultaneously in a loading plot, which can be interpreted like a normal PCA loading plot. Thus, features of which the loadings are closely together in the loading plot and far away from the origin are highly correlated in the common part of the data.

Figure 7.2 shows the simultaneous component analysis results of selected features of positive mode LC-MS data and transcriptomics data. The score plot shows a clear separation between the erp2 group and the remainder of the genotypes on PC1 while also some separation is observed between erp1 and the other genotypes. The colors identify the different genotypes while the treatment (water/Pathogen) and time (6h/12h) settings are not indicated. There is much variation within the erp2 group on PC1. Because these scores represent the common variation in both the metabolomics and transcriptomics data sets, the loadings of both sets can be interpreted directly to each other. There is a cluster of metabolomics features on the left side of the loading plot that is related to the erp2 group. The identified metabolic features (indicated by a blue asterisk) have very low loading values. This could be expected as they were also not able to distinguish between the genotypes.

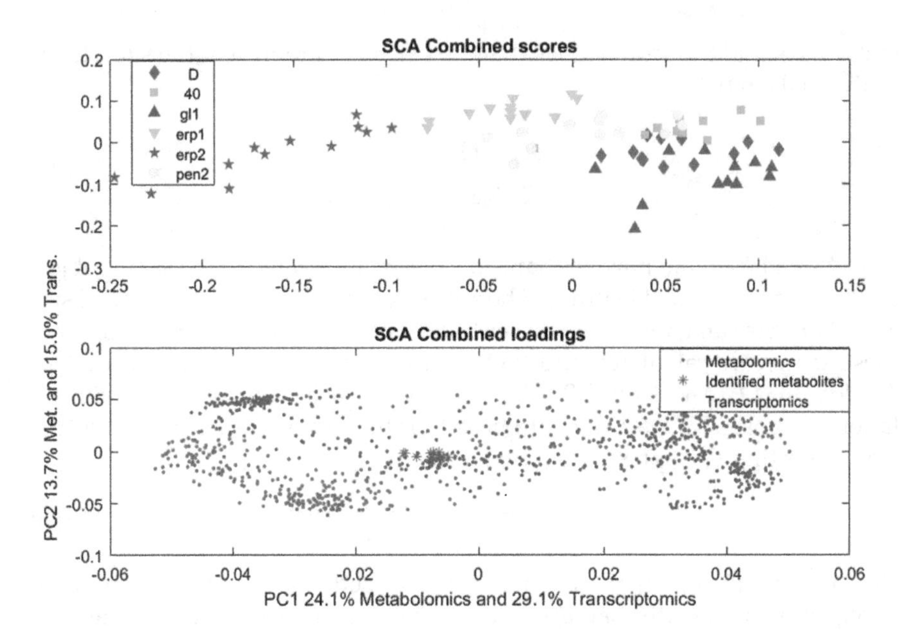

FIGURE 7.2
Global scores and global loading plots of the Arabidopsis experiment. The top figure shows the global (common) scores indicating some differences between the various genotypes. The bottom plot shows the metabolomics features (blue) and transcriptomics features (red) that were selected. The metabolites that were identified are indicated with blue asterisks. PC1 explains 24.1% of the metabolomics data while 29.1% is explained of the transcriptomics data. PC2 explains 13.7% of metabolomics data and 15.0% of the transcriptomics data.

Methods for Common and Distinct Information

The data fusion methods discussed until this point focus on the overlap between two data sets, also called the common part. This common part in different data sets describes systematic variation between the objects that is similar in both sets of data. In addition, there can also be systematic variation in one data set that is not found in the other. Such information can be very informative and should not be ignored. Therefore, methods have been developed that decompose the two data sets into three parts, a common part that overlaps with the other data set, a distinctive part which cannot be found in the other sets and a residual part, which is not systematic. OnPLS, JIVE (Joint and Individual Variation Explained) and DISCO-SCA (Distinct and Common-SCA) are three methods that make such a decomposition. If there are two data sets $\boldsymbol{X}_1(N \times M_1)$ and $X_2(N \times M_2)$ that have the same objects measured at different platforms leading to M_1 and M_2 variables respectively, then the general model for common and distinctive information can be written as follows.

$$\boldsymbol{X}_1 = \boldsymbol{C}_1 + \boldsymbol{D}_1 + \boldsymbol{E}_1 \quad \boldsymbol{X}_2 = \boldsymbol{C}_2 + \boldsymbol{D}_2 + \boldsymbol{E}_2 \tag{7.5}$$

Here $C_1(N \times M_1)$ and $C_2(N \times M_2)$ refer to the common parts, $D_1(N \times M_1)$ and $D_2(N \times M_2)$ to the distinctive parts and $E_1(N \times M_1)$ and $E_2(N \times M_2)$ to the residual error for both data sets. Both the common part and the distinctive parts are modeled with a few components to describe the systematic variation. The common parts of the two data sets can be described by the same scores $T_C(N \times R_C)$. Thus the variation between the objects is exactly the same for the common parts, but it is represented differently in the data sets because of the different features in the different data sets. The distinctive part of each set is modeled with distinctive scores $T_{D1}(N \times R_{D2})$ and $T_{D2}(N \times R_{D2})$. Note that the number of distinctive components can be different for each set and can even be 0 for sets if no systematic distinctive variation is present. Finally, the part of each data set that is not systematic is left in the residuals $E_1(N \times M_1)$ and $E_2(N \times M_2)$.

$$X_1 = T_C P_{C1}^T + T_{D1} P_{D1}^T + E_1 \quad X_2 = T_C P_{C2}^T + T_{D2} P_{D2}^T + E_2 \tag{7.6}$$

Equation 7.6 is the general model valid for the common and distinctive methods; however, the various methods described below may have small differences compared to this general model.

DISCO-SCA

DISCO-SCA, the first model we will discuss, was first applied in a cross-cultural study of emotions (Schouteden et al. 2013, 2014). The method starts with a simultaneous component analysis (SCA) model of all data blocks, which is the model described in Equation (7.4). The number of components used in this model equals the sum of the common and distinctive components. The SCA scores and loadings, however, are not yet separated into common and distinct as SCA models all systematic variation simultaneously. The main idea of the method is to rotate the SCA loadings toward a predefined structure that distinguishes between common loadings and distinctive loadings for each data block. The score matrix is then also rotated such that the modeled part of the original data does not change. An example of a target loading is:

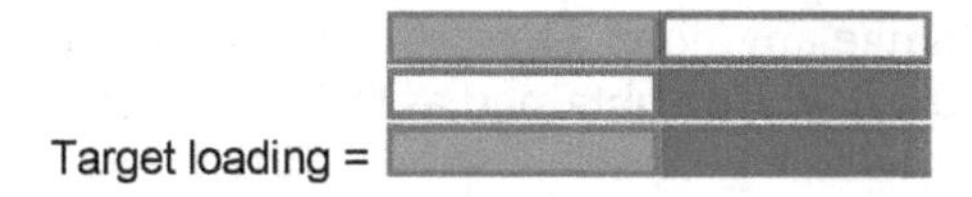

Here the target loading consists of three rows (three PC's) and two columns (blue representing block X_1 and red representing block X_2). The first loading (first row) represents a loading distinctive for block 1 (note that the red loading block is empty representing all loadings values to be zero, while the blue loading block is filled, and these loadings are allowed to be non-zero). This component describes variation only present in block 1 and not in block 2, as the loading values of the second block are assumed to be zero. The second row represents a loading distinctive for block 2 (here the part for $\boldsymbol{X}_1$ is empty) while the last row is the common loading, which describes variation that is present in both blocks (now both blocks for $\boldsymbol{X}_1$ and $\boldsymbol{X}_2$ are filled and are allowed to have nonzero values). Thus the rotation makes the contribution of block 2 to the first component as small as possible and the contribution of block 1 to the second component as small as possible. The rotation forces the loadings of the SCA to become similar to the target matrix. In most cases, the exact target is not reached. In the example, this means that the loadings of the first PC for the second block are not exactly zero, but they are small—the separation between common and distinctive is not perfect. The last row of the loadings matrix refers to a component that is common, as all loadings of all blocks are expected to contribute to the description of the data.

JIVE

The Joint and Individual Variation Explained (JIVE) method introduced by Lock et al. (Lock et al. 2013; O'Connell and Lock 2016) is also based on a simultaneous component analysis of both sets similar to Equation (7.4). The distinctive parts ($\boldsymbol{D}_1$ and $\boldsymbol{D}_2$) are estimated separately and iteratively. First the remainder of each block (R_1 and R_2) is calculated by deflating the common information from the original data sets:

$$\boldsymbol{R}_1 = \boldsymbol{X}_1 - \boldsymbol{T}_C\boldsymbol{P}_{C1}^T \quad \boldsymbol{R}_2 = \boldsymbol{X}_2 - \boldsymbol{T}_C\boldsymbol{P}_{C2}^T \tag{7.7}$$

Then, from the remainder of each block, ($\boldsymbol{R}_1$ and $\boldsymbol{R}_2$), distinctive principal components are calculated in such a way that they are orthogonal to the common information. $\boldsymbol{D}_1 = \boldsymbol{T}_{D1}\boldsymbol{P}_{D1}^T$ and $\boldsymbol{D}_2 = \boldsymbol{T}_{D2}\boldsymbol{P}_{D2}^T$ are low rank estimations of $\boldsymbol{X}_1$ and $\boldsymbol{X}_2$ composed of the distinctive components. The orthogonality is important as it allows the separation of the total variation into the variation explained by the common parts and distinctive parts. Furthermore, the interpretation of the common part is not confused by the distinctive parts and otherwise. The JIVE approach iterates between estimating the common part and the distinctive parts until the method converges and the scores and loadings do not change anymore. Thus after $\boldsymbol{D}_1$ and $\boldsymbol{D}_2$ have been obtained, new estimates for $\boldsymbol{C}_1$ and $\boldsymbol{C}_2$ are obtained as follows. First the distinctive parts are removed from the data

$$\boldsymbol{S}_1 = \boldsymbol{X}_1 - \boldsymbol{T}_{D1}\boldsymbol{P}_{D1}^T \quad \boldsymbol{S}_2 = \boldsymbol{X}_2 - \boldsymbol{T}_{D2}\boldsymbol{P}_{D2}^T \tag{7.8}$$

The new common parts can be estimated using $[\boldsymbol{S}_1 \mid \boldsymbol{S}_2] = \boldsymbol{T}_C\left[\boldsymbol{P}_{C1}^T \mid \boldsymbol{P}_{C2}^T\right] + [\boldsymbol{R}_1 \mid \boldsymbol{R}_2]$. The new estimates for $\boldsymbol{C}_1 = \boldsymbol{T}_C\boldsymbol{P}_{C1}^T$ and $\boldsymbol{C}_2 = \boldsymbol{T}_C\boldsymbol{P}_{C2}^T$ are obtained. The number of common components is the same for all blocks, but the number of distinctive components can be selected differently for each block.

OnPLS

O2PLS was developed for two blocks and later extended to OnPLS for studies with more than 2 blocks of data (Bylesjo et al. 2007; Lofstedt and Trygg 2011). In contrast to DISCO and JIVE that use a component model of the concatenated data sets, OnPLS starts with a singular value decomposition (SVD) of the covariance matrix ($\boldsymbol{X}_1^T\boldsymbol{X}_2$) for an analysis of the common variation. This SVD gives the loadings $\boldsymbol{P}_{C1}$ and $\boldsymbol{P}_{C2}$ for each of the blocks to obtain the common part of the data.

$$[\boldsymbol{P}_{C1}, \boldsymbol{D}, \boldsymbol{P}_{C2}] = SVD(\boldsymbol{X}_1^T\boldsymbol{X}_2) \tag{7.9}$$

$$\boldsymbol{T}_{C1} = \boldsymbol{X}_1\boldsymbol{P}_{C1}$$

$$\boldsymbol{T}_{C2} = \boldsymbol{X}_2\boldsymbol{P}_{C2} \tag{7.10}$$

Here it is clear that the common scores are not the same for the two blocks. Actually $\boldsymbol{T}_{C1}$ contains the variation of $\boldsymbol{X}_1$ to optimally predict the variation in X_2 and vice versa. This is clearly a different definition of what is common compared to the other methods.

The common scores between $\boldsymbol{X}_1$ and $\boldsymbol{X}_2$ described by $\boldsymbol{T}_{C1}$ and $\boldsymbol{T}_{C2}$, could still contain variation that is independent from the other block. Therefore, directions in $\boldsymbol{T}_{C1}$ that are orthogonal to $\boldsymbol{X}_2$ (called $\boldsymbol{T}_{D1}$) are obtained, and also directions in $\boldsymbol{T}_{C2}$ that are orthogonal to $\boldsymbol{X}_1$ (called $\boldsymbol{T}_{D2}$) are obtained. These orthogonal directions form the distinctive part of the model.

$$\boldsymbol{P}_{D1} = \boldsymbol{X}_1{}^T\boldsymbol{T}_{D1}(\boldsymbol{T}_{D1}{}^T\boldsymbol{T}_{D1})^{-1} \tag{7.11}$$

$$\boldsymbol{P}_{D2} = \boldsymbol{X}_2{}^T\boldsymbol{T}_{D2}(\boldsymbol{T}_{D2}{}^T\boldsymbol{T}_{D2})^{-1}$$

A thorough and detailed overview of all steps in the O2PLS algorithm can be found in a recent paper by Stocchero et al. (2018).

Interpretation of Common and Distinctive Methods

One of the advantages of data fusion methods is that features of different blocks can be interpreted simultaneously in the loading plot of the common part. But also the distinctive part can lead to an improved interpretation. Contrary to the common part, which is only valid for the total data, and not for each block separately, the distinctive part should be interpreted per block. Important is that the distinctive part is orthogonal to the common part, so that they do not describe the same information, and that the variance is separated over the two parts of the data.

The distinctive scores and loadings can also be interpreted as in a normal PCA model. However, the distinctive information should always be interpreted on top of the common part. For the total interpretation of a single block the common part and the distinctive part should be combined in such a way that the explained variation in the common and distinctive part is also taken into account. It is possible that the common part is only present in a small amount of the block (this can be observed by the amount of variation explained by the common part in that block).

Another advantage of the common and distinctive models is that the distinctive components clean up the data such that the common part can be better estimated. This makes interpretation of the common part easier as the distinctive variation is removed and cannot disturb interpretation.

Orthogonality Aspects

The separation of the total variation into separate parts for the common, distinctive and residual parts depends on the orthogonality properties of the different blocks. This orthogonality depends on how the blocks are estimated and this differs per method. Within a block, the common and distinctive parts are orthogonal for all methods, but the residuals are not orthogonal to the common part for JIVE and OnPLS. This means that part of the variation of the residuals is already described by the common part. Thus, for these methods:

$$\|\boldsymbol{X}_1\|^2 \neq \|\boldsymbol{C}_1\|^2 + \|\boldsymbol{D}_1\|^2 + \|\boldsymbol{E}_1\|^2$$

In an ANOVA context we are used to the situation that our model explains a certain amount of variance, and the residuals explain the rest of the variance. Here, this is not the case. The variances of all the blocks do not add up to the total variance in the original data. The distinctive parts for the different data sets are not orthogonal to each other for both JIVE and OnPLS. For OnPLS they are also not orthogonal to the common part of the other block. If we do want to explain the variances in such a way that they add up to the total variance of the data X, type III partial explained sum of squares for the residuals could be used, i.e., the sum of squares of the residuals that are not yet explained by the other parts (Stanimirova et al. 2011).

Generally, it seems that DISCO is the most restrictive model with many orthogonality restrictions between the different blocks, while OnPLS is the least restrictive. The similarity between DISCO and JIVE is a consequence of the use of SCA in both methods.

Results

In this example we will look at the metabolic and transcriptional response of *Arabidopsis thaliana* wild type and mutants to *Phytophtora infestans* with different data fusion methods that are able to separate the common variation in the metabolomics and transcriptomics data from the variation that is distinctive for each set. We will use two data fusion methods, JIVE and OnPLS to look for similarities and differences between these two methods.

The model selection procedure (estimation of the number of components for the common and distinctive part in the model) is different for each of the methods. To be able to compare the scores and loadings of the methods, the same number of common and distinctive components were used in all cases. Two common components were selected and also two distinct components for each data set.

Figure 7.3 shows the common part of the JIVE model. The scores plot (top) shows a nice separation between the different genotypes. The large variation of the erp2 group, which was present only in the transcriptomics data, is not present anymore in the common part score plot, because it is not common variation (as it is not present in the metabolomics block). Therefore, the common 1st component also only explains 15% of the variation of the transcriptomics block, which is much lower than 29% in the simultaneous component model. You will see below that this large variation in the erp2 group is specific for the transcriptomics data and thus appears in the transcriptomics distinctive variation. The result now is a nice separation of most groups in the common part of the JIVE model. Also, here the loading plot

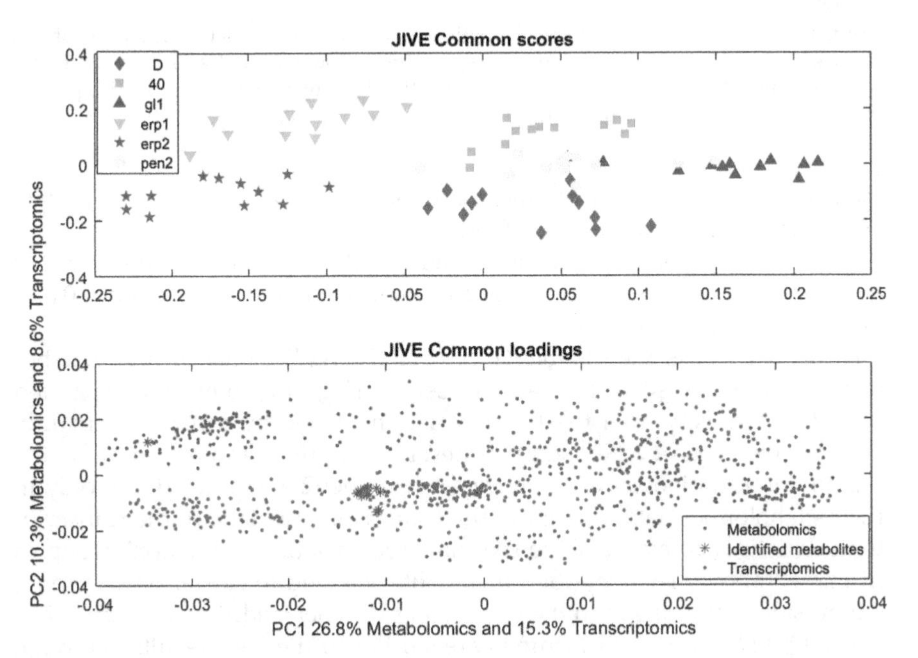

FIGURE 7.3
The top figure shows the common scores of metabolomics and transcriptomics data using JIVE. These two common components describe 37.1% of the metabolomics and 23.9% of the transcriptomics data. The bottom plot shows the common loadings of metabolomics and transcriptomics data. Indicated with asterisk are the identified metabolite features that were not able to distinguish between the genotypes.

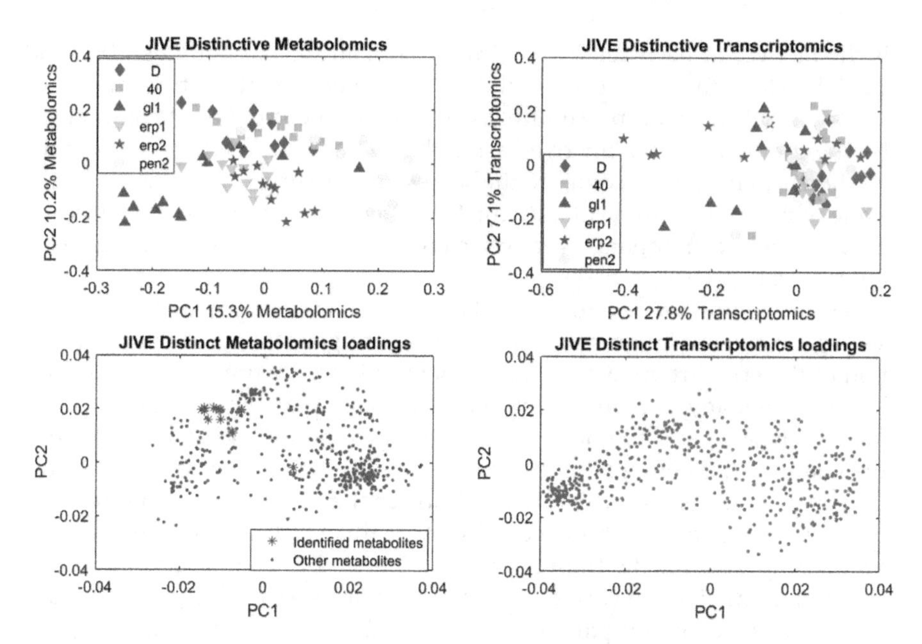

FIGURE 7.4
Distinct part of the JIVE model of metabolomics and transcriptomics data. The top row shows the distinctive scores for the metabolomics and transcriptomics part and the bottom row shows the distinct loadings. The amount of variation explained by these distinctive parts is indicated at the labels of the score plots.

metabolomics and transcriptomics features can directly be compared as they are related to the same score profiles.

The metabolite loadings plot again shows a cluster at the left side related to the erp2 and erpl genotypes. Again the identified loadings are close to 0 and hardly take part in the model.

Figure 7.4 shows the distinctive part of the JIVE model. In the metabolomics distinctive score plot we can see pen2 group somewhat separated from the rest. Now that the main variation caused by differences mainly due to separation of erp2 and gl1 is described in the common variation, the remainder variation shows more clearly that pen2 is separated, but only in the metabolomics data and not in the transcriptomic data. In the transcriptomics distinct component, we see the large variation of the erp2 group as discussed above. The identified metabolites (*) tend to cluster on the upper left outside of the cluster in the metabolites distinct loading plot (Figure 7.5).

The OnPLS scores and loadings are similar to the JIVE results, although OnPLS has different common loadings for each block. Therefore it is possible that the common part of the transcriptomics data now is better explained than with the JIVE model (Figure 7.5).

In the distinctive part of the OnPLS model again the pen2 group is somewhat separated from the rest, while again in the distinct transcriptomics part

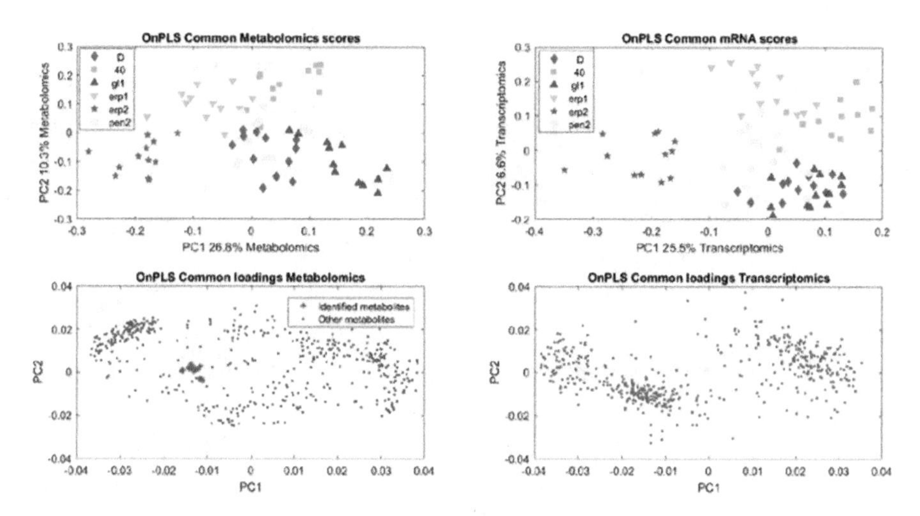

FIGURE 7.5
OnPLS common part of the Arabidopsis study.

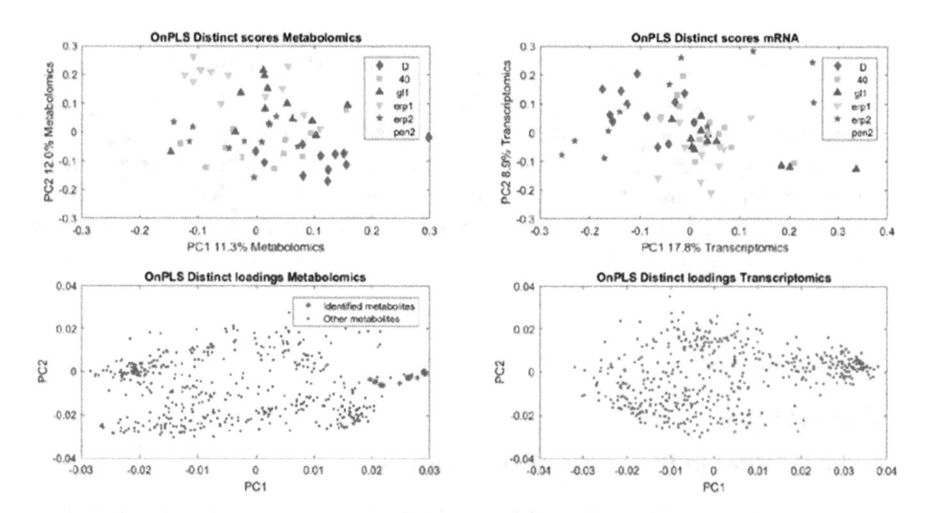

FIGURE 7.6
OnPLS distinct part of the Arabidopsis study.

we see the large variation in the erp2 group. Now the identified metabolomics are clearly different from the other metabolites in the distinct metabolomics loadings (Figure 7.6).

Finally, the DISCO results are again similar to the previous models. Because there is only one set of common components for both blocks again the transcriptomics part is not well explained in the common part of the model. The large variation in the erp2 group is visible in the transcriptomics distinct part, while the pen2 group is less separated from the other groups

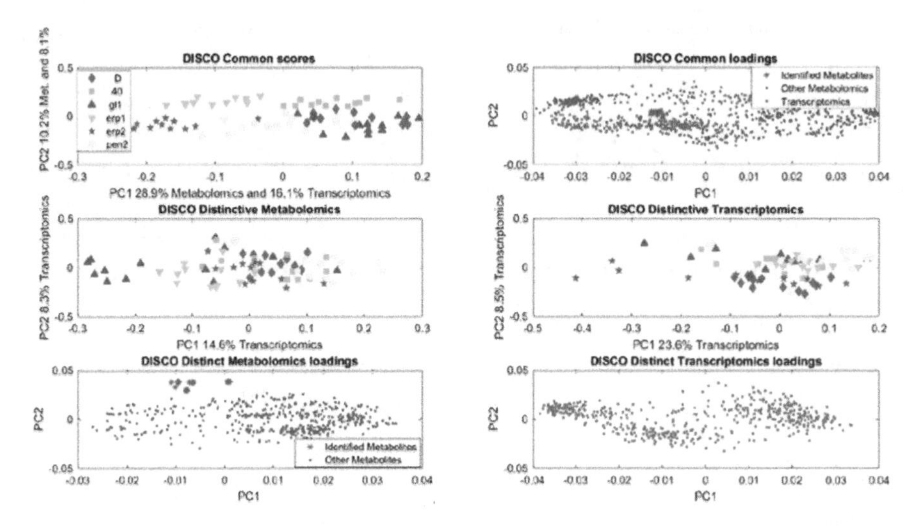

FIGURE 7.7
DISCO results of the Arabidopsis study.

in the distinct metabolomics part than in the OnPLS model. Here also the identified metabolites loadings are clearly separated from the other loadings in the distinctive metabolomics part of the DISCO model (Figure 7.7).

Concluding, we see that although there are slight changes between the three models that separate common variation from distinct variation, the major conclusions are similar. Note that this was an explorative view on the data, and the leads have to be followed up with more dedicated analysis.

Discussion

General Issues in SCA Type of Data Fusion

There are several important issues in the SCA type of models that need to be addressed in order to make an appropriate interpretation of the results possible. The mathematical criteria applied in the different methods make sure that each of these methods gives a different view on the data. None of these views are wrong; however, they highlight different features in the data.

Orthogonality is helpful in separating variance and it makes estimation of the components easier. However, in terms of biological interpretation it can become cumbersome. In biology, treatment effects are usually not orthogonal. If orthogonality is forced by the method, then the estimated effects may be a distorted version of the true effect, which will make interpretation more difficult.

Not only the selected method, but also the selected data pre-processing gives different views to the data. Block scaling, autoscaling, normalization all transform the data in different ways. As the SCA type of methods described above are aimed to describe variance, the interpretation depends on how the data is treated before data fusion. All the methods are based on mathematical properties such as maximum variation explained or maximum correlation. These mathematical concepts make sure that in mathematical terms the model is optimal. What this means for biological terms is still not fully understood. Some general guidelines to pre-processing are given below.

Sample Alignment

The most common application of data fusion methods is when multiple platforms are used to measure a number of samples. In data fusion the methods use correlations between variables of different sets. It is therefore mandatory that the samples on the same row in each data set are comparable. This seems logical, but in case of replicates can become complex. As each platform usually has its own sample workup, technical replicates cannot be used in data fusion. The relationship of replicate 1 in platform A is not the same for both replicates in platform B. Therefore, data fusion of technical replicates has to be performed on the averages over technical replicates. This still leads to potential problems when the number of technical replicates is not constant over all samples—typically, only a few samples are repeatedly measured to get an idea of the technical variation, and their averages will have smaller errors than the unreplicated sample. In contrast, biological replicates can be used directly in data fusion.

Block Scaling

Another issue in data fusion is block scaling. As stated above, we assume that the variables of each block are already centered and scaled by the standard deviation. The number of features in each block, however, can be rather different. As most data fusion methods (just as a normal PCA) focus on explaining variance for the blocks, differences in variance per block can have a large effect on the final results. Just as auto scaling is often applied to a single block for explorative analysis, in data fusion it is common to scale the variance of each block to a constant value. In this way, each block is given the same importance, irrespective of its size. The consequence is that an important single variable in a large block is less likely to have a large contribution in the final model than a similar variable from a smaller block. Prior knowledge about the system can help finding a better approach of giving weights to each of the blocks in the data fusion model.

Model Complexity

For the SCA models, and especially for the common and distinctive models, model complexity is an important issue. How many components are being explored in the model? For the common and distinctive component models such a decision has to be made for each distinctive block and also for the common block, and these numbers are not independent. Several approaches are used in literature. Moreover, each of the methods discussed above has its own approach, which makes comparison difficult. For the SCA models, the number of components is not so critical as one usually explores the first few components only. For the common and distinctive component models it is more critical: when more distinctive components are selected, this usually corresponds to fewer common components, and vice versa.

Conclusion

Data fusion methods can be used to explore data in situations where different platforms have been used to measure the same set of samples. To combine these data sets and to learn from the interactions between the sets, several data fusion methods have been discussed that, besides the common part of the variation in the data, also model the parts distinctive for each data set. Correlation based methods visualize the correlation between features of different block using correlation maps or by correlation networks. SCA based methods calculate scores and loadings to visualize similarities between the samples and the features in the different blocks. By exploring the data using score and loading plots of the common and distinctive information, one can gain an improved understanding of the particular sources of variation within the data. Note that the methods shown in this chapter are all unsupervised; they are used for exploration (like a normal PCA) and not explicitly for prediction or for finding differences between the mutants.

The differences between the models are due to the different mathematical properties (e.g., orthogonality constraints, applied algorithm) used to calculate the model. They give somewhat different views of the data, which are all correct, but one might be more useful in one occasion while another may be more useful in another application. At this moment it is not clear which mathematical properties lead to the most insightful biological view, and this has to be assessed on a case-by-case basis.

Experimental

Data Pre-Processing

The microarray data from the MTBLS18 experiment was processed and normalized with the R-package *simpleaffy* (Wilson and Miller 2005), which resulted in a 72 × 22810 transcriptomics data set. The transcriptomics data was normalized using Robust Multi-array Average. Feature selection was performed by only allowing features that were different between the various genotypes. For the final analysis, 519 features were selected.

The LC-MS data consist of positive and negative mode data sets, with 5931 and 1612 features, respectively. Only the positive mode data set was used in this example. First a log scaling was applied as commonly measurement error variance increases with intensity and log scaling helps reducing the heteroscedastic variance. Feature selection for metabolomics was performed by allowing only features that were able to distinguish between the different genotypes. Furthermore, 15 features that had identification were added. In total, 493 positive mode features were selected for an explorative analysis of the data fusion methods.

References

Bylesjo, M., D. Eriksson, M. Kusano, T. Moritz, and J. Trygg. 2007. "Data Integration in Plant Biology: The O2PLS Method for Combined Modeling of Transcript and Metabolite Data." *Plant Journal* 52 (6): 1181–1191.

Camacho, D., A. de la Fuente, and P. Mendes. 2005. "The Origin of Correlations in Metabolomics Data." *Metabolomics* 1 (1): 53–63.

Cloarec, O. et al. 2005. "Statistical Total Correlation Spectroscopy: An Exploratory Approach for Latent Biomarker Identification from Metabolic H-1 NMR Data Sets." *Analytical Chemistry* 77 (5): 1282–1289.

Crockford, D.J., A.D. Maher, K.R. Ahmadi, A. Barrett, R.S. Plumb, I.D. Wilson, and J.K. Nicholson. 2008. "H-1 NMR and UPLC-MSE Statistical Heterospectroscopy: Characterization of Drug Metabolites (Xenometabolome) in Epidemiological Studies." *Analytical Chemistry* 80 (18): 6835–6844.

Gomez-Cabrero, D., I. Abugessaisa, D. Maier, A. Teschendorff, M. Merkenschlager, A. Gisel, E. Ballestar, E. Bongcam-Rudloff, A. Conesa, and J. Tegner. 2014. "Data Integration in the Era of Omics: Current and Future Challenges." *Bmc Systems Biology* 8: I1.

Kamburov, A., R. Cavill, T.M.D. Ebbels, R. H., and H.C. Keun. 2011. "Integrated Pathway-Level Analysis of Transcriptomics and Metabolomics Data with IMPaLA." *Bioinformatics* 27 (20): 2917–2918.

Lock, E.F., K.A. Hoadley, J. S. Marron, and A.B. Nobel. 2013. "Joint and Individual Variation Explained (Jive) for Integrated Analysis of Multiple Data Types." *Annals of Applied Statistics* 7 (1): 523–542.

Lofstedt, T. and J. Trygg. 2011. "OnPLS-a Novel Multiblock Method for the Modelling of Predictive and Orthogonal Variation." *Journal of Chemometrics* 25 (8): 441–455.

Noda, I. 1993. "Generalized 2-Dimensional Correlation Method Applicable to Infrared, Raman, and Other Types of Spectroscopy." *Applied Spectroscopy* 47 (9): 1329–1336.

Noda, I., A.E. Dowrey, C. Marcott, G.M. Story, and Y. Ozaki. 2000. "Generalized Two-Dimensional Correlation Spectroscopy." *Applied Spectroscopy* 54 (7): 236A–248A.

O'Connell, M.J. and E.F. Lock. 2016. "R.JIVE for Exploration of Multi-Source Molecular Data." *Bioinformatics* 32 (18): 2877–2879.

Pavlopoulos, G.A., A.-L. Wegener, and R. Schneider. 2008. "A Survey of Visualization Tools for Biological Network." *Biodata Mining* 1: UNSP 12.

Richards, S.E., M.-E. Dumas, J.M. Fonville, T.M.D. Ebbels, E. Holmes, and J.K. Nicholson. 2010. "Intra- and Inter-Omic Fusion of Metabolic Profiling Data in a Systems Biology Framework." *Chemometrics and Intelligent Laboratory Systems* 104 (1): 121–131.

Schouteden, M., K. Van Deun, S. Pattyn, and I. Van Mechelen. 2013. "SCA with Rotation to Distinguish Common and Distinctive Information in Linked Data." *Behavior Research Methods* 45 (3): 822–833.

Schouteden, M., K. Van Deun, T.F. Wilderjans, and I. Van Mechelen. 2014. "Performing DISCO-SCA to Search for Distinctive and Common Information in Linked Data." *Behavior Research Methods* 46 (2): 576–587.

Smilde, A.K., J.A. Westerhuis, and S. de Jong. 2003. "A Framework for Sequential Multiblock Component Methods." *Journal of Chemometrics* 17 (6): 323–337.

Smolinska, A. et al. 2012. "Simultaneous Analysis of Plasma and CSF by NMR and Hierarchical Models Fusion." *Analytical and Bioanalytical Chemistry* 403 (4): 947–959.

Stanimirova, I., K. Michalik, Z. Drzazga, H. Trzeciak, P.D. Wentzell, and B. Walczak. 2011. "Interpretation of Analysis of Variance Models using Principal Component Analysis to Assess the Effect of a Maternal Anticancer Treatment on the Mineralization of Rat Bones." *Analytica Chimica Acta* 689 (1): 1–7.

Steuer, R. 2006. "On the Analysis and Interpretation of Correlations in Metabolomic Data." *Briefings in Bioinformatics* 7 (2): 151–158.

Stocchero, M., S. Riccadonna and P. Franceschi. 2018. "Projection to Latent Structures with Orthogonal Constraints for Metabolomics Data." *Journal of Chemometrics* 32 (5): e2987.

Van Deun, K., A.K. Smilde, M.J. van der Werf, H.A.L. Kiers, and I. Van Mechelen. 2009. "A Structured Overview of Simultaneous Component Based Data Integration." *Bmc Bioinformatics* 10: 246.

Westerhuis, J.A., T. Kourti, and J.F. MacGregor. 1998. "Analysis of Multiblock and Hierarchical PCA and PLS Models." *Journal of Chemometrics* 12 (5): 301–321.

Wilson, C.L. and C.J. Miller. 2005. "Simpleaffy: A BioConductor Package for Affymetrix Quality Control and Data Analysis." *Bioinformatics* 21 (18): 3683–3685.

Xia, J. and D.S. Wishart. 2010a. "MetPA: A Web-Based Metabolomics Tool for Pathway Analysis and Visualization." *Bioinformatics* 26 (18): 2342–2344.

Xia, J. and D.S. Wishart. 2010b. "MSEA: A Web-Based Tool to Identify Biologically Meaningful Patterns in Quantitative Metabolomic Data." *Nucleic Acids Research* 38: W71–W77.

Xia, J. and D.S. Wishart. 2011. "Web-Based Inference of Biological Patterns, Functions and Pathways from Metabolomic Data using MetaboAnalyst." *Nature Protocols* 6 (6): 743–760.

8

Genome-Scale Metabolic Networks

Clément Frainay and Fabien Jourdan

CONTENTS

Introduction 177
Genome-Scale Reconstruction of Metabolic Networks 178
  Reconstruction Principle 178
  Databases 181
  Systems Biology Markup Language (SBML) Format for Genome-Scale Metabolic Networks Description 183
  Genome-Scale Reconstruction Limitations 184
  Metabolome-Based *Ab Initio* Network Reconstruction 185
Pathway Mapping 187
  From Metabolites to Pathways 187
  Pathway Enrichment 188
Graph Modeling: Looking for Sub-Networks Based on Metabolomics Data 189
  Graph Models 190
  Extracting Relevant Sub-Graphs from Global Metabolic Network 192
Network Visualization 194
Conclusion 196
References 196

Introduction

Metabolomics experiments and subsequent data analyses often yield a list of metabolites with concentrations changing under environmental or genetic stress. However, this list does not provide by itself a mechanistic interpretation of the perturbation, that is, the enzymatic reactions leading to these metabolite concentration changes. This biochemical context is provided by the organism metabolic network (the union of all metabolic pathways), also called the Genome-Scale Metabolic Network, since it is based on genomic information and aims to cover the whole metabolism. The aim of this chapter is to present how these networks are built and how metabolites can be mapped into metabolic pathways/networks. Once this mapping is performed, we will show how it is possible, using a specific mathematical formalism called graph modeling, to identify metabolic paths between metabolites of interest. Finally, we will present how network visualization is achieved and what are the benefits it provides.

Networks are data structures commonly used in many application fields since they allow describing connectivity (edges) between elements (nodes). Currently, the most studied networks are the ones modeling social interactions since they allow gaining new insights on communities and global sociological behavior (Trpevski et al. 2010; Schich et al. 2014). In the field of biology, networks are successfully used at different levels: the organism level (e.g., trophic networks) (Thébault and Fontaine 2010), the cellular level (e.g., neuronal networks) (Bullmore and Sporns 2009), the molecular level (e.g., protein-protein interaction networks, metabolic networks) (Albert 2005; Barabási et al. 2011) and even the intracellular level (e.g., De Brujin networks for next generation sequencing data analysis) (Compeau et al. 2011). Networks have the double advantage of providing a visual representation of relational data as well as allowing the use of mathematical algorithms to make sense out of the thousands of connections they may contain.

This chapter will focus on networks made of connections between substrates and products involved in the same biochemical reaction. Nevertheless, the reader has to be aware that other networks are used in the field of metabolomics as well. For instance, the correlation networks mentioned in Chapter 6 connect metabolites (or features) with coordinated changes in intensities. These are usually not modeling reactions (Camacho et al. 2005). MS/MS networks, mostly called molecular networks, connect features if they share common substructures (Watrous et al. 2012). It is important to note that even if they are not addressed in this chapter, some of the algorithms that will be presented here can be applied to these other networks too.

Genome-Scale Reconstruction of Metabolic Networks

Reconstruction Principle

The main aim of Genome-Scale Metabolic Network reconstruction is to assemble all the knowledge about an organism metabolism into a single database which can then be used for modeling purposes. Hence, reconstruction constitutes the first step of network-based analysis of metabolomics data, and quality of this reconstruction may have impacts on the results obtained using computational methods.

Genome-Scale Metabolic Network reconstruction is based on genomic and bibliomic data and represents the biochemical, genetic and genomic knowledge base for the target organism (Reed et al. 2006; Watrous et al. 2012). However, up to date, it is still infeasible to automatically reconstruct a high-quality metabolic network without going through a manual curation process. This network annotation consists in continuous back and forth between experimental validation and model update (Feist et al. 2009).

A metabolic network is made of the union of all reactions that turn substrate compounds into products. Contrary to metabolic pathways focusing on specific anabolic or catabolic functions, a metabolic network considers all of these simultaneously. Most of the reactions it contains are catalyzed by enzymes which are proteins coded for by genes. There is thus a clear relation between the genome and the metabolome, often called GPRs (Gene-Protein-Reaction). Such a metabolic network is therefore a relevant framework to interpret multi-omics data related to metabolism, since transcriptomic or proteomic data can be mapped on reactions while metabolomics data can be mapped on metabolites.

Various methods and software products are available to assemble a genome-scale network, but most of them follow the same semi-automatic protocol (Thiele and Palsson 2010). The input data is the annotated genome of the organism. This genomic information is then used to retrieve in databases, by sequence homology, which genes code for an enzyme that will achieve the particular metabolic function of interest. The individual reconstruction steps are described below (see Figure 8.1 for an overview).

Automatic reconstruction. The first step in genome-scale reconstruction is the automated generation of a draft reconstruction, which is based on the genome annotation and biochemical databases of the organism under study.

The main objective of the automatic reconstruction is listing the metabolic reactions which can be associated with the corresponding metabolic genes by using enzyme commission (EC) identifier (Bairoch 2000) or biochemical reaction databases such as BRENDA (Bairoch 2000; Schomburg et al. 2013) or KEGG (Kanehisa et al. 2014). The aims of this process are: (i) to link metabolic genes with their corresponding encoded enzymes and (ii) to determine the stoichiometric relationships of metabolic reactions with the corresponding reagents and products (metabolites and/or cofactors). The relationship

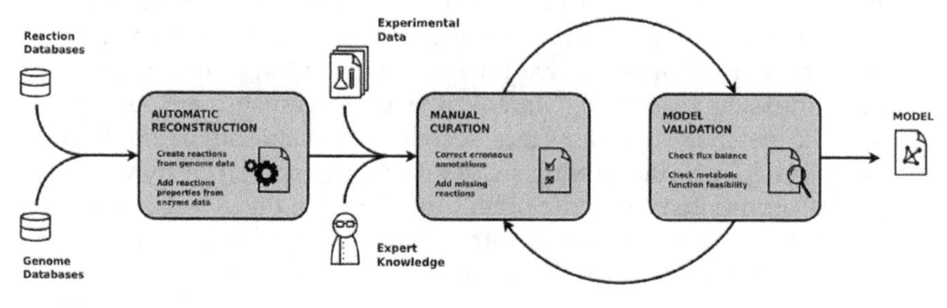

FIGURE 8.1
Genome-Scale Metabolic Network reconstruction pipeline.

between genes-proteins and metabolic reactions is encoded by the GPR associations which are boolean logical expressions: isoenzymes catalyzing the same reaction have an "OR" relationship (only one of the genes that encode the different isoenzymes is required to enable the reaction), and the complexes that catalyze a reaction have an "AND" relation (all the genes that encode the different complex subunits are necessary to have the reaction active) (Duarte et al. 2007). These associations enable the systematic mapping of transcriptomics or proteomics data to the corresponding reactions.

The other challenge in automatic network reconstruction for an organism with cellular compartments is to assign reactions and metabolites to these intra-cellular bodies (e.g., which reactions are taking place in mitochondria). The sub-cellular localization in which metabolic reactions occur is determined from protein location data or can be derived from the presence of certain motifs in a protein sequence (Thiele and Palsson 2010). Additionally, genome-scale metabolic models use artificial reactions to predict or impose certain phenotypic conditions on the mathematical model. Some of the most common artificial reactions are the exchange reactions that define the overall rate of nutrients consumption or production, or the biomass reaction defining the ratio at which biomass constituents (nucleic acids, lipids, proteins, etc.) are produced. Nevertheless, localization is less of a concern in metabolomics since metabolites are usually measured without information on their localization.

This automatic reconstruction can be achieved by using software tools such as Pathway Tools (Karp et al. 2015) or metaSHARK (Pinney et al. 2005) in combination with databases such as GOLD (Pagani et al. 2012) (https://gold.jgi.doe.gov/) or NCBI Entrez Gene databases (http://www.ncbi.nlm.nih.gov/gene) (Maglott et al. 2005). If the genome is not sufficiently annotated, putative annotations can be identified by using tools like Blast (Camacho et al. 2009) or Inparanoid (Sonnhammer and Östlund 2015).

Manual curation. Once the automated reconstruction is finished, it is necessary to manually curate this draft reconstruction. In fact, this reconstruction will generate false positive predictions (reactions that are added but not active for the organism) and false negative ones (missing reactions). Moreover, if the organism can perform specific metabolic functions (e.g., secondary metabolism in plants or fungi), they may be missing in the reconstruction since they cannot be propagated from any other organism. Finally, a large part of the genes are not annotated with a function. This is, e.g., true for more than one third for *Escherichia coli* K-12 MG1655 (Díaz-Mejía et al. 2009). The main objective of curation is to identify and correct incomplete or erroneous annotations, and add reactions that occur spontaneously, through a literature search and other databases, and to remove gaps and metabolites that cannot be produced or consumed (Francke et al. 2005; Duarte et al. 2007).

Model validation. The next step is to evaluate and validate the curated reconstruction by using mathematical tools (Schellenberger et al. 2011) that will be detailed in Chapter 9. One validation process is to evaluate if the model is stoichiometrically balanced with respect to the quantitative evaluation of biomass precursor production and growth rate. Another one is to check if no gaps (i.e., missing reactions) are remaining, which would result in a disconnected network. Candidate reactions for gap filling can be suggested at this stage for the next manual curation iteration. Finally, validation of the model can also be achieved by checking if known metabolic functions can be performed using the organism's metabolic network (e.g., does the network allow to create essential metabolites?). Other methods, such as inparanoid (Sonnhammer and Östlund 2015), assess the quality of the resulting model by comparison with validated reconstruction of phylogenetically close organisms, based on sequence orthology.

Iterations between validation and curation. Consequently, metabolic reconstruction requires a long and tedious manual curation—validation iterations that can last few months from years to get a reliable network, meaning that predictions (like cell growth) based on this network will fit most of experimental data (the risk of overfitting is not yet discussed in the field). This level of curation is not provided in the KEGG database. On the other hand, BioCyc is classifying reconstructions based on their level of curation: tier 1 (highly curated), tier 2 (under curation) and tier 3 (automatically reconstructed). In addition to the curation quality, another issue is that, for the same organism, several reconstructions can exist—for *Arabidopsis Thaliana*, e.g., at least five different reconstructions are available/known (Jourdan 2013). The challenge is then to select the most appropriate reconstruction from these. In a metabolomics experiment, one criterion can be the availability in the model of unambiguous identifiers that could help mapping identified compounds into the network (see the section "From Metabolites to Pathways" for further discussion).

Databases

Metabolic reactions for a large range of organisms have been gathered through metabolic network reconstruction process and made available in several databases (Box 8.1). We will here focus mainly on KEGG and BioCyc since they cover many organisms and they come with a large variety of tools. Both KEGG and BioCyc provide information on reactions and related proteins and genes. Nevertheless, a BioCyc-based database, e.g., EcoCyc for *E. coli*

BOX 8.1 BASIC DEFINITIONS IN GENOME-SCALE METABOLIC NETWORK RECONSTRUCTIONS

- **Metabolic reaction**: A metabolic reaction transforms metabolites (substrates) into other metabolites (products). Most reactions in metabolic networks are catalyzed by enzymes while some can be spontaneous.
- **EC number**: Enzyme Commission (EC) numbers classify enzymes based on their reaction activity. It is a four digit number (e.g., 5.4.2.2 codes for phosphoglucomutase, which is the first reaction of the glycolysis). Each digit provides information on the enzyme metabolic activity (e.g., 5 as a first digit means that an internal rearrangement of the molecule will occur).
- **GPR**: Gene Protein Reaction (GPR) is the association of the genes that can code for proteins that are involved in a given reaction. It allows linking enzymatic data with genomics ones.
- **Metabolic pathway**: A metabolic pathway is a set of metabolic reactions involved in a particular metabolic function (e.g., glycolysis). Generally, pathways have input compounds and output ones. Pathway definition is subjective and may vary from one database to another.
- **Genome-scale metabolic network**: A genome-scale metabolic network aims at gathering all metabolic reactions an organism can achieve. This list is generally based on the genome annotation of the organism.

(Keseler et al. 2013) or TrypanoCyc for *Trypanosoma brucei* (Shameer et al. 2015), will only contain reactions that are associated to a gene or used to complete a pathway (pathway holes). KEGG, on the other hand, is providing generic metabolic maps and will highlight feasible reactions from a selected organism by using different colored reactions associated to genes. Figure 8.2, e.g., is presenting glycolysis/gluconeogenesis for KEGG and HumanCyc (Romero and Pedro 2012). In KEGG, reactions colored in green will have a GPR association while those with no color do not. Another important difference is the definition of pathways which vary from one database to the other. All the reactions displayed in glycolysis/gluconeogenesis KEGG map are split into more than four pathways in HumanCyc (see the right side of Figure 8.2; Rapoport-Luebering glycolytic shunt and lactate fermentation are not represented on the HumanCyc side for clarity). This highlights a weakness in pathway-based analysis, since it relies on very subjective definitions (Ginsburg 2009; Keseler et al. 2013).

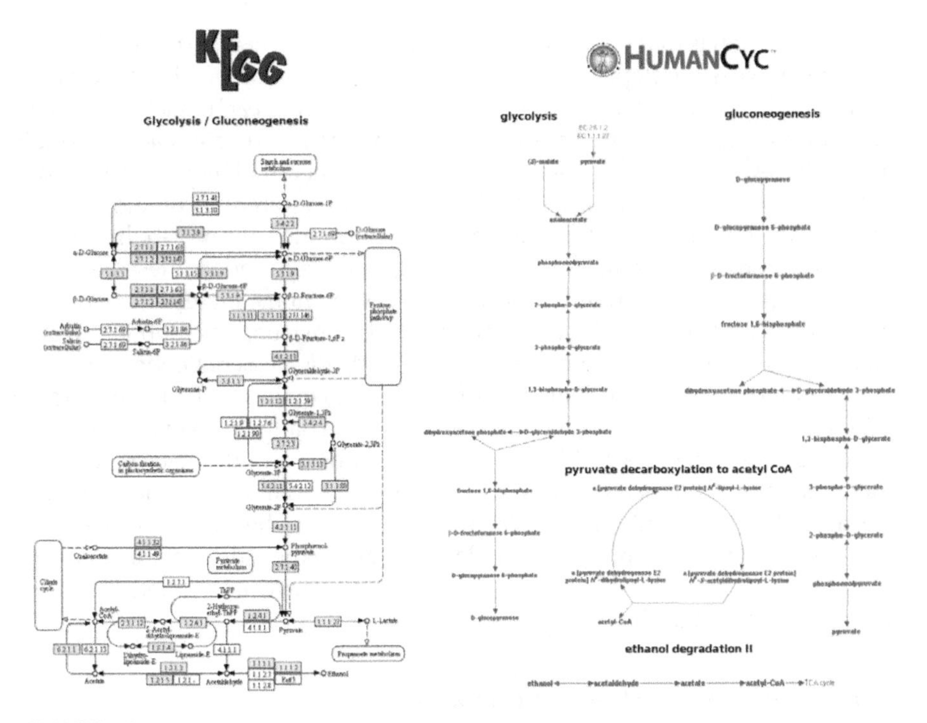

FIGURE 8.2
Glycolysis/gluconeogenesis in KEGG (left) and corresponding pathways in HumanCyc (right).

Both BioCyc and KEGG provide programmatic ways to access data through web services. For instance, using a single URL it is possible to get information on a gene or on a metabolite. In addition, both databases are available in text files. However, KEGG FTP repository usage requires a license.

Sharing knowledge on metabolism can also be achieved by storing information in flat files. In order to formalize this knowledge exchange, it has been necessary to standardize the files describing metabolic networks. The format getting current consensus, called SBML (Systems Biology Markup Language), is described in the following section.

Systems Biology Markup Language (SBML) Format for Genome-Scale Metabolic Networks Description

A convenient way to exchange and use all the information on a metabolic network consists of gathering all reactions and metabolites into a single file. That is the aim of SBML (Systems Biology Markup Language) files (Hucka et al. 2003). This XML-based format provides information on cellular compartments, metabolites, reactions proteins and genes (see sbml.org for SBML specifications). SBML is also designed to be enriched with mathematical information on reactions (kinetic parameters, stoichiometry). The advantage of this standardization is that these files can be handled by several software without any pre-processing step. Finally, non-formalized sections called "NOTES" may contain information about the level of curation of each reaction. Thiele and Palsson suggest scores ranging from 0 (typically a reaction added to fill a gap) to level 4 (a reaction experimentally validated) (Thiele and Palsson 2010). Also other meta-information, such as article references, ontology tags, or, relevant for the field of metabolomics, information on metabolites such as cross references to databases like ChEBI, may be found in these NOTES. These single files can often be found as supplementary material of articles describing Genome-Scale Metabolic Network reconstructions. Publicly available databases such as Biomodels (Li et al. 2010), BIGG (Schellenberger et al. 2010) or MetExplore (Cottret et al. 2010) also aim to make these files available.

SBML is an ongoing project and versions are changing. Nevertheless, it is possible to check if a SBML file is valid or not using the "SBML validator" provided by the consortium maintaining this standard. Other formats are available to describe molecular networks. Nevertheless, since they are designed to describe other molecular processes they do not necessarily encode all metabolic information, e.g., CellML (Lloyd and Geoff 2013) or BioPAX (Demir et al. 2010).

Genome-Scale Reconstruction Limitations

Genome-based reconstruction aims at providing an overview of all metabolic functions an organism can perform. Nevertheless, each tissue or cell may express these capacities in different ways. Some parts of the metabolism may not even be active. Muscle and hepatic cells, e.g., show different anabolic and catabolic activities. To be more specific, methods have been developed to use functional data (transcriptomics or proteomics) on reactions to extract the active parts in a particular condition from the global network (Shlomi et al. 2008; Jerby et al. 2010). This kind of approach has been successfully used in cancer research to decipher which parts of the metabolism are specific for cancerous cell lines, with the aim to identify potential oncogenes or oncometabolites (Prensner and Chinnaiyan 2011).

Metabolic networks currently contain a few thousands of reactions (e.g., 7440 for the *H. sapiens* reconstruction published in 2013 (Thiele et al., 2013)). Note that the size of these lists of reactions and metabolites tends to increase with the ongoing annotation of metabolic networks (13 543 reactions in the recently published new release of human metabolic network (Brunk et al. 2018)). Even if most reactions are well defined, some parts of the metabolism are mentioned using generic reactions and compounds, in particular for lipids. Figure 8.3 shows the generic reaction for phosphatidylcholines synthesis. The R1 and R2 parts (radicals) can be replaced with various carbon chains, thus defining a family of molecular structures. This generic definition can raise issues when using the network to map metabolomics data since it will not be possible to distinguish in the network two compounds of the same

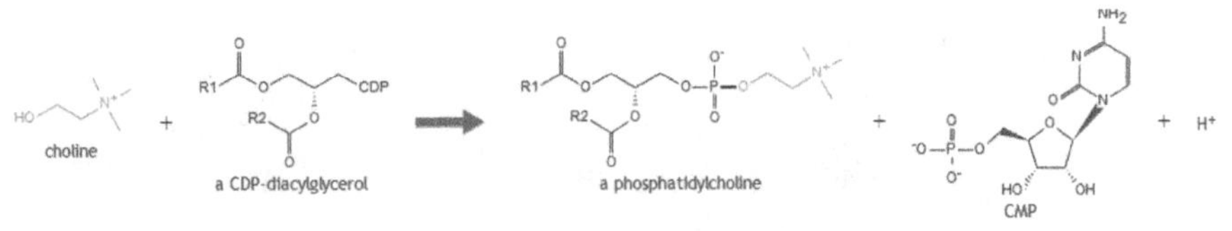

FIGURE 8.3
Generic synthesis reaction of phosphatidylcholines in MetaCyc.

class (e.g., two sphingolipids). Another example is the use of the network to perform atom tracing: not all carbons are described. Finally, it is not possible to balance these equations since for some molecules the number of atoms is not precisely defined.

One study proposed to expand the human metabolic network reconstruction Recon2 (Smallbone 2013) by integrating lipids. To do so, they assigned 16 carbons to these R parts of the molecule (as suggested in Duarte et al. 2007) and for triglycerides they chose most abundant families: myristate (14:0), palmitate (16:0), palmitoleate (16:1), stearate (18:0), oleate (18:1), linoleate (18:2), α-linoleate (18:3) and γ-linoleate (18:3). The size of the network has increased massively, going from 7440 reactions to 71,159 and from 5063 compounds to 13,940. Moreover, 562 reactions that were not balanced in Recon2 were now balanced thanks to adding these precise descriptions of substrates and products.

Finally, no genome is fully annotated, and some metabolic functions remain undefined. This limitation implies that some part of the metabolome cannot be predicted and consequently, some metabolites detected using metabolomics cannot be mapped in the network. That is the reason why methods were developed to build networks based on metabolomics data, in particular from high-resolution MS data. This approach will be described in the next section.

Metabolome-Based *Ab Initio* Network Reconstruction

High resolution mass spectrometry (HRMS) allows detecting masses with a very small ppm (parts-per-million) error. This is quite convenient to assign molecular formulas to peaks. Nevertheless, peaks are often considered independently, while they are in reality often the inputs and outputs of biochemical reactions. In fact, the product of a reaction is a substrate to which atoms had been added or removed through the action of an enzyme. For example, if a reaction corresponds to the loss of an alanine (with formula C_3H_5NO and exact mass 71.0371138), then the mass difference between substrate and product should be equal (up to few ppms) to the mass of alanine. This structural change can be translated into a mass difference between the substrate and the product. This mass shift, also called transformation function, can appear several times in the network since a single enzyme can act on several substrates (see Figure 8.4).

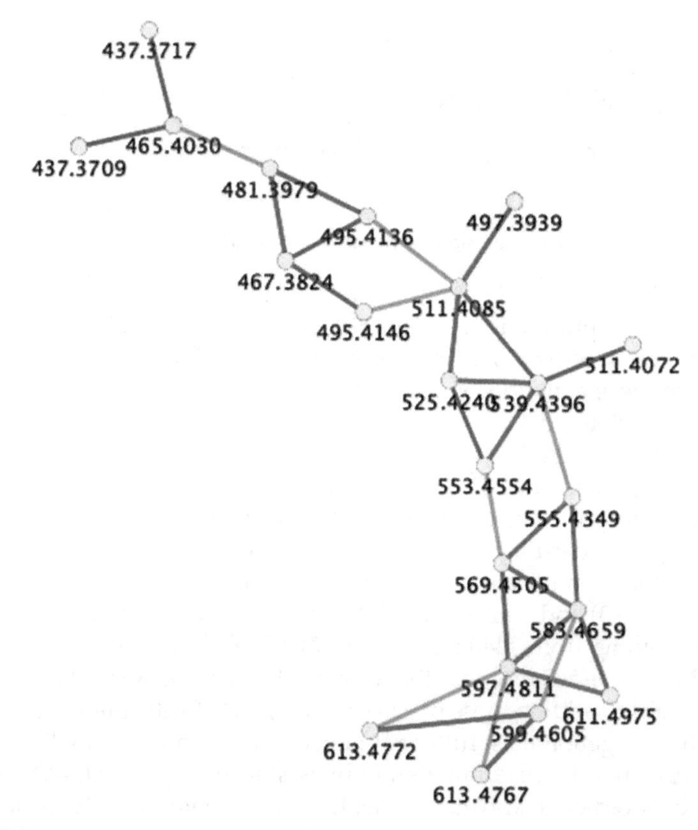

FIGURE 8.4
Network created using MetaNetter. (Jourdan, F. et al., *Bioinformatics*, 24, 143–45, 2008.) Each node corresponds to a mass obtained in HRMS. Three mass differences are used in this example. They correspond to hydroxylation (green), ethylation (blue) and methylation (red).

Based on this observation, it is possible to define a list of potential transformations that can occur in the network (Breitling et al. 2006). Then the reconstruction algorithm takes each combination of two peaks in the peak list and checks if their mass difference corresponds to a transformation in the list. If this is the case, the two peaks (nodes) are connected by a link (edge) labeled by the transformation.

This method may generate some false positives. The first reason is that metabolomics does not allow to associate a peak to a cellular compartment. It can therefore happen that the method predicts a reaction that is not possible since the two metabolites never appear simultaneously in the same compartment. The other drawback is that HRMS is not covering the entire metabolome and may thus miss some intermediary compounds (false negatives). Compounds may also be missed when a reaction is very fast, meaning that the concentration of the substrate will be below the detection threshold.

The *ab initio* approach nicely complements genome-based reconstruction since it is based on the functional observation of the metabolism. It is also used to help in metabolite identification since mass transformations correspond to well-defined formula modifications. One can then propagate annotations in the network based on these transformations (Silva et al. 2014).

When quantitative or semi- quantitative data are available, it is possible to use statistical approaches to infer network connections between metabolites or peaks (as detailed in Chapter 6). For instance, the Pearson correlation coefficient will model a linear relation between chemical entities (Steuer 2006). Nevertheless, the Pearson correlation will fail in predicting reactions where there is a nonlinear relationship between concentrations of substrates and products.

Another approach consists of using the chemical similarity between compounds, based on the observation that biochemical reactions tend to convert compounds with closely related chemical structures. Chemical similarity can be computed using chemical fingerprints. A chemical fingerprint is a numerical vector where each position refers to a chemical substructure. A value is attributed to each position given the presence or absence (or even the number of occurrences) of the substructure in the considered compound. The similarity of two fingerprints can then be computed using, e.g., the Tanimoto score (Barupal et al. 2012). Compound pairs with the highest similarities can be turned into edges, enriching metabolic networks built from reactions data from databases such as KEGG. This method helps overcoming the poor coverage of metabolomics data set mapping onto classical reactions databases, which can be imputed to enzyme promiscuity. This algorithm is implemented in the MetaMapp tool (Barupal et al. 2012).

The aim of metabolic network based analysis of metabolomics data is to focus on sub-parts of the network which are strongly related to the metabolites belonging to the identified fingerprint. Detecting those modulated subnetworks can be achieved using two kinds of approaches: pathway based and graph based. The first one will rely on the division of the network into predefined metabolic pathways corresponding to functions and the second one will interpret data in a more holistic way by using the entire network. These two approaches are presented in the following sections.

Pathway Mapping

From Metabolites to Pathways

Metabolic pathways divide the metabolism into functional modules which gather up to a few tens of reactions and metabolites (e.g., glycolysis). By organizing metabolites and/or genes into pathways, it is possible to list which ones contain monitored biological entities. Then, each pathway can be visualized highlighting identified metabolites. This pathway painting pipeline can also be applied to reactions when data on proteins or transcripts are available.

The main challenge before performing pathway mapping consists of finding for each compound of interest the corresponding identifiers in the pathway database. For instance, the KEGG identifier for glucose is C00031 while it is GLC for BioCyc. The name of a compound can be ambiguous, as most of them carry several synonyms. If database identifiers are not provided, finding them from names can therefore become a difficult task even if some methods were developed (Bernard et al. 2014). To overcome this issue the InChI project provides an unambiguous compound identifier system based on chemical structure (Heller et al. 2013). Several tools provide identifiers translations services, internally using InChI (or the InChI-Key hashed version) to link entries from various databases (Wohlgemuth et al. 2010; Chambers et al. 2013). Without any InChIs or well-known database identifiers, those kinds of tools can still perform a name-based search using fuzzy string matching and synonyms maps. However, it may return several possibilities for a given name and thus still requires a manual check.

Pathway Enrichment

Metabolites identified using metabolomics can be mapped in several metabolic pathways. The challenge then is to decide which pathways require to be further investigated. This cannot only be decided based on the number of metabolites mapped among the metabolites of the pathway (pathway coverage). In fact, mapping three metabolites in a pathway containing a tenth of metabolites is less interesting than a pathway containing few metabolites with two of them mapped. Distinguishing these two configurations is the aim of Pathway Enrichment. Hence Pathway Enrichment computes a p-value, telling if the pathway coverage obtained with an experimental list of metabolites is significantly higher than a coverage obtained with a random list of metabolites. This approach is often used in transcriptomics analysis to detect, based on expression data, which biological processes are significantly modulated during the experimental protocol (Subramanian et al. 2005). In Xia and Wishart 2010, the authors also propose to combine this pathway enrichment with a topological property of metabolic pathways. This measure, called centrality, is high when pathways are in between many metabolic processes and low when they are less central.

Pathway enrichment analysis (or metabolite set enrichment analysis) usually refers to statistical methods known as over-representation analysis (ORA), applied to metabolomics data. Those methods are used to evaluate if a predefined set of elements (metabolites of a given pathway or associated with a given disease) is represented more than expected in a sample (the list of differential metabolites).

One example is the Fisher's Exact Test. It's used on a contingency table with four entries:

- The number of differential metabolites found in the target pathway
- The number of differential metabolites not found in the target pathway
- The number of metabolites in the target pathway but not in the differential metabolites list
- The number of metabolites neither in the differential metabolites list nor in the pathway

The basic idea is completely analogous to the statistical testing scheme summarized in Chapter 6: one computes the probability of obtaining the same overlap (or a better one) between the list and the target pathway when doing this randomly. For example, a pathway which contains many differential metabolites can be discarded if its size is so large (compared to the size of the whole network) that we could get the same number of hits by randomly picking metabolites from the model. The strength of this approach is that exact probabilities can be computed using a hypergeometric distribution, and that no assumptions are needed. Fisher's exact test should be performed for each pathway of the model. A threshold is used on the p-values to identify pathways that are significantly overrepresented among our differential metabolites. This threshold should define the acceptable risk of obtaining a false positive result; however, by running the test several times on the same data set, we are more likely to encounter a false-positive result, and a multiple-testing approach is needed (see Chapter 6). More advanced methods allow to take into account quantitative data associated to metabolites, such as Quantitative Enrichment Analysis (QEA) (Xia and Wishart 2010).

Metabolic pathways provide information on metabolic functions, but they give a fragmented view of the metabolism. In particular, if the metabolic modulation under study is spanning several pathways it will be hard to identify this process using pathway mapping. This is the reason why, especially when dealing with global untargeted data, it is necessary to consider the metabolic network gathering all the pathways into a single model.

Graph Modeling: Looking for Sub-Networks Based on Metabolomics Data

Genome-scale reconstructions offer a tremendous tool to interpret metabolomic data by gathering and organizing knowledge about metabolism into a single object. However, the vast amount of information provided and the heterogeneity of sources make the linking of those models to the metabolomic data far from trivial. The following section will address the problem of bridging data to genome-scale reconstruction, as well as some methods to help refine the information in metabolic reconstruction to the most relevant part according to the data.

Graph Models

A network contains the information on the relationship between elements of a system. In order to mine its structure (topology), it is necessary to turn this database of connections into a formal representation called a graph. Graphs are mathematical objects containing nodes (or vertices) representing elements of a system, linked by edges representing relationships. This formalism allows exploiting a large panel of algorithms and topological measures from the mathematical field known as graph theory. These algorithms can give biological insights about global properties of the metabolism (Albert 2005), the role of a given metabolite or reaction in the network (Rahman and Schomburg 2006), identify indirect relations between metabolites or reactions (Frainay and Jourdan 2016).

The challenge when translating metabolic networks (list of reactions) into graphs (single mathematical object) is to find which biological entities are mapped to nodes and what is the biochemical meaning of the connections between them. Several options are possible (Lacroix et al. 2008; Cottret and Jourdan 2010):

- Compound graph: nodes are metabolites and two metabolites are connected by an edge if they are involved in the same reaction as substrate and product. Consequently, a reaction is represented by several edges (one for each pair of substrate-product). See Figure 8.5a
- Reaction graph: nodes are reactions and two reactions are connected if one is producing a compound which is a substrate of the other one. See Figure 8.5b
- Bipartite graph: it contains two kinds of nodes, reactions and metabolites. A metabolite is linked to a reaction if it is the substrate or product of the reaction. See Figure 8.5c
- Hypergraphs: it is a particular class of graphs where edges (called hyperedges) can connect several nodes. For metabolic networks, hyperedges connect a set of substrates to a set of products. See Figure 8.5d

Thus, the first choice to be made when building a metabolic graph consists of using the relevant graph model. Several aspects have to be taken into account. The first one depends on the data. If the available data only focus on metabolites, a compound graph may be sufficient. If data is monitored both on reactions and metabolites, a bipartite graph may be more suitable.

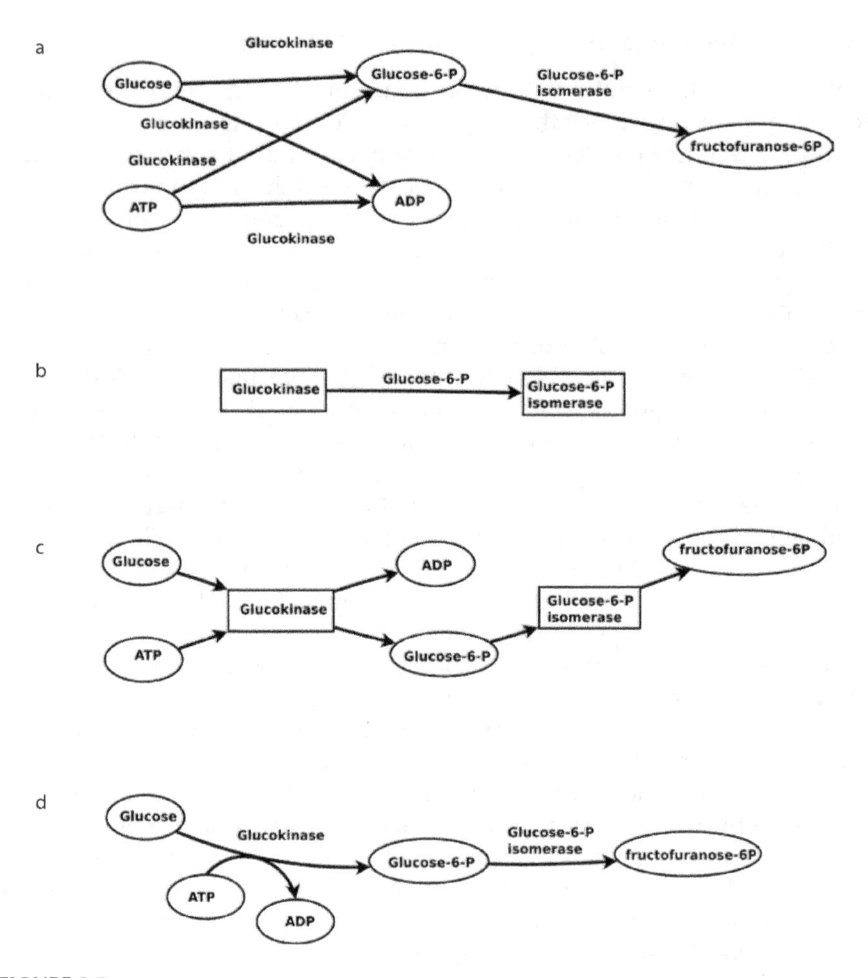

FIGURE 8.5
Metabolic network graph representation. (a) Compound graph. (b) Reaction graph. (c) Bipartite graph. (d) Hypergraph.

The best option is to use hypergraphs, but unfortunately there are far fewer algorithms available for this data structure than the ones available for simple graphs (Pearcy et al. 2016). It is important to note that for the same list of reactions, topology of the corresponding graph may vary according to the modeling choice (see Figure 8.5).

One important topological feature of graphs is the degree of a node. This value corresponds to the number of edges connected to a node (glucokinase degree on Figure 8.5c is four). If the graph is directed then there will be an indegree (number of edges pointing to this node; two for glucokinase on Figure 8.5c) and an outdegree (number of edges leaving this node; two for glucokinase on Figure 8.5c).

Extracting Relevant Sub-Graphs from Global Metabolic Network

Untargeted metabolomics will detect nodes in the graph that may belong to several pathways. Since the graph allows considering the metabolism as a single object, it is thus possible to identify cascades of reactions spanning more than one metabolic pathway. To do so, one of the main approaches currently used consists of looking into the metabolic graph for paths between each pair of metabolites of interest. A path in a graph is a sequence of edges (or nodes) allowing to connect a source node to a target one. Since metabolic graphs contain thousands of nodes, there are many alternatives to go from one metabolite of interest to the other. For instance, using the KEGG human metabolic network there exist more than 500,000 different paths to connect glucose and pyruvate (Küffner et al. 2000). Among these paths, the shortest one is going from glucose to glucokinase, ADP, Pyruvate kinase and then reaching pyruvate. This path is not relevant on a biochemical point of view (we expect a path similar to the well characterized glycolysis). Undoubtedly, it is also the case for most of the 500,000 paths. The aim of the algorithms described in this section is to focus on the most relevant paths in this huge solution space. It will be achieved by including biochemical rules into the path search algorithms.

The first option consists of looking for a path that will go through the minimal number of steps (Arita 2000). The parsimony assumption underlying this choice is based on the idea that evolution may have preferentially oriented metabolism toward short catabolic and anabolic processes. Nevertheless, the shortest paths are often not biologically relevant (c.f. the glucose to pyruvate shortest path). In fact, these paths often involve ubiquitous cofactors like water that are involved in many reactions and thus behave as shortcuts in the graph (Faust et al. 2011; Frainay and Jourdan 2016). The main challenge in metabolic path search is thus to avoid these side compounds during the routing. A side compound can be defined as a molecule that plays an auxiliary role in a biochemical reaction, such as a proton donor or acceptor or an energy carrier for example.

One solution to avoid side compounds consists of defining a list of metabolites that will be ignored during path computation. While for some of them this definition is relatively obvious (water, CO_2, etc.), it is hard and time consuming to define an exhaustive list. One way to perform this selection can be based on the degree of nodes (van Helden et al. 2002). Nodes (metabolites) involved in many reactions are often called hubs, which will have a high degree (see Table 8.1). By removing all highly connected nodes, it is then possible to remove these hubs. The issue then is to define a threshold value that may vary from one network to another. Moreover, a side-compound can occasionally act as a main one, and their removal may lead to the loss of relevant links. For example, removing ATP from the graph will avoid some irrelevant paths; however, it will prevent from retrieving the dATP nucleotide biosynthesis.

TABLE 8.1

20th Highest Degree Compounds in Recon 2 Version 2. Degree Computed as Number of Neighbors in Metabolic Network

Compound	Degree
Proton	1271
Water	982
Dioxygen	390
Nicotinamide adenine dinucleotide phosphate reduced	355
Nicotinamide adenine dinucleotide phosphate	339
Nicotinamide adenine dinucleotide reduced	287
Nicotinamide adenine dinucleotide	278
Adenosine triphosphate	269
Coenzyme A	252
Hydrogen phosphate	202
Adenosine diphosphate	182
Acetyl-CoA	171
Uridine diphosphate	167
Diphosphate	151
Hydrogen peroxide	145
Carbon dioxide	135
Adenosine monophosphate	127
Flavin adenine dinucleotide	118
Flavin adenine dinucleotide reduced	99
L-glutamate	90

To avoid this subjective definition, the notion of lightest paths had been introduced. The idea is to compute a path which goes through a set of nodes with a minimal degree sum (Croes et al. 2006). These paths may be longer than shortest path ones, but they tend to avoid ubiquitous compounds. Another approach based on lightest path had been proposed to avoid side compounds using chemical structure (Box 8.2). The chemical similarity can be computed using the Tanimoto coefficient on chemical fingerprints. Using the chemical similarity as weight in a compound graph tends to obtain the lightest path that avoids side compounds (Rahman et al. 2005).

A path search uses a pair of nodes as input, a starting point and an ending one. Metabolomics results usually highlight a larger list of metabolites and provide no clues about which are "sources" and which are "targets." Other methods have been proposed to handle larger list of metabolites, such as Steiner Tree computation (Faust et al. 2010), Random Walks (Dupont et al. 2006) or Metabolic Stories (Milreu et al. 2014).

BOX 8.2 BASIC DEFINITIONS FOR GRAPH MODELLING

- **Node**: A node is a fundamental element (e.g., a metabolite) of a network.
- **Edge**: An edge is a set of two nodes. If the edge is directed (it can then be called an arc), the first node will be the source while the second one will be the target.
- **Graph**: A graph is composed of a set of nodes and a set of edges.
- **Path**: A path is a sequence of edges in a graph allowing to connect two nodes. Length of a path is the number of edges belonging to this path.
- **Cycle**: A cycle is a path which can connect a node to itself.
- **Tree**: A tree is a graph with no cycle.
- **Degree**: Degree of a node is the number of edges involving this node. If the graph is directed, then this number is divided into out-degree (number of edges for which the node is the source) and in-degree (number of edges for which the node is the target).

Network Visualization

While graph algorithms can extract meaningful parts of the network and help in building new hypotheses, obtained results require visual inspection to get interpreted. Interacting with a network composed of thousands of tightly connected nodes has become an important challenge in the field of metabolic network analysis (Dinkla and Westenberg 2012). Specific data-visualization techniques for graph drawing are required to enable a full exploitation of results gained from metabolic networks.

Drawing a graph requires computing coordinates in two (or three) dimensions for all nodes. Then edges are drawn as lines or bent lines (in that case it is necessary to compute coordinates of bending points). The quality of a drawing has been defined in the community by a series of esthetical criteria: limiting the number of edge crossings, avoiding node overlap, spreading nodes in the representation space (Purchase 1997). Satisfying all these criteria at once for a large graph is, in the general case, not tractable. That is the reason why heuristics (algorithms giving a good, but not necessarily optimal solution) had been developed (Purchase 1997). The most common family of graph drawing algorithms are force-based methods (Fruchterman and Reingold 1991; Herman et al. 2000). The idea underlying these methods is to consider the graph as a physical system, for instance associating nodes to steel balls and edges to springs. Then, using a simulation guided by physical forces (e.g., Hook's law), it is possible to compute an equilibrium state that will be selected as a suitable visualization. The fact that most of these methods are based on simulation implies that they may not always give the same result (non-deterministic algorithms). Thus, the same graph drawn twice may result in two different images while the graph is the same. Moreover, since generally most methods consider all edges as identical in terms of their spring force, the length of edges in the drawing is not proportional to any biologically relevant data.

When a graph shows a hierarchical structure (some roots and leaves), it is possible to use dedicated algorithms called hierarchical drawing. These methods will organize nodes by layer in order to minimize the number of edge crossing between layers. They are suitable for small graphs but will give not readable results for large ones. For metabolic graphs they have the advantage to highlight directionality better than force-based methods (Figure 8.6).

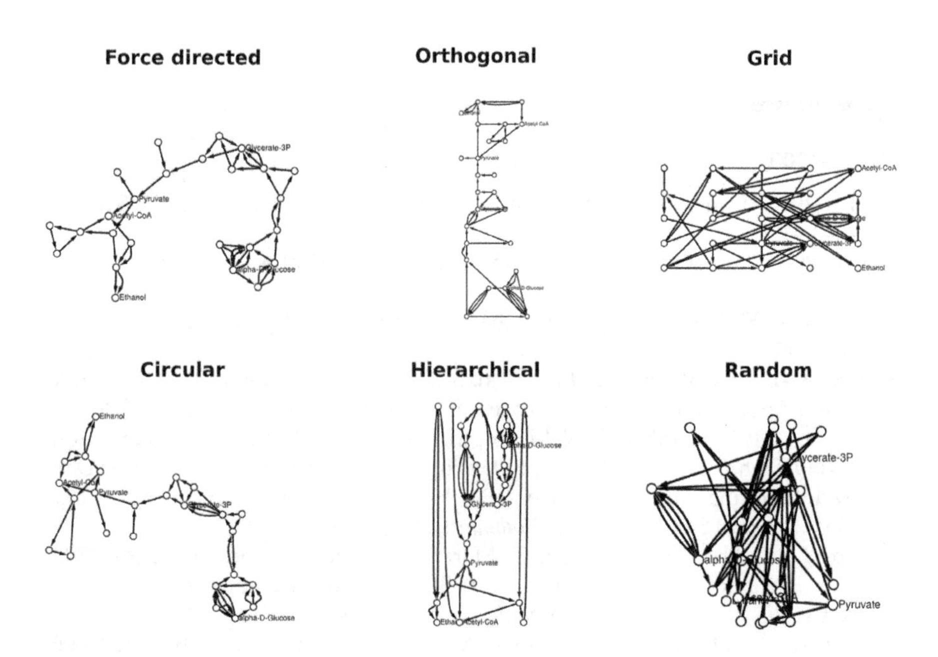

FIGURE 8.6
KEGG Glycolysis/Gluconeogenesis pathway map represented with different graph drawing algorithms. (Visualization made using Cytoscape (From Shannon, P. et al., *Genome Research* 13, 2498–2504, 2003.))

Conclusion

Network- and graph-based modeling provide ways to represent metabolomics data in the context of a complex web of biochemical reactions. The technology allows generating hypotheses on potential sub-networks involved in the response to an environmental stress or a genetic modification. Solutions proposed by the algorithms strongly rely on the quality of the reconstructed network. The bioinformatics research community is putting a lot of effort in order to generate networks of better quality by using omics data.

Visualization is a powerful tool to mine complex data since it exploits human visual system capabilities. Once again it is important to be aware of the various drawing algorithms, their limits and what kind of conclusions can/cannot be raised.

This chapter has presented a descriptive computational approach where metabolomics data are mainly used in a boolean manner (e.g., being a biomarker or not). The next chapter will show how networks can be used to predict growth and impact of genetic modifications using constraint-based modeling.

References

Albert, R. 2005. "Scale-Free Networks in Cell Biology." *Journal of Cell Science* 118 (Pt 21): 4947–4957.

Arita, M. 2000. "Metabolic Reconstruction Using Shortest Paths." *Simulation Practice and Theory* 8 (1–2): 109–125.

Bairoch, A. 2000. "The ENZYME Database in 2000." *Nucleic Acids Research* 28 (1): 304–305.

Barabási, A. -L., N. Gulbahce, and J. Loscalzo. 2011. "Network Medicine: A Network-Based Approach to Human Disease." *Nature Reviews Genetics* 12 (1): 56–68.

Barupal, D. K., P. K. Haldiya, G. Wohlgemuth, T. Kind, S. L. Kothari, K. E. Pinkerton, and O. Fiehn. 2012. "MetaMapp: Mapping and Visualizing Metabolomic Data by Integrating Information from Biochemical Pathways and Chemical and Mass Spectral Similarity." *BMC Bioinformatics* 13: 99.

Bernard, T., A. Bridge, A. Morgat, S. Moretti, I. Xenarios, and M. Pagni. 2014. "Reconciliation of Metabolites and Biochemical Reactions for Metabolic Networks." *Briefings in Bioinformatics* 15 (1): 123–135.

Breitling, R., S. Ritchie, D. Goodenowe, M. L. Stewart, and M. P. Barrett. 2006. "*Ab Initio* Prediction of Metabolic Networks Using Fourier Transform Mass Spectrometry Data." *Metabolomics: Official Journal of the Metabolomic Society* 2 (3): 155–164.

Brunk, E. et al. 2018. "Recon3D Enables a Three-Dimensional View of Gene Variation in Human Metabolism." *Nature Biotechnology* 36 (3): 272–281.

Bullmore, E., and O. Sporns. 2009. "Complex Brain Networks: Graph Theoretical Analysis of Structural and Functional Systems." *Nature Reviews. Neuroscience* 10 (3): 186–198.

Camacho, C., C. Christiam, C. George, A. Vahram, M. Ning, P. Jason, B. Kevin, and Thomas L. Madden. 2009. "BLAST: Architecture and Applications." *BMC Bioinformatics* 10 (1): 421.

Camacho, D., C. Diogo, A. de la Fuente, and M. Pedro. 2005. "The Origin of Correlations in Metabolomics Data." *Metabolomics: Official Journal of the Metabolomic Society* 1 (1): 53–63.

Chambers, J., M. Davies, A. Gaulton, A. Hersey, S. Velankar, R. Petryszak, J. Hastings, L. Bellis, S. McGlinchey, and J. P. Overington. 2013. "UniChem: A Unified Chemical Structure Cross-Referencing and Identifier Tracking System." *Journal of Cheminformatics* 5 (1): 3.

Compeau, P. E. C., P. A. Pevzner, and T. Glenn. 2011. "How to Apply de Bruijn Graphs to Genome Assembly." *Nature Biotechnology* 29 (11): 987–991.

Cottret, L., and F. Jourdan. 2010. "Graph Methods for the Investigation of Metabolic Networks in Parasitology." *Parasitology* 137 (9): 1393–1407.

Cottret, L., D. Wildridge, F. Vinson, M. P. Barrett, H. Charles, M. -F. Sagot, and F. Jourdan. 2010. "MetExplore: A Web Server to Link Metabolomic Experiments and Genome-Scale Metabolic Networks." *Nucleic Acids Research* 38 (Web Server issue): W132–W137.

Croes, D., F. Couche, S. J. Wodak, and J. van Helden. 2006. "Inferring Meaningful Pathways in Weighted Metabolic Networks." *Journal of Molecular Biology* 356 (1): 222–236.

Demir, E. et al. 2010. "The BioPAX Community Standard for Pathway Data Sharing." *Nature Biotechnology* 28 (9): 935–942.

Díaz-Mejía, J. J., M. Babu, and A. Emili. 2009. "Computational and Experimental Approaches to Chart the *Escherichia Coli* Cell-Envelope-Associated Proteome and Interactome." *FEMS Microbiology Reviews* 33 (1): 66–97.

Dinkla, K., and M. A. Westenberg. 2012. "Network Visualization in Cell Biology." *Tsinghua Science and Technology* 17 (4): 365–382.

Duarte, N. C., S. A. Becker, N. Jamshidi, I. Thiele, M. L. Mo, T. D. Vo, R. Srivas, and B. Ø. Palsson. 2007. "Global Reconstruction of the Human Metabolic Network Based on Genomic and Bibliomic Data." *Proceedings of the National Academy of Sciences of the United States of America* 104 (6): 1777–1782.

Dupont, P., J. Callut, G. Dooms, J. -N. Monette, Y. Deville, and B. P. Sainte. 2006. "Relevant Subgraph Extraction from Random Walks in a Graph." *Universite Catholique de Louvain, UCL/INGI, Number RR* 7. Available at: https://www.info.ucl.ac.be/~yde/Papers/kwalks_2006.pdf. [Accessed March 12, 2019].

Faust, K., D. Croes, and J. van Helden. 2011. "Prediction of Metabolic Pathways from Genome-Scale Metabolic Networks." *Bio Systems* 105 (2): 109–121.

Faust, K., P. Dupont, J. Callut, and J. van Helden. 2010. "Pathway Discovery in Metabolic Networks by Subgraph Extraction." *Bioinformatics* 26 (9): 1211–1218.

Feist, A. M., M. J. Herrgård, I. Thiele, J. L. Reed, and B. Ø. Palsson. 2009. "Reconstruction of Biochemical Networks in Microorganisms." *Nature Reviews Microbiology* 7 (2): 129–143.

Frainay, C., and F. Jourdan. 2016. "Computational Methods to Identify Metabolic Subnetworks Based on Metabolomic Profiles." *Briefings in Bioinformatics*, January. doi:10.1093/bib/bbv115.

Francke, C., F. Christof, R. J. Siezen, and T. Bas. 2005. "Reconstructing the Metabolic Network of a Bacterium from Its Genome." *Trends in Microbiology* 13 (11): 550–558.

Fruchterman, T. M. J., and E. M. Reingold. 1991. "Graph Drawing by Force-Directed Placement." *Software: Practice & Experience* 21 (11): 1129–1164.

Ginsburg, H. 2009. "Caveat Emptor: Limitations of the Automated Reconstruction of Metabolic Pathways in Plasmodium." *Trends in Parasitology* 25 (1): 37–43.

Helden, J. van, L. Wernisch, D. Gilbert, and S. J. Wodak. 2002. "Graph-Based Analysis of Metabolic Networks." *Ernst Schering Research Foundation Workshop* 38: 245–274.

Heller, S., A. McNaught, S. Stein, D. Tchekhovskoi, and I. Pletnev. 2013. "InChI—The Worldwide Chemical Structure Identifier Standard." *Journal of Cheminformatics* 5 (1): 7.

Herman, I., G. Melancon, and M. S. Marshall. 2000. "Graph Visualization and Navigation in Information Visualization: A Survey." *IEEE Transactions on Visualization and Computer Graphics* 6 (1): 24–43.

Hucka, M. et al. 2003. "The Systems Biology Markup Language (SBML): A Medium for Representation and Exchange of Biochemical Network Models." *Bioinformatics* 19 (4): 524–531.

Jerby, L., T. Shlomi, and E. Ruppin. 2010. "Computational Reconstruction of Tissue-Specific Metabolic Models: Application to Human Liver Metabolism." *Molecular Systems Biology* 6 (September): 401.

Jourdan, F., R. Breitling, M. P. Barrett, and D. Gilbert. 2008. "MetaNetter: Inference and Visualization of High-Resolution Metabolomic Networks." *Bioinformatics* 24 (1): 143–145.

Jourdan, F. 2013. "Qualitative Modelling of Metabolic Networks." In *Advances in Botanical Research*. Elsevier, ISSN: 0065-2296, pp. 557–591.

Kanehisa, M., S. Goto, Y. Sato, M. Kawashima, M. Furumichi, and M. Tanabe. 2014. "Data, Information, Knowledge and Principle: Back to Metabolism in KEGG." *Nucleic Acids Research* 42 (Database issue): D199–D205.

Karp, P. D. et al. 2015. "Pathway Tools Version 19.0 Update: Software for Pathway/genome Informatics and Systems Biology." *Briefings in Bioinformatics* 17 (5): 877–890.

Keseler, I. M. et al. 2013. "EcoCyc: Fusing Model Organism Databases with Systems Biology." *Nucleic Acids Research* 41 (Database issue): D605–D612.

Küffner, R., R. Zimmer, and T. Lengauer. 2000. "Pathway Analysis in Metabolic Databases via Differential Metabolic Display (DMD)." *Bioinformatics* 16 (9): 825–836.

Lacroix, V., L. Cottret, P. Thébault, and M. -F. Sagot. 2008. "An Introduction to Metabolic Networks and Their Structural Analysis." *IEEE/ACM Transactions on Computational Biology and Bioinformatics/IEEE, ACM* 5 (4): 594–617.

Li, C. et al. 2010. "BioModels Database: An Enhanced, Curated and Annotated Resource for Published Quantitative Kinetic Models." *BMC Systems Biology* 4 (June): 92.

Lloyd, C. M., and N. Geoff. 2013. "CellML Model Curation." In *Encyclopedia of Systems Biology*, edited by W. Dubitzky, O. Wolkenhauer, H. Yokota, K.-H. Cho. New York: Springer, pp. 372–375.

Maglott, D., J. Ostell, K. D. Pruitt, and T. Tatusova. 2005. "Entrez Gene: Gene-Centered Information at NCBI." *Nucleic Acids Research* 33 (Database issue): D54–D58.

Milreu, P. V. et al. 2014. "Telling Metabolic Stories to Explore Metabolomics Data: A Case Study on the Yeast Response to Cadmium Exposure." *Bioinformatics* 30 (1): 61–70.

Pagani, I., K. Liolios, J. Jansson, I-Min A. Chen, T. Smirnova, B. Nosrat, V. M. Markowitz, and N. C. Kyrpides. 2012. "The Genomes OnLine Database (GOLD) v.4: Status of Genomic and Metagenomic Projects and Their Associated Metadata." *Nucleic Acids Research* 40 (Database issue): D571–D579.

Pearcy, N., N. Chuzhanova, and J. J. Crofts. 2016. "Complexity and Robustness in Hypernetwork Models of Metabolism." *Journal of Theoretical Biology* 406: 99–104.

Pinney, J. W., M. W. Shirley, G. A. McConkey, and D. R. Westhead. 2005. "metaSHARK: Software for Automated Metabolic Network Prediction from DNA Sequence and Its Application to the Genomes of Plasmodium Falciparum and Eimeria Tenella." *Nucleic Acids Research* 33 (4): 1399–1409.

Prensner, J. R., and A. M. Chinnaiyan. 2011. "Metabolism Unhinged: IDH Mutations in Cancer." *Nature Medicine* 17 (3): 291–293.

Purchase, H. 1997. "Which Aesthetic Has the Greatest Effect on Human Understanding?" In *Lecture Notes in Computer Science*, edited by G. Di Battista. New York: Springer, pp. 248–261.

Rahman, S. A., and D. Schomburg. 2006. "Observing Local and Global Properties of Metabolic Pathways: 'Load Points' and 'Choke Points' in the Metabolic Networks." *Bioinformatics* 22 (14): 1767–1774.

Rahman, S. A., P. Advani, R. Schunk, R. Schrader, and D. Schomburg. 2005. "Metabolic Pathway Analysis Web Service (Pathway Hunter Tool at CUBIC)." *Bioinformatics* 21 (7): 1189–1193.

Reed, J. L., F. Iman, T. Ines, and B. O. Palsson. 2006. "Towards Multidimensional Genome Annotation." *Nature Reviews. Genetics* 7 (2): 130–141.

Romero, P., and R. Pedro. 2012. "The HumanCyc Pathway-Genome Database and Pathway Tools Software as Tools for Imaging and Analyzing Metabolomics Data." In *Methods in Pharmacology and Toxicology*, edited by T. M. Fan, A. Lane, and R. Higashi, Totowa, NJ: Humana Press, pp. 419–438.

Schellenberger, J., J. O. Park, T. M. Conrad, and B. Ø. Palsson. 2010. "BiGG: A Biochemical Genetic and Genomic Knowledgebase of Large Scale Metabolic Reconstructions." *BMC Bioinformatics* 11 (1): 213.

Schellenberger, J. et al. 2011. "Quantitative Prediction of Cellular Metabolism with Constraint-Based Models: The COBRA Toolbox v2.0." *Nature Protocols* 6 (9): 1290–1307.

Schich, M., C. Song, Y.-Y. Ahn, A. Mirsky, M. Martino, A.-L. Barabasi, and D. Helbing. 2014. "A Network Framework of Cultural History." *Science* 345 (6196): 558–562.

Schomburg, I. et al. 2013. "BRENDA in 2013: Integrated Reactions, Kinetic Data, Enzyme Function Data, Improved Disease Classification: New Options and Contents in BRENDA." *Nucleic Acids Research* 41 (Database issue): D764–D772.

Shameer, S. et al. 2015. "TrypanoCyc: A Community-Led Biochemical Pathways Database for Trypanosoma Brucei." *Nucleic Acids Research* 43 (Database issue): D637–D644.

Shannon, P., A. Markiel, O. Ozier, N. S. Baliga, J. T. Wang, D. Ramage, N. Amin, B. Schwikowski, and T. Ideker. 2003. "Cytoscape: A Software Environment for Integrated Models of Biomolecular Interaction Networks." *Genome Research* 13 (11): 2498–2504.

Shlomi, T., M. N. C., M. J. Herrgård, B. Ø. Palsson, and E. Ruppin. 2008. "Network-Based Prediction of Human Tissue-Specific Metabolism." *Nature Biotechnology* 26 (9): 1003–1010.

Silva, R. R., F. Jourdan, D. M. Salvanha, F. Letisse, E. L. Jamin, S. Guidetti-Gonzalez, C. A. Labate, and R. Z. N. Vêncio. 2014. "ProbMetab: An R Package for Bayesian Probabilistic Annotation of LC-MS-Based Metabolomics." *Bioinformatics* 30 (9): 1336–1337.

Smallbone, K. 2013. "Striking a Balance with Recon 2.1." arXiv [q-bio. MN]. arXiv. Available at: http://arxiv.org/abs/1311.5696. [Accessed March 12, 2019].

Sonnhammer, E. L. L., and G. Östlund. 2015. "InParanoid 8: Orthology Analysis between 273 Proteomes, Mostly Eukaryotic." *Nucleic Acids Research* 43 (Database issue): D234–D239.

Steuer, R. 2006. "Review: On the Analysis and Interpretation of Correlations in Metabolomic Data." *Briefings in Bioinformatics* 7 (2): 151–158.

Subramanian, A. et al. 2005. "Gene Set Enrichment Analysis: A Knowledge-Based Approach for Interpreting Genome-Wide Expression Profiles." *Proceedings of the National Academy of Sciences of the United States of America* 102 (43): 15545–15550.

Thébault, E., and C. Fontaine. 2010. "Stability of Ecological Communities and the Architecture of Mutualistic and Trophic Networks." *Science* 329 (5993): 853–856.

Thiele, I., and B. Ø. Palsson. 2010. "A Protocol for Generating a High-Quality Genome-Scale Metabolic Reconstruction." *Nature Protocols* 5 (1): 93–121.

Thiele, I. et al. 2013. "A Community-Driven Global Reconstruction of Human Metabolism." *Nature Biotechnology* 31 (5): 419–425.

Trpevski, D., W. K. S. Tang, and L. Kocarev. 2010. "Model for Rumor Spreading over Networks." *Physical Review E, Statistical, Nonlinear, and Soft Matter Physics* 81 (5 Pt 2): 056102.

Watrous, J. et al. 2012. "Mass Spectral Molecular Networking of Living Microbial Colonies." *Proceedings of the National Academy of Sciences of the United States of America* 109 (26): E1743–E1752.

Wohlgemuth, G., P. K. Haldiya, E. Willighagen, T. Kind, and O. Fiehn. 2010. "The Chemical Translation Service—A Web-Based Tool to Improve Standardization of Metabolomic Reports." *Bioinformatics* 26 (20): 2647–2648.

Xia, J., and D. S. Wishart. 2010. "MSEA: A Web-Based Tool to Identify Biologically Meaningful Patterns in Quantitative Metabolomic Data." *Nucleic Acids Research* 38 (Web Server issue): W71–W77.

9

Metabolic Flux

Igor Marín de Mas and Marta Cascante

CONTENTS

Model-Driven Strategies in Metabolic Network Analysis 201
Dynamic Flux-Map Computation and Kinetic Modeling 204
  Mathematical Basis of Kinetic Modeling 204
  Dynamic Metabolic Modeling and ^{13}C Fluxomics 206
  Working Example of Kinetic Modeling Application 208
Large-Scale Flux-Map Computation and Constraint-Based Modeling 211
  Mathematical Basis of Constraint-Based Modeling 211
  Metabolic Flux-Map Analysis Enriched by Omics Data Integration...... 213
  Working Example of Constraint-Based Modeling Application 216
Applications of Metabolic Flux Modeling 218
  Drug Discovery 218
  Bioengineering 221
  Evolution 222
Future Challenges of Metabolic Flux Modeling 223
  Challenges in Dynamic Flux-Map Computation and Kinetic Modeling 223
  Challenges in Large-Scale Flux-Map Computation and Constraint-Based Modeling 225
  Large-Scale ^{13}C-Constrained FBA 226
  Genome-Scale Kinetic Models 227
References 229

Model-Driven Strategies in Metabolic Network Analysis

As discussed in previous chapters, data-driven methods are extremely efficient to extract knowledge from a massive amount of data. However, in some situations these approaches are not suitable to analyze biological phenomena. For example, the amount of experimental data required to determine patterns, correlations and mechanisms is not available or it is necessary to add new knowledge extracted from the literature in order to constrain the space of possible solutions. Typically, this information is introduced in the form of models. A model is an abstraction of a biological system representing the interactions and interdependencies between the components of a given system (metabolites, genes, proteins, etc.).

These models can be represented in the form of graphs. However, these representations usually cannot represent all the dynamic behavior of the biological systems. Moreover, handling large networks in the form of graphs is cumbersome. Computational models, on the other hand, provide a precise mathematical representation of knowledge allowing to interpret and evaluate measured data, analyze a system's behavior (e.g., identify important parts for a particular behavior, etc.) and generate and test hypotheses. Additionally, these computational tools allow us to improve the consistency between models and data in an iterative fashion (Fisher and Henzinger 2007). Consequently, mathematical modeling has become an essential tool for the comprehensive understanding of cell metabolism and its interactions with environmental and process conditions.

A large variety of mathematical models to represent different aspects of the cellular metabolism is available. Since metabolism provides the closest link to the macroscopic phenotype, models that somehow represent the metabolic processes occurring within the cell are especially useful to study different aspects of cellular biology. However, the study of the dynamic behavior of metabolic processes requires further considerations such as the rate of turnover of metabolites through a network of reactions (metabolic flux). The study of metabolic fluxes or fluxomics (by analogy to other omic-types analysis such a as proteomics or transcriptomics) is a discipline that gathers the set of experimental and computational tools enabling to determine the metabolic reactions fluxes (Mardinoglu et al. 2013). Fluxomics integrates in vivo or in vitro measurements of metabolic fluxes with mathematical models to allow the determination of absolute flux through a metabolic network. This systems-biology approach has been widely used for a large variety of applications involving the study of the mechanisms underlying diseases with strong metabolic components such as cancer or chronic diseases (Mardinoglu et al. 2013), the evolutionary adaptation due to environmental perturbations (Papp et al. 2011) and the design of strains to optimize cellular processes in order to maximize the production of certain substances (Matsuda et al. 2011). Based on how the metabolic reactions are formalized in such models, one can distinguish two main types of metabolic modeling approaches:

1. kinetic modeling and
2. constraints-based modeling.

Kinetic modeling defines the processes that represent metabolic networks using kinetic rate law equations; these equations are defined by a set of kinetic parameters allowing the study of the dynamic behavior of the system. This approach can be based on label-free or label metabolomics (typically ^{13}C). Kinetic modeling permits a thorough characterization of metabolic fluxes. However, the paucity of kinetic parameters limits this approach to relatively small metabolic network models. On the other hand, constraints-based modeling defines the metabolic systems based on the mass balance between the substrates and the products of metabolic reactions (stoichiometry) and their reversibility (thermodynamics). This approach requires less parametrization than the kinetic modeling which makes it less effective to define specific metabolic fluxes but makes it especially suitable for the study of large metabolic network systems in a holistic manner.

In the following, we summarize the most relevant model-driven approaches to infer the metabolic flux map of a given metabolic network, as well as their application in a variety of studies. Finally, we will discuss some novel strategies to integrate both approaches and future challenges that need to be overcome (Box 9.1).

BOX 9.1 BASIC DEFINITIONS IN METABOLIC FLUX

Metabolic flux: The rate of biotransformation of substrates to products through a metabolic reaction. It is typically defined as units/mass of substrate transformed to products per unit of time (e.g., micromole/minute; mg/kg/minute).

Stoichometry: Is the relation of relative quantities between substrates and products in a chemical reaction. This relation is based on the law of conservation of mass. Thus, the total mass of the substrates equals the total mass of the products.

Thermodynamic: Refers to the quantity of energy necessary to catalize a given metabolic reaction (to transform substrate/s into the corresponding product/s). In biochemistry, thermodynamic defines if the metabolic flux throughout a given reaction goes only from substrates to products (irreversible reaction) or if the products can be also transformed into substrates (reversible reaction).

Kinetic: Is associated to enzymatic biochemical reactions and refers to the enzymatic parameters and equations that define the flux rate (velocity) through a given reaction.

Dynamic Flux-Map Computation and Kinetic Modeling

Biological systems are not static entities. They change over time in response to a variety of perturbations. Kinetic modeling methods aim to integrate the corresponding data sets, allowing one to infer the kinetics and dynamics of the reactions between all the chemical entities in a cell (Resat et al. 2009). Thus, one can represent a metabolic network that incorporates the stoichiometry and thermodynamics of the system (in particular, the direction of metabolic reactions) as well as detailed knowledge of the enzyme regulation in the form of kinetic rate laws and their associated parameter values to formulate kinetic models that can accurately capture the dynamic response of the metabolic network fluxes (Chakrabarti et al. 2013).

Mathematical Basis of Kinetic Modeling

The mathematical formalization is performed by representing the metabolic concentrations in a system of ordinary differential equations (ODEs) that can be solved by Flux Balance Analysis (FBA). Thus, the concentration of the ith metabolite in the model is described by a variable S_i. As a result of mass balances (no matter can appear or disappear), the change of this variable over time (dS_i/dt) is given by the sum of the rates of the enzymes synthesizing the metabolite minus the sum of the rates of the enzymes utilizing the metabolite as is expressed in the following equation:

$$\frac{dS_i}{dt} = R_{input} - R_{output} \tag{9.1}$$

Here R_{input} and R_{output} are the set of reactions in which the ith metabolite is a product or a substrate, respectively. In kinetic modeling approaches, the rate of each enzymatic reaction can be described by kinetic law equations. The following equation represents an example of a Michaelis-Menten kinetic equation, one of the best-known models of enzyme kinetics:

$$R_m = \frac{V_{max} \times S_i}{K_m + S_i} \tag{9.2}$$

This equation represents the mth reaction that uses the ith metabolite as substrate. Here S_i is the concentration of the ith metabolite that is a substrate of the reaction, V_{max} and K_m are the kinetic parameters of the equation. V_{max} represents the maximum rate achieved by the system, at maximum substrate concentrations (saturating conditions); K_m is the substrate concentration at which the reaction rate is half of V_{max}.

This process yields a system of ODEs in which dS_i/dt is on one side and the metabolite-dependent rate laws are on the other side of the equations. With this system of differential equation, the metabolic network can be simulated, and by solving the system of ODEs the steady state can be calculated, in which all reaction rates and metabolite concentrations are constant (Figure 9.1). Figure 9.1 shows how to express the information embedded in a metabolic network and the kinetic law equations into a system of ODEs and finally in a stoichiometric matrix that can capture the dynamics of the metabolism. Based on the kinetic laws governing the metabolic fluxes through the reactions (Figure 9.1b and c) and the stoichiometric and thermodynamic information embedded in the network one can construct a system of ODEs that

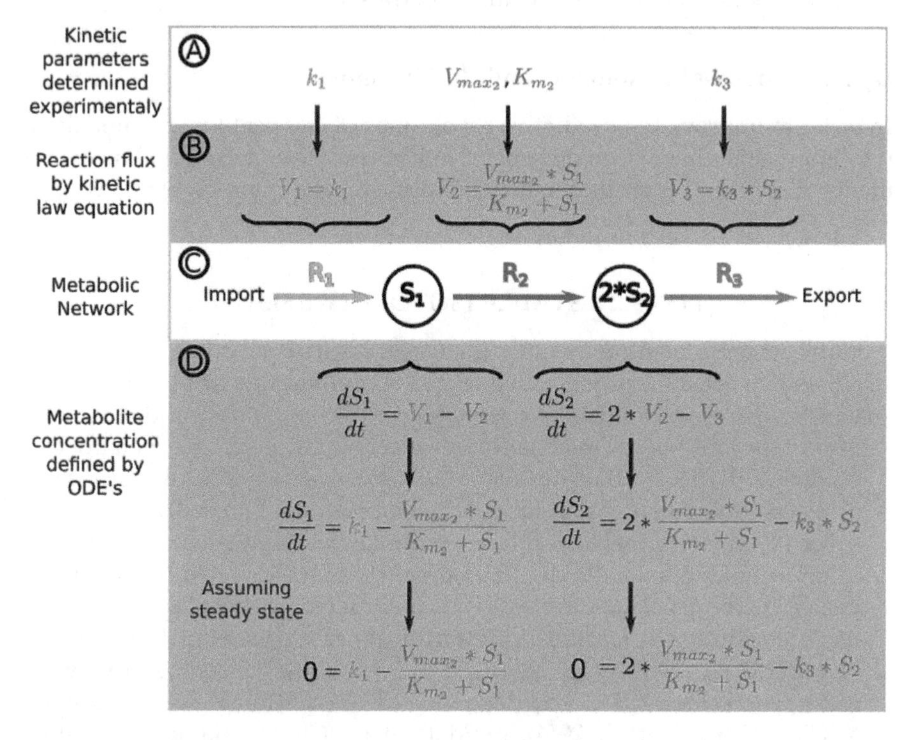

FIGURE 9.1
Kinetic metabolic model. (a) Toy network representing a metabolic process in which the nodes are the metabolites involved in the process and the arrows the biochemical reactions that transform the substrates into the final products. Metabolite S_1 is imported into the toy network by reaction R_1; the reaction R_2 produces two molecules of S_2 from one molecule of S_1; finally, the reaction R_3 exports the metabolite S_2 outside the network. (b) Equations describing the metabolic fluxes; V_1 is the flux of reaction R_1 and represents a linear equation that depends on the kinetic parameter k_1; V_2 is the flux of reaction R_2 that is based on a Michaelis-Menten kinetic equation; finally V_3 represents the metabolic flux through reaction R_3—it is based on a linear equation that depends on the kinetic parameter k_3 and the concentration of the metabolite S_3. (c) Kinetic parameter that determines the metabolic fluxes through the reaction in the network. (d) System of ODEs that summarizes the kinetic information from B and the stoichiometric and thermodynamic information embed into the metabolic network. This system of ODEs determines the metabolic concentrations of the metabolites S_1 and S_2. Transforming the system of ODEs to a stoichiometric matrix regarding defines the concentrations.

defines how the metabolic concentrations varies along time (Figure 9.1d). This system of ODEs summarizes the stoichiometric relation between the metabolites and reactions. If we are analyzing the system at steady state, then $dS_i/dt = 0$ because, by definition, the metabolic concentrations are constant. Thus, integrating metabolite concentration measurements and a previous knowledge of kinetic parameters governing metabolic reaction dynamics into this system of ODEs, one can infer the metabolic flux map of a given pathway. This method is widely used in systems biology. Many software packages have been developed to simulate biological networks based on these kinetic models (Alves et al. 2006) (Box 9.2).

Dynamic Metabolic Modeling and ^{13}C Fluxomics

In order to improve the predictive capabilities of classical kinetic modeling, new approaches have been developed to incorporate experimental measurements allowing a better understanding of systems dynamics. Stable isotope

BOX 9.2 STABLE ISOTOPE LABELING

Stable isotopic labeling involves the use of an artificial culturing nutrient source or environment, such as glucose, glutamine or CO_2 that are labeled using non-radioactive isotopes (e.g., ^{13}C or ^{15}N) *t*, and it is an important tool for tracing the flux of reactions through the metabolic pathways. Different bioinformatic tools exist for LC-MS and GC-MS stable isotopic labeling non-targeted metabolomics data analysis.

For LC-MS, two main workflows or strategies have been developed in order to automatically detect isotopologues (i.e., labeled metabolite peaks) in the labeled samples in an untargeted way. Both strategies use algorithms for feature detection to determine metabolite peaks and perform retention-time alignment in liquid chromatography/mass spectrometry (LC-MS) data. One of these strategies, shared by X^{13}CMS (Huang et al. 2014), mzMatch-ISO (Chokkathukalam et al. 2013) and MetExtract (Bueschl et al. 2012), use labeled samples to track stable isotopes by iterating over all mzRT features using the theoretical mass difference between the light and heavy isotopes, such as ^{12}C and ^{13}C. A second workflow implemented in the computational tool geoRge (Capellades et al. 2016) uses unlabeled and labeled biologically equivalent samples to compare isotopic distributions in the mass spectra. Isotopically enriched compounds change their isotopic distribution as compared to unlabeled compounds. This is directly reflected in a number of new *m/z* peaks and higher intensity peaks in the mass spectra of labeled samples relative to the unlabeled equivalents, in consequence being distinguishable features by statistical testing.

tracing using [1,2–^{13}C2]-glucose as a source of carbon has been described as a very powerful tool for metabolic flux profiling (Vizán et al. 2009). The metabolite labeling proceeds during cell incubation with substrates containing stable isotope tracer, usually ^{13}C. The specific pattern of various ^{13}C isotopic isomers (isotopomers- positional isomers- or isotopologues-mass isomers-) measured using mass spectrometry or nuclear magnetic resonance techniques characterize the distribution of metabolic fluxes in the cells under the studied conditions. To evaluate the flux distribution from the measured isotopomer distribution, special software tools have been developed (Nargund and Sriram 2014).

However, for a long time the detailed analysis of isotopic isomer distribution was restricted to an isotopic steady state (Wiechert 2001). The analysis of metabolic fluxes not restricted by isotopic steady state is very important even if in the experiment metabolic fluxes do not change. Although intracellular metabolites could reach steady state within minutes, intracellular stores, such as glycogen, amino acids or lipids intensively exchange with intermediates of the metabolism and delay the time necessary for establishing an isotopic steady state. These internal stores and external metabolites could be far from isotopic steady state during hours of labeling experiment. Of course, there is always the possibility of measuring the labeling of such stores and applying classical ^{13}C-FBA for the "fast" intermediates of the central metabolism, assuming that they are in quasi-steady state. The simulation and comparison of such "slow" variables provide additional restrictions that helps to evaluate the fluxes. Moreover, there is another reason for using non-stationary analysis based on a kinetic model of the considered pathways: it has the potential to perform a more profound analysis of kinetic characteristics and regulation in the pathway, if enough experimental data are available. Such advantages have stimulated the development of other bioinformatic tools for non-stationary flux analysis. For example, the software package Isodyn has been developed to perform ^{13}C non-stationary flux analysis (Selivanov et al. 2010). This software simulates ^{13}C redistribution in metabolites by automatically constructing and solving large systems of differential equations for isotopologues. The basis of Isodyn is a kinetic model of a metabolic pathway. Such a model simulates the metabolic fluxes, which transfer the ^{13}C label from externally supplied substrates to the intermediates and products of the cellular metabolism. Using the metabolic fluxes simulated by the kinetic model, one specific module of Isodyn simulates the distribution of the labels, which can be compared with the experimental distribution. By adjusting the parameters of the kinetic model to fit the experimental label distribution, Isodyn defines the characteristics of the analyzed metabolic pathway. Currently, the stable isotope tracing of metabolites has been used to, e.g., identify the adaptive changes of fluxes in humans in normal and diseased states (Fan et al. 2005), in isolated cells (Amaral et al. 2010), cancer cell cultures (Selivanov et al. 2010) and organisms such as fungi or yeast (Jouhten et al. 2009).

Deterministic ODE-based enzyme-kinetic models have the longest history in the area of metabolic pathway modeling. These approaches, and in particular those that permit to perform non-stationary ^{13}C-FBA, offer a highly curated representation of the metabolic and regulatory processes and are able to predict specific metabolic fluxes with high precision. However, the local complexity of kinetic models leads to the situation that, for reasons of feasibility, usually only single pathways or even just some reactions are modeled. This restricts the kinetic models to relatively small networks: small metabolic networks can be described with high precision, but large networks like Genome-Scale Metabolic Models (GSMM) cannot (Box 9.3).

Working Example of Kinetic Modeling Application

The following example illustrates how kinetic modeling methods can be used to study how metabolite concentrations and metabolic fluxes dynamically evolve along time. To this aim we use a simple metabolic network model that represents the upper part of glycolysis (Figure 9.2a). This model considers the transport of glucose from media to cytosol and its metabolization to trioses.

The activity of each reaction is described by a kinetic rate equation (Figure 9.2b). These equations are functions of kinetic parameters (variables starting by *K* or *V* in the kinetic equations) and the concentration of substrates and/or products (variables in brackets in the kinetic equations) that regulates the metabolic flux through each metabolic reaction. Thus, the transport of glucose from media to cytosol (V_{glc} transport) is inhibited by the

BOX 9.3 BASIC DEFINITIONS IN KINETIC MODELING

Flux balance analysis: Is a CBM for simulating metabolism in metabolic network models.

Ordinary differential equation: Is a differential equation containing one or more functions of one independent variable and its derivatives. In an FBA problem, an ODE defines the evolution of a given metabolite concentration along the time and the functions that it contains define metabolic fluxes that either consume (influx) or produce (outflux) the metabolite.

Linear equation: Is an algebraic equation defining constant or the product of a constant and a single variable (first power).

Stable isotope tracing: Is based experimental and computational techniques developed to determine how the label of a substrate enriched with stable isotopes (i.e., ^{13}C) is propagated through a metabolic network and hence defines a metabolic flux profile.

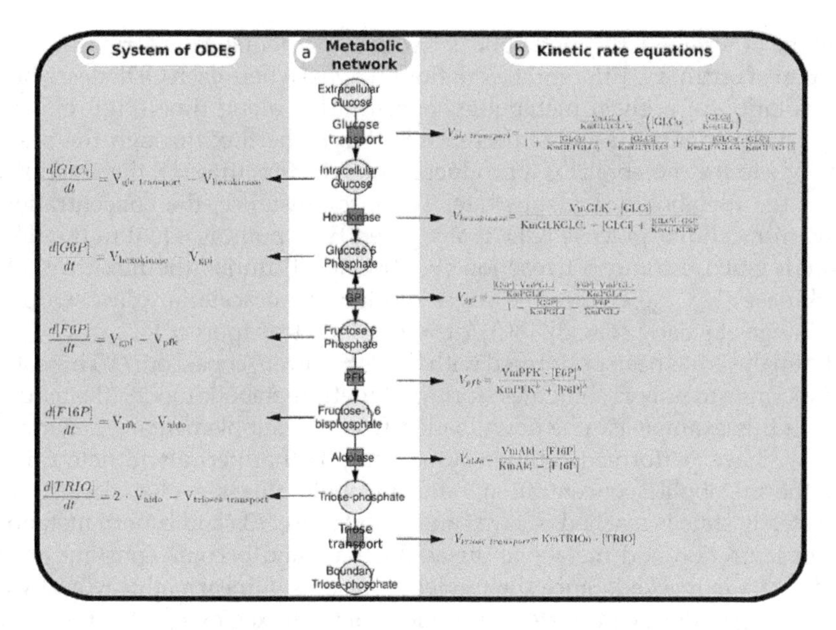

FIGURE 9.2
Metabolic network representing the upper part of glycolysis and its mathematical formalization as a kinetic model. (a) Metabolic network representing the transport of glucose to cytosol, its metabolization to trioses and the triose flux through the rest of the metabolism not considered in the model. Here the metabolites are represented as circles and the reaction as arrows. Glucose-6-phosphate isomerase reaction is abbreviated as GPI and phosphofructokinase as PFK. Boundary triose-phosphate represents the pool of triose phosphate used by other metabolic processes not considered in the model. (b) Kinetic rate equations: equations that describe the kinetics of each reaction. The reactions represented by kinetic rate equations are: hexokinase ($V_{hexokinase}$), glucose transport (V_{glc} transport), aldolase (V_{aldo}), phosphofructokinase (V_{pfk}), glucose-6-phosphate isomerase (V_{gpi}) and triose phosphate transport to other metabolic processes (V_{triose} transport). The metabolic concentrations are represented as follows: Glucose extracellular: [GLCo]; Glucose intracellular: [GLCi]; Glucose-6-phosphate: [G6P]; Fructose-6-phosphate: [F6P]; Fructose-1,6-bisphosphate: [F16P] and triose-phosphate: [TRIO]. The kinetic parameters are represented as variables starting by *K* or *V* and their values are described in the metabolic model stored as an sbml file in the supplementary material. (c) Mathematical formalization of the metabolic network by a system of ODEs where each ODE defines the evolution of a given metabolite concentration along time.

intracellular concentration of glucose, the hexokinase activity ($V_{hexokinase}$) is defined by noncompetitive inhibition kinetics, a reversible Michaelis-Menten kinetic describes the enzymatic activity of glucose-6-phosphate isomerase (V_{gpi}), phosphofructokinase (V_{pfk}) is defined by a Hill Cooperative kinetic equation while a Henry-Michaelis-Menten equation describes the aldolase enzymatic activity (V_{aldo}). Finally, triose transport (V_{triose} transport) is a fictitious reaction linking our model with other metabolic processes not considered here. The kinetic parameters are generally determined experimentally and are embed into the metabolic model that can be retrieved as an sbml file from the supplementary material.

On the other hand, the metabolic network is mathematically formalized as a system of ordinary differential equations (ODEs) where each ODE describes the evolution of a given metabolite concentration along time (Figure 9.2c). Each ODE is defined as the difference between the flux through the reactions where the metabolite is a product minus the flux through the reactions where the metabolite is a substrate. Then, for instance, the concentration of the intracellular glucose [GLCi] at a given time point is equal to the flux through glucose transport reaction (V_{glc} transport) minus the flux through hexokinase ($V_{hexokinase}$). Since the concentration of intracellular glucose does not change at steady state d[GLCi]/dt = 0 and V_{glc} transport = $V_{hexokinase}$.

The analysis has been performed with COPASI (http://copasi.org/). This software permits to import sbml files storing a kinetic metabolic model (the model used in this example can be downloaded from the supplementary material). Here we have performed a time-course analysis that permits to determine how the metabolite concentrations and metabolic fluxes evolve along time until steady state is reached. As we can see in Figure 9.3a and b, both metabolite concentration and metabolic fluxes fluctuate and become constant once steady state is reached. Since the model represents a linear pathway and we are working under steady state assumption, all the fluxes, except the transport of trioses, converge at the same metabolic flux value at steady state. This is because the reaction previous to the transport of trioses (catalyzed by aldolase) split a molecule of fructose-1,6 bisphosphate (a six-carbon molecule) into two molecules of triose (a three-carbon molecule). Thus, based on this 1:2 stoichiometry, all the subsequent biochemical reactions in a linear pathway will have exactly the double metabolic flux than the previous reactions at steady state. The results of this analysis as well as the sbml file with the metabolic model and a COPASI file containing the metabolic model to perform different analyses can be downloaded from the supplementary material.

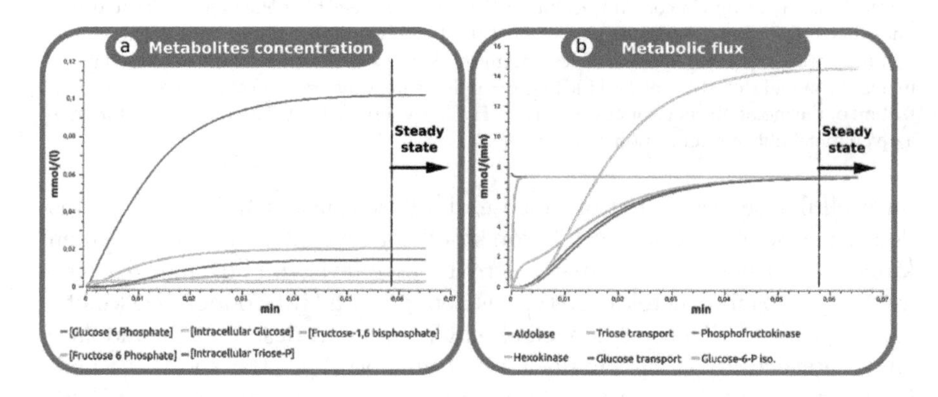

FIGURE 9.3
Graphical representation of a time-course analysis that represent the evolution of metabolite concentration (a) and metabolic fluxes (b) until steady state is reached. Metabolite concentrations are in mmol/L and metabolic fluxes are represented in mmol/min

Large-Scale Flux-Map Computation and Constraint-Based Modeling

Constraint-based modeling approaches permit the study of metabolic systems at large-scale (Imam et al. 2015). In this sense, Genome-Scale Metabolic Models (GSMMs) are emerging as a potential solution to decipher the molecular mechanisms underlying metabolism in a holistic manner (Lazar and Birnbaum 2012).

This systems-biology tool represents the metabolic reactions encoded by the genome, summarizing the information known about the metabolism of an organism. GSMMs are built based on the literature and databases using algorithms developed for these purposes (Caspi et al. 2010). Over a hundred GSMMs have been built for different species, ranging from Archaea to Mammals (Thiele et al. 2013). They are widely used to study the metabolic mechanisms underlying complex phenotypes (diseases, evolution, adaptation to environment, etc.). This enables the integration of the increasing amount of "omic" data generated by the different high-throughput technologies with a high potential in a variety of applications, involving studies on evolution (Pál et al. 2006), metabolic engineering (Park et al. 2009), genome annotation (Kumar and Maranas 2009) or drug discovery (Kim et al. 2011). Here, we discuss computational approaches to analyze and integrate "omics" data into these large-scale metabolic network models (Figure 9.4).

Mathematical Basis of Constraint-Based Modeling

GSMMs define the stoichiometric details for the set of known reactions in a given organism, which can contain up to thousands of metabolic reactions. As was previously mentioned, kinetic modeling, based on ordinary differential equation (ODE) is not suitable to analyze large scale networks (Stelling et al. 2002), but constraint-based models (CBMs) are (Heinemann et al. 2005).

Genome-scale constraint-based metabolic models can be used to predict or describe cellular behaviors, such as growth rates, uptake/secretion rate or intracellular fluxes (Matthew et al. 2009). Flux Balance Analysis (FBA) is one of the most widely used CBMs for the study of metabolic networks. The FBA approach uses the fluxes through the metabolic and transport reactions, and as model parameters the stoichiometry of the metabolic reactions, ATP requirements, biomass composition and the lower and upper boundaries for individual fluxes.

At the first step of FBA, the metabolic reactions are mathematically represented in the form of a numerical matrix containing the stoichiometric coefficients of each reaction (the stoichiometric matrix). Reactions are represented in the columns and metabolites in the rows (Figure 9.5c).

The stoichiometric matrix is mathematically represented by employing a mass action formalism: $dC/Dt = S.v.$ (Figure 9.5b), where S is the stoichiometric matrix, v and C are vectors of reaction fluxes and metabolites, respectively,

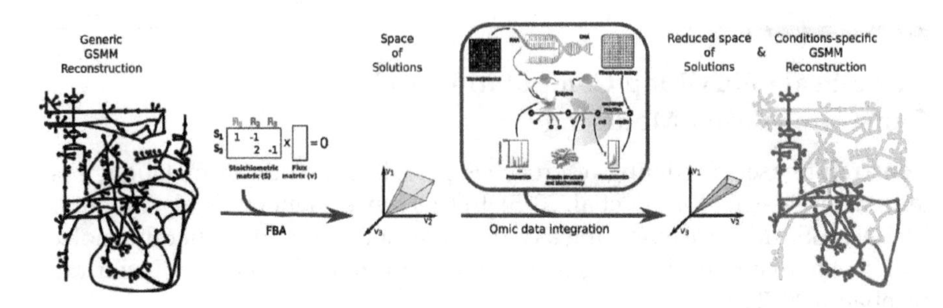

FIGURE 9.4
Genome-Scale Metabolic Models (GSMMs) analysis by Flux Balance Analysis (FBA) assuming steady state (ss) B. GSMMs can be used as a platform to integrate and combine omic data from multiple layers and further constrain the space of feasible flux solutions. In these models, metabolomics data can be associated with metabolites; while genomics, transcriptomics and proteomics can be associated with metabolic reactions, these associations are established through gene-protein-reaction associations (GPR). The phenotypic assays can constrain properties of the network such as growth rate under certain experimental conditions. By integrating omic data into a GSMM we can determine either disease-specific biomarkers or drug-targets and reconstruct condition-specific GSMM.

and t is time. In order to narrow the space of feasible flux solutions, it is necessary to impose constraints which are fundamentally represented in two ways:

1. Steady-state mass-balance constraints: These constraints are imposed by the stoichiometry and network topology of the metabolic network (Stelling et al. 2002). Steady-state assumptions also impose constraints on the system. By definition, at steady state, the concentration of a certain metabolite is constant, consequently $dC/Dt = 0$ and $S.v = 0$. These considerations ensure that for each metabolite in the network the net production rate equals the net consumption rate (Figure 9.5b and c).
2. Inequalities that impose bounds on the system: For every reaction, upper and lower bounds can be imposed, defining maximum and minimum allowable fluxes. These restrictions are based on reaction reversibility (irreversible fluxes have a zero lower bound) and/or measured rates (e.g., metabolite uptake/secretion rates), allowing to define the environmental conditions in a given simulation such as nutrient or O_2 availability.

In addition, it is necessary to define a phenotype in the form of a biological objective that is relevant to the problem being studied, the objective function. In many studies, this objective function is related to growth rate prediction and, in CBM approaches, it is defined by an artificial biomass production reaction. The design of this artificial reaction is based on experimental measurements of biomass composition and is unique for each species.

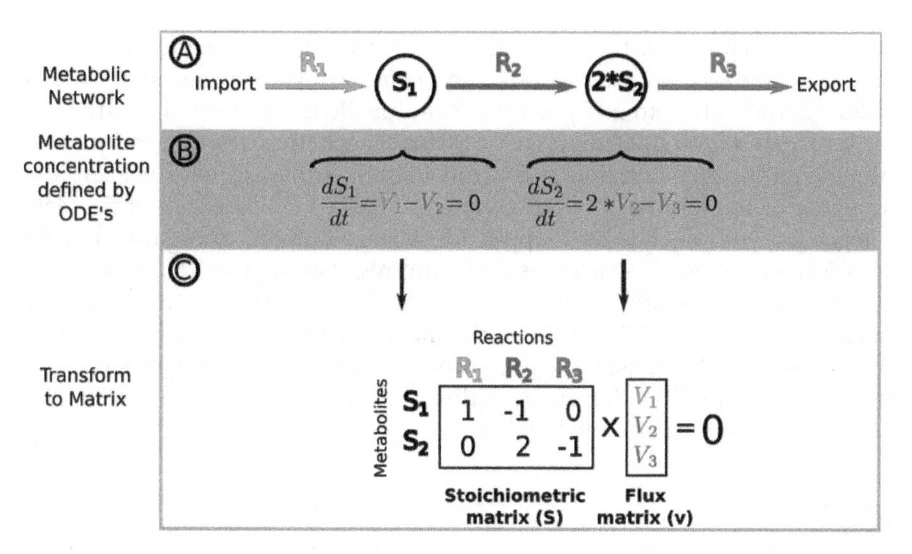

FIGURE 9.5
(a) Toy network representing a metabolic process in which the nodes are the metabolites involved in the process and the arrows the biochemical reactions that transform the substrates into the final products. Metabolite S_1 is imported into the toy network by reaction R_1; the reaction R_2 produces two molecules of S_2 from one molecule of S_1; finally, the reaction R_3 exports the metabolite S_2 outside the network. (b) System of ODEs describing the metabolic fluxes; V_1 is the flux of reaction R_1, V_2 is the flux of reaction R_2 and V_3 represents the metabolic flux through reaction R_3. This system of ODEs determines the metabolic concentrations of the metabolites S_1 and S_2. (c) Transforming the system of ODEs to a stoichiometric matrix and a flux matrix that define the concentration of the metabolites. Here steady state is imposed ($S \times v = 0$).

Typically, the objective function is the maximization of growth rate that can be accomplished by calculating the set of metabolic fluxes that result in the maximum flux through the biomass production reaction. This approach is widely used in the simulation of micro-organism or cancer cell metabolism. The objective function can be adapted to the specific case and can integrate more than one biological aspect. Then one can, e.g., minimize overall flux through the network while maximizing the biomass production (maximizing enzymatic efficiency). However, it is not always obvious what the objective should be, especially in multicellular organisms (Shlomi et al. 2008).

Finally, based on the stoichiometric matrix, a system of linear equations is defined, which is solved by applying FBA (Schmidt et al. 2013) while simultaneously satisfying the previously imposed constraint and optimizing an objective function (Babaei et al. 2014).

Metabolic Flux-Map Analysis Enriched by Omics Data Integration

High-throughput technologies have provided substantial amounts of quantitative omics data across a variety of scales in the scope of systems biology (e.g., gene or protein expressions, metabolite concentrations, metabolic fluxes, etc.). However, the extraction of "knowledge" from this increasing amount of omics data has proven to be a complex task (Palsson and Zengler 2010). GSMMs offer an adequate platform for the integration of a variety of omics data (Palsson 2002) that is used to further constrain the non-uniqueness of constraint-based solution space and thereby enhance the precision and accuracy of model prediction (Lewis et al. 2012). GSMM-based omics data integration analysis has been applied to a large number of organisms with a variety of applications such as deducing regulatory rules (Kümmel et al. 2006), data visualization (Duarte et al. 2007), constructing tissue-specific models (Gille et al. 2010), multi-cellular modeling (Bordbar et al. 2010) or network medicine (Barabási et al. 2011). The following table highlights some of the most relevant FBA-driven algorithms recently developed to incorporate experimental omic data into GSMMs:

Name	Input	Description
iMAT (Shlomi et al. 2008)	Gene expression data	Assigns each gene a discrete expression state (low, moderate or high), seeks to maximize the number of reactions consistent with their expression state.
E-Flux (Colijn et al. 2009)	Gene expression data	Applies FBA constraining the maximum flux through reactions catalyzed by genes with expression levels below a certain threshold.
MADE (Jensen and Papin 2011)	Gene expression data from two or more conditions	Calculates changes in gene expression across conditions and classifies each gene into increasing, decreasing or constant expression and seeks the network structure that better reproduces these changes while applying FBA to the different conditions.
PROM (Chandrasekaran and Price 2010)	Gene expression data from multiple conditions, Regulatory Network Structure	Calculates the probability of gene being highly expressed if a transcription factor (TF) is active or inactive based on gene expression data in multiple conditions. Uses such probabilities to constrain the upper bound of reactions in FBA.
Gonçalves et al. (2012)	Relative expression of genes in a given strain compared to the wild type	Gene expression data is used to constrain the upper bound of fluxes in FBA.
mCADRE (Wang et al. 2012)	Multiple gene expression data form the tissue/cell type, Metabolomics	Gives each reaction a ubiquity score based on gene expression data. Those with higher score are considered core reactions. Non-core

(*Continued*)

Name	Input	Description
		reactions are ranked based on ubiquity score and connectivity based evidence. The algorithm seeks to remove all non-core reactions, starting from those ranked at the bottom with reactions being removed only if upon removal the core set of reactions remains consistent and the model is capable of producing key metabolites.
GIM^{3}E (Schmidt et al. 2013)	Gene expression data, Metabolomics	Builds a network that satisfies an objective function while penalizing the inclusion of reactions catalyzed by genes with expression below a certain threshold. Can be constrained to produce metabolites that have been detected experimentally.
INIT (Agren et al. 2012)	Gene expression data (particularly adapted to work with data from the human proteome atlas), Metabolomics	Seeks to build a model prioritizing the addition of reactions with strong evidence of their presence based on gene expression data. Can be forced to produce metabolites that have been detected experimentally.
MBA (Jerby et al. 2010)	Tissue-specific molecular data sources (Transcriptome, Proteome, Metabolme, Bibliome etc.)	Uses tissue specific data to identify high and moderate probability core reactions. Seeks to build a network with all the high probability core reactions, the maximum moderate probability core reactions and the non-core reaction required to prevent gaps.
Fastcore (Galhardo et al. 2014)	Tissue-specific molecular data sources (Transcriptome, Proteome, Metabolme, Bibliome, etc.)	Uses tissue specific data to identify a set of core reactions. Seeks to build a network that contains the minimum set of additional reactions necessary to build a feasible metabolic network.
FARM (Dreyfuss et al. 2013)	Experimental KO phenotypes (growth/no-growth phenotype), Bibliome, Gene expression data	Combination of three algorithms: (1) limed-FBA which applies FBA linearly accounting for metabolite dilution; (2) One Prune which removes blocked reactions; (3) CROP, gives a score to each reaction based on available evidence, uses this score as guide to remove reactions when the models fail to predict a no growth phenotype. When the models fail to reproduce a growth phenotype, adds reactions from a reactions database.

The algorithms in the table above represent some of the most widely used approaches allowing the integration of the increasing amount of complex and heterogeneous omic data into comprehensive reconstruction of metabolism. This integration enhances the predictive capabilities of metabolic models with important implications in a wide range of fields, some of which will be exposed below (Box 9.4).

Working Example of Constraint-Based Modeling Application

The following example illustrates how constraint-based modeling methods can be used to estimate a space of feasible metabolic fluxes describing the behavior of a given metabolic network model. To this aim we use a simple metabolic network model depicted in Figure 9.6a. Here A, B and C are metabolites; int_1, int_2, int_3 and int_4 are intracellular reactions and ex_1, ex_2 and ex_3 are exchange reactions between cell and media.

Next, the graphical representation is mathematically formalized in a matrix form where S is the stoichiometric matrix and v is a vector of fluxes (Figure 9.6b). The S matrix represents the stoichiometric relation between metabolites (rows) and reactions (columns). Then, for instance, reaction int_1 transforms one

BOX 9.4 BASIC DEFINITIONS IN CONSTRAINT-BASED MODELING

Genome-scale metabolic model: Is a comprehensive reconstruction gathering all biochemical reactions encoded by an organism genome. These reconstructions define the associated metabolites and genes as well as the stoichiometry and thermodynamics for each reaction.

Stoichiometry matrix: Is a mathematical representation of the relative relation between substrates and products in a given reaction.

Objective function: Defines a phenotype (i.e., biomass production or a given metabolite production) that is mathematically formalized in optimization problems such as FBA. Then, for instance, a possible objective function would be the maximization of the biomass production.

Omic data integration: Defines a variety of computational techniques developed to integrate the increasing amount of data generated by the different high-throughput technologies to infer the metabolic flux profile in a GSMM analysis.

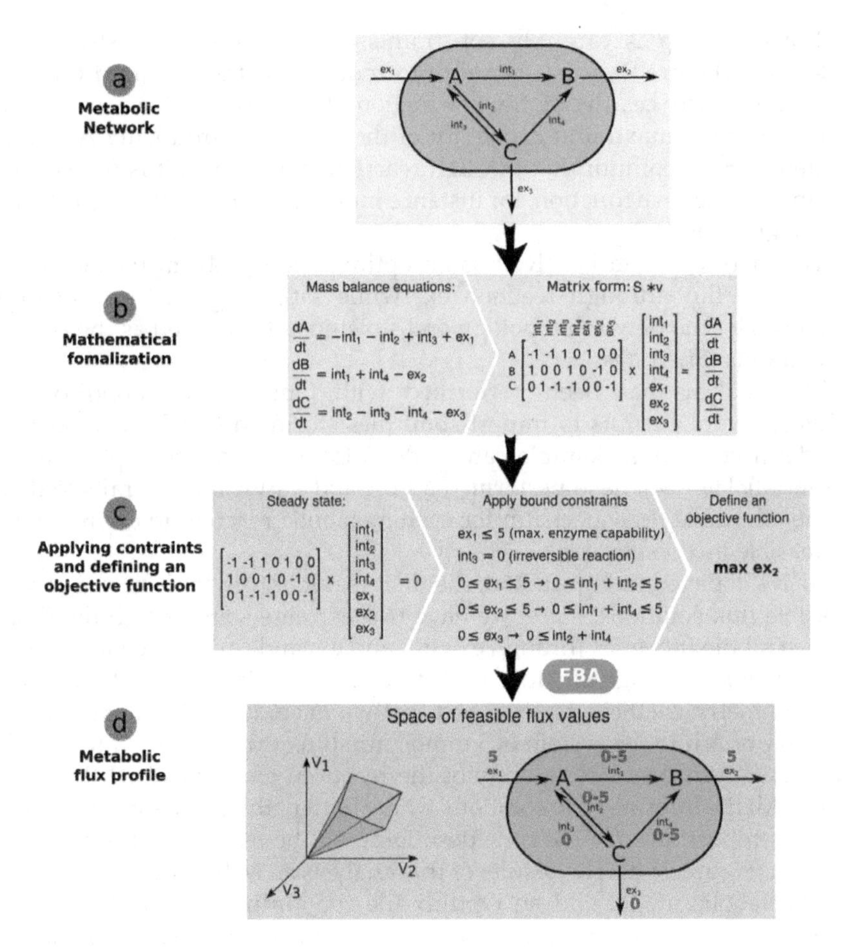

FIGURE 9.6
Pipeline of how apply CBM to determine the feasible space of metabolic fluxes of a metabolic network. (a) Metabolic network where A, B and C are metabolites; int_1, int_2, int_3 and int_4 are intracellular reactions and ex_1, ex_2 and ex_3 are exchange reactions between cell and media. (b) Mathematical formalization of the metabolic network into a matrix form. (c) Impose constraints based on steady state assumption and on previous knowledge such as reaction irreversibility or maximum enzyme capability (impose a maximum flux through reaction ex_1 of 5 mmol/min). Finally, it is defined as the maximization of the flux through reaction ex_2 as objective function. (d) As a result of the FBA, a spectrum of feasible flux solutions for each metabolic reaction (in mmol/min and highlighted in red).

molecule of A into one molecule of B. Consequently, the stoichiometric coefficient of A in this reaction is −1 (because it is a substrate), B has a stoichiometric coefficient of 1 (product) and for C it's 0 (not related with this reaction). Thus, the vector of stoichiometric coefficients for reaction int_1 is −1, 1, 0 which is represented in the left most column in the S matrix (Figure 9.6b).

The next step is to apply constraints, first by assuming steady state $S \times v = 0$ (Figure 9.6c). We can also introduce constraints based on previous knowledge; i.e., the biotransformation of A into C is irreversible, thus int_3 is 0 or the maximum capability of the enzyme introducing A into the system is 5 (maximum flux through reaction ex_1). Finally, it is necessary to define an objective function, for instance maximize the flux through reaction ex_2 (Figure 9.6c).

The matrix system is solved as an optimization problem trying to maximize the flux through reaction ex_2 while satisfying all the constraints imposed by the network topology and stoichiometry as well as the additive constraints (FBA).

The analysis has been performed with Optflux (www.optflux.org/). This software permits to import sbml files storing a metabolic model (the model used in this example can be downloaded from the supplementary material). Here we have performed a FBA and a FVA that permits to determine allowable flux spectrum for each metabolic reaction to define a space of feasible flux solutions.

As it's depicted in Figure 9.6d, the final result of this analysis is a space of feasible flux solutions. Here, we have two extreme solutions: (i) the first in which all the flux goes through ex_1, int_1 and ex_2 and (ii) the second where all the flux goes through ex_1, int_2, int_4 and ex_2. Taking into account that all the stoichiometry coefficients are 1 and that we have defined a maximum flux of entry of A into the system of 5 mmol/min (flux through ex_1), it results that the maximum flux through any of the reactions cannot exceed this upper limit. All the intermediate solutions where both paths carry some flux in different proportions make a cone describing all the feasible solutions for our system (Figure 9.6d). The results of this analysis as well as the sbml file with the metabolic model and an Optflux file containing the metabolic model to perform different analyses can be downloaded from the supplementary material.

Applications of Metabolic Flux Modeling

Drug Discovery

The elucidation of diverse metabolic alterations for the identification of biomarkers and novel drug targets have been increased in the last years. An increasing number of methods and algorithms have been recently developed to integrate disease-specific omic data into GSMMs. It has enabled us to gain further biological and mechanistic understanding of how the diseases with a strong metabolic component benefit from metabolic modifications (Palsson 2002). This model-driven approach allows the discovery of potential biomarkers and drug targets (Oberhardt et al. 2010).

Biomarkers: The identification of new biomarkers is of major importance to biomedical research for early diagnosis and monitoring treatments efficiently. The identification of disease-specific biomarkers is possible due to aberrant metabolism observed in such diseases that alters the profile of absorption and nutrients secretion. In the clinical practice, it is achieved using different methods. Then, for instance, in cancer, it is achieved by detecting increased glucose uptake via F-deoxyglucose positron emission tomography (PET) and other metabolites such as ^{11}C-acetate, ^{11}C-methionine or ^{18}F-labeled amino acids. However, it is still challenging to develop non-invasive and cost-effective methods for the identification of metabolic biomarkers in the biofluids (Barabási et al. 2011) for clinical diagnosis.

As described above, the modeling of disease-specific metabolism is based on integrating high-throughput data into a GSMM reconstruction analysis. Omic data of clinical samples (mainly transcriptomics data) can be used to infer the exchange rates of different metabolites for each individual sample via GSMM analysis (alterations in exchange reactions in the model). Thus, those metabolites that significantly differ between two clinical groups in their exchange rates are then considered as potential biomarkers.

Constraint-based GSMMs approach has been applied to predict biomarkers for inborn errors of metabolism (IEM). Here the loss of functionality was done via "in silico" specific metabolic gene KOs simulations (Gille et al. 2010).

However, despite the accurate level of predictions, this method is not suitable to identify biomarkers in more complex diseases, where the metabolic abnormalities result from more complex and elaborative genetic and epigenetic alterations that alter the expression of a variety of disease-associated isoenzymes. In order to bridge this gap, other approaches have been developed to predict metabolic biomarkers in cancer. Metabolic phenotypic analysis (MPA) method uses GPR association to integrate transcriptomics and proteomics data within a GSMM to infer metabolic phenotypes (Jerby et al. 2010). MPA was used to study breast cancer metabolism including biomarker predictions. These predictions that included amino acid and choline-containing metabolites are supported by a number of experimental evidences (Lewis et al. 2012). Other approaches such as flux variability analysis (Murabito et al. 2009) or sampling analysis (Schellenberger and Palsson 2009) are also suitable to predict metabolic biomarker candidates by integrating omic data into a GSMM.

Drug target: The specific metabolic abnormalities observed in metabolic diseases, such as cancer, diabetes or chronic diseases among others, are in the basis of novel drug discovery. These differences can be used as drug targets to attack specific weaknesses of the diseased cell and hence compromising its viability, but not that of non-diseased cells (Duggal et al. 2013). Based on this rationale, several approaches have been developed that consider different aspects of the disease metabolism for the discovery of new drug targets:

Antimetabolite: One of the most common drugs are antimetabolites. An antimetabolite is structurally similar to a certain metabolite, but it cannot be used to produce any physiologically important molecule. Antimetabolite-based drugs act on key enzymes preventing the use of endogenous metabolites, resulting in the disruption of the robustness of diseased cells and reduction or suppression of cell growth. For example, antimetabolites such as antifolates or anti purines mimic folic acid and purines (Duggal et al. 2013).

The GSMMs approach can be used to systematically simulate the effect of potential antimetabolites in drug discovery research. To achieve this, methods such as tINIT (Task-driven Integrative Network Inference for Tissues) algorithm have been developed. This method has been used to (i) reconstruct personalized GSMMs for six Hepatocellular carcinoma (HCC) patients based on proteomics data and the Human Metabolic Reaction database (HMR) (Mardinoglu and Nielsen 2012) and (ii) identify anticancer drugs that are structural analogs to targeted metabolites (antimetabolites). tINIT algorithm was able to identify 101 antimetabolites, 22 of which are already used in cancer therapies, and the remaining can be considered as new potential anticancer drugs.

Synthetic lethal: The genetic lesions occurring in diseases with a genetic background are associated with dependencies that are specific to these lesions and absent in non-diseased cells. Two genes are considered "synthetic lethal" if the isolated mutation on either of them is compatible with the cell viability but simultaneous mutation is lethal (Conde-Pueyo et al. 2009). Analogously, two genes are considered to interact in a "synthetic sick" fashion if simultaneous mutation reduces cell fitness below a certain threshold without being lethal (Conde-Pueyo et al. 2009). Enzymes encoded by genes that are in synthetic lethal or sick interactions with known, non-druggable disease-driving mutations represent ideal drug targets. This approach has two main advantages: first, we can indirectly target non-druggable disease-promoting lesions by inhibiting druggable synthetic lethal interactors, and secondly, we can achieve a high selectivity by exploiting true synthetic lethal interactions for a specific disease therapy. Silencing or inhibiting the protein product of a gene that is synthetic lethal with a given mutated disease gene will selectively kill diseased cells while it is tolerated by healthy cells that lack the disease cell-specific mutation.

GSMMs provide an excellent tool for the systematic simulation of specific pairs of gene knock-out (KO) to unveil those combinations that compromise the viability of diseased cells (synthetic lethal). By definition, gene KO is simulated by giving value zero to gene expression and the effect of gene deletion is transferred to the metabolic reaction level by GPR association. Thus, for instance, the flux through a reaction that is associated only to a knocked-out gene is zero. If the reaction is catalyzed by isoenzymes or complexes, the effect of a gene deletion is more complex.

However, predicting the metabolic state of a cell after a gene knockout is a challenging task because after the gene knock-out, the system evolves to a new steady-state that tends to be as close as possible to the original steady-state (Barbash and Lorigan 2007). To overcome these difficulties several algorithms have been developed. For example, MOMA algorithm minimizes the Euclidean norm of flux differences between metabolic state of the knockout compared with the wild type (Ren et al. 2013). ROOM method minimizes the total number of significant flux changes from the wild type flux distribution (Ren et al. 2013). In other words, MOMA minimizes the changes in the overall flux distribution while ROOM minimizes the number of fluxes to be modified after the gene knockout.

Bioengineering

GSMM-enabled studies of microorganisms have direct application to metabolic engineering. This approach aims to optimize genetic and regulatory processes within cells to increase the production of a certain substance. The whole-genome sequencing together with detailed biochemical and enzymatic data (e.g., bibliome data) on microbial metabolism has led to the reconstruction of detailed metabolic networks at the genome-scale that can be used to perform computational and quantitative queries to answer various questions about the capabilities of organisms and their likely phenotypic states (Orth et al. 2010) as well as to determine the minimal set of interventions (gene modifications, media restrictions, etc.) necessary to achieve a given phenotype (e.g., maximize a metabolite production). In this sense, bacteria metabolic GSMMs have been used in a variety of metabolic engineering studies that has matured strain design from academic to industrial.

Through metabolic engineering, the native biochemical pathways of bacteria can be manipulated and optimized to more efficiently produce industrial and therapeutically relevant compounds (Lee et al. 2012). In particular, the *E. coli* GSMM has guided metabolic engineers toward the production of a variety of compounds, including alcohols, amino acids, organic acids, etc. (Ranganathan et al. 2010). GSMMs are used for the rational strain design allowing the prediction of cellular phenotypes from a systems level using genomic, regulatory, stoichiometric and kinetic knowledge. This approach permits to identify engineering strategies that can be lately implemented on in vivo models. These strategies include: (i) gene over- and underexpression (Fowler et al. 2009), (ii) gene deletions (Fong et al. 2005), (iii) identification of bottlenecks or competing pathways by mapping high-throughput data into Genome-scale metabolic network models (Lee et al. 2007), (iv) integration of non-native pathways integration for production of compounds that are either not natively found or only synthesized in minute concentrations by the studied microorganism (Yim et al. 2011) and (v) couple the growth of different bacterial strains to target product overproduction (Kim and Reed 2010) (growth-coupled strain).

An increasing number of model-driven based algorithms have been developed to identify engineering strategies (Kim and Reed 2010). Following are highlighted some of the most interesting recent findings. For instance, studies using *E. coli* GSMM predicted a high flavonone production by strategically knocking out genes; it was experimentally validated demonstrating that the production of the redox carrier (NADPH) to drive the heterologous flavanone catalyst was increased while the optimal redox potential of the cell was maintained (Fowler et al. 2009). Other studies in *E. coli* have determined a gene that needed to be upregulated and tune the expression level of this gene by performing GSMM-based analysis of a deleterious overexpression event (Lee et al. 2012). GSMM-based approaches also have permitted to find engineering strategies allowing to enhance by several orders of magnitude the production of the nonnative metabolite 1,4-butanediol in *E. coli* (Yim et al. 2011). It was achieved by rewiring the host cell (activation/inhibition of native and nonnative pathways) to force that the production of the compound was the only means by which the cell could maintain a redox balance and grow anaerobically (Yim et al. 2011). These results evidence that the analysis of genetic alterations at the systems level allows to predict the adaptation to the loss of functionalities associated to gene deletion by activating compensatory pathways. This knowledge can be used to design novel engineering strategies that couple cellular goals to target compound overproduction with important implications in the industry.

Evolution

Demands imposed by environmental conditions modify the genomic content and phenotypic landscape that are constantly adapting. This evolution occurs via alterations in gene expression, enzyme capacity or enzyme kinetics, addition of new reactions by horizontal gene transfer, gene duplication, and by loss-of-function or gain-of-function mutations that eliminate or incorporate individual reactions to the metabolic network. GSMMs provide an adequate mapping tool between genotype and phenotype by incorporating GPR associations that permit to test evolutionary hypotheses and determine the molecular mechanism underlying these changes.

This approach has been employed to study bacterial evolution showing a set of conserved common reactions in organisms with similar lifestyles (Pál et al. 2006) that suggest the existence of a common enzymatic machinery to metabolize specific carbon sources. Other studies have shown that, even being the genes deletion in mutagenesis a random process, the order in which genes are lost follows a coordinated and consistent pattern that can be accounted mostly by a GSMM and is consistent with available phylogenetic data (Yizhak et al. 2011). As is previously mentioned *E. coli* GSMM is one of the most curated metabolic reconstructions, which makes it suitable to be used as a scaffold on which similar bacterial strains can be reconstructed, and their divergent evolutions understood (e.g., five *E. coli* strain-specific GSMM (Baumler et al. 2011)). These studies have shown that the horizontal gene transfer is the dominant evolutionary mechanism in bacteria (Pál et al. 2006), which is closely related with environment-specific genes (Pál et al. 2006).

Future Challenges of Metabolic Flux Modeling

Metabolism represents the essence of how cells interact with their environment to provide themselves with energy and the essential building blocks for life. In this section, we have presented two main approaches to address the study of metabolism (constraint-based and kinetic modeling). In the following we will expose some caveats and limitations of these approaches that need to be overcome, as well as novel strategies aiming to combine constraint-based and kinetic modeling to improve our knowledge of the metabolic processes underlying cell behavior.

Challenges in Dynamic Flux-Map Computation and Kinetic Modeling

Analysis based on kinetic modeling approaches, enhanced by incorporating ^{13}C-labeling information from ^{13}C-tracer experiments, provides a powerful tool to interrogate the metabolism of a cell (Marin de Mas et al. 2011). However, the local complexity of kinetic models, in spite to describe metabolic networks with high precision, restricts this approach to systems with relatively small size. Several factors limit the optimum resolution of flux parameters in kinetic modeling and more specifically in those studies based on ^{13}C metabolomics:

1. The significant costs of the custom-synthesized labeling patterns which means that the commonly used labeled sugars and simple organic acids (e.g., acetate) are the only affordable substrates used in the majority of tracer experiments

2. The relatively small number of groups of compounds that are currently accessible for labeling measurements via MS or NMR (mainly amino acids from proteins or sugars from carbohydrates)
3. The design of the metabolic model and the experiments in order to choose the combination of labeling substrates and pathways to study (set of metabolic reactions) allowing an optimal resolution of the fluxes in a network. In other words, not all the labeling substrates provide the same information of all the metabolic pathways. Thus, choosing the right labeling substrate that permits the deepest analysis of the path under study is not always obvious due to the non-linearity and complexity of metabolic networks.

While the first limitation is related with the availability and the price of tracer compounds and the second concerns limitations in the current analytical techniques used in tracer-based metabolomics, the third can be addressed by applying computational solutions. To this aim two different strategies has been recently developed:

1. A priori rational tracer selection and experiment design based on elementary metabolite units (EMU) decomposition (Crown and Antoniewicz 2012). This method identifies the minimum amount of information needed to simulate isotopic labeling within a reaction network and has enabled the identification of two optimal novel tracers on a network model of mammalian metabolism.
2. Elementary flux modes at the carbon level to integrate tracer-based experimental data into constraint-based models (elementary carbon mode). This approach represents the non-linear tracer-based metabolomics data in a linear fashion, which makes this information suitable to restrict the steady-state solution space (Pey et al. 2011).

In spite the efforts spent in developing novel computational tools, there is still a lack of methods to analyze the dynamics of metabolic networks based on tracer-based metabolomics. In particular, the approaches described above omit the reversible reactions in the network in order to keep the problem linear. Thus, these approaches provide a given set of target analytes rather than predicting the optimal combination of analytes considering all the possible substrates to resolve a set of flux parameters. Further improvements in sampling metabolites, to separate metabolite pools from different compartments or from different cell types in tissues or microbial communities, should permit the individual labeling to become accessible and could be used for ^{13}C-fluxomics. In addition, since the less likely to achieve the metabolic and isotopic steady state the more complex the system becomes, the development and improvement of dynamic approaches is required (Marin de Mas et al. 2011).

Challenges in Large-Scale Flux-Map Computation and Constraint-Based Modeling

Up to now a large number of constraint-based algorithms have been developed for genome-scale metabolic models. Most of the studies have focused on the metabolism of microorganisms due to their smaller and simpler metabolic networks and the availability of experimental data. Thus, for instance, recent studies on artificial microbial ecosystems have demonstrated the potential of GSMM approach to study synergies in heterogeneous cellular communities (Ye et al. 2014). This approach may be of special interest in the study of diseases where a metabolic cooperation is reported such as cancer, which would permit to unveil the mechanisms underlying the cooperation between tumoral and stromal cells (Hulit et al. 2013), as well as between intratumoral subpopulations (Peinado et al. 2007). The study of the metabolic coupling between different cellular populations as potential drug targets can be achieved by reconstructing an artificial disease-specific microenvironment by using GSMM approach. Another interesting approach in drug discovery is the metabolic phenotypic analysis (MPA) (Jerby et al. 2012). This algorithm has been used to determine the metabolic state of different patients with breast cancer, providing generic and subtype predictions of potential metabolic biomarkers. However, there is a need for novel methods applicable when the similarity in gene expression pattern between samples is high.

Constraint-based analysis on GSMM has been widely used to develop novel engineering strategies with important applications in industry. This approach has made it possible to design strains to improve metabolite production by exploring the native and nonnative genetic space in a variety of microorganisms such as *E. coli*. However, a strong understanding of metabolic biochemistry is a prerequisite for successful strain design. Thus, the results in these studies often lead to nonviable and suboptimal phenotypes, or they are not predicted at all as is the case of the in vivo performance of nonnative genes and proteins (Yim et al. 2011). This is mainly due to the fact that GSMM does not account for detailed enzyme and isoenzyme kinetics, translational regulation, posttranscriptional and posttranslational modifications or optimal codon usage (Alper et al. 2005). Hence, it is necessary to carefully select which predicted knockout strain designs will be constructed in vivo. This fact limits the application of many potential engineering strategies, which cannot be addressed with the current generation of GSMM. Thus, more complete biochemical information is necessary to be considered in strains design studies in order to consider cellular aspects beyond those accounted in the current GSMM reconstructions (e.g., expression, regulation, and enzyme kinetics, etc.). Additionally, in order to improve the development of novel strain mutant with non-native functionalities, further advancements in pathway finding procedures are required, allowing the identification of non-native heterologous pathways and techniques to optimize the expression of nonnative enzymes. GPR associations allow

GSMMs to explore a vast region of genotypic space, enabling the study of cellular evolution specially in bacteria. However, GSMMs are currently limited to metabolic genes and do not account for regulation of metabolism during evolution. Several studies have demonstrated the utility of large-scale constraint-based approaches modeling horizontal gene transfer and gene loss (Pál et al. 2006) which is supported by studies on comparative genomics (Yizhak et al. 2011). However, the current state of the art of GSMMs are limited in determining the precise genes and their exact loss, the location of mutations in the genome, and predicting their effect on the physiology of the organism. In addition, large strain-specific portions of the genomes still remain uncharacterized, which hinders the comparison of the evolutionary trajectories of different bacterial strains.

GSMM-based analysis on adaptive evolution will be nourished from studies on adaptive laboratory evolution experiments. This experimental procedure introduces a selection pressure in a controlled environmental setting and permits the characterization of the genetic changes occurring in an organism during the process of adaption (Conrad et al. 2011). This experimental approach in combination with GSMM-based analysis will provide a tool for understanding the mutations associated to the adaptive process. Finally, a further extension of GSMM methodology is required for characterizing promiscuity of known enzymes. Enzyme promiscuity refers to the flexibility of the enzyme catalytic units in recognizing substrates different from the canonical ones. Accounting for enzyme promiscuity will open the door for a more detailed examination of unknown cellular processes and will correct pitfalls and errors in gene, essentially predictions (Guzmán et al. 2015).

Large-Scale ^{13}C-Constrained FBA

A number of FBA-driven algorithms have been developed recently, making it possible to use ^{13}C-based metabolomics in a high-throughput fashion in large-scale studies of cellular metabolism (Winter and Krömer 2013). The integration of ^{13}C metabolomics in large-scale metabolic networks is mainly focused on the study of microorganism metabolism and provides a better characterization of complex metabolic networks, allowing the correlation of the whole metabolic flux profile with specific phenotypes (Winter and Krömer 2013). For instance, a large-scale ^{13}C-constrained FBA approach has been used to study the transcriptional regulation of the respiration on four different carbon sources in *Saccharomyces cerevisiae* (Fendt and Sauer 2010). It was found that glucose or mannose as carbon sources induce fermentation, an intermediate degree of respiration is achieved using galactose and the use of pyruvate as the source of carbon induces a full respiration (Fendt and Sauer 2010). Furthermore, by integrating proteomic data with known regulatory networks the transcription factors were identified that affect respiration. These were validated experimentally using strains with specific genetic mutations. In another study Van Rijsewijk et al. (2011) aimed to identify the transcriptional mechanisms that regulate the metabolic flux profile associated to aerobic growth in *E. coli.* It was carried out on knockout mutant strains for 81 transcriptions and 10 sigma and anti-sigma factors known to directly or indirectly control central metabolic enzymes using glucose (induces fermentation) or galactose (induces respiration) as the carbon source. Here, flux ratios and intracellular fluxes were quantified by applying large-scale ^{13}C-constrained FBA that enabled to identify nine transcription factors that control flux distribution during growth on glucose and one with galactose (Crp dependent of cAMP) (Van Rijsewijk et al. 2011).

These results can be partially extrapolated to yeast which network regulating growth under glucose conditions is similar but not the network regulating galactose metabolism which is highly different between these two organisms. However, these results cannot be applied to other organisms such as *S. cerevisiae,* where only a few reactions in the central carbon metabolism are transcriptionally regulated (Bordel et al. 2010).

Overall, these results clearly demonstrated the utility of large-scale ^{13}C fluxomics approaches to analyze metabolic fluxes and perform comparative analysis between organisms, and on the other hand evidences the need to further study the energy metabolism, particularly as a similar number of transcription factors target the TCA cycle, 36 for *E. coli* (Keseler et al. 2005) and 35 for *S. cerevisiae* (Monteiro et al. 2008).

In addition, it is suggested that flux distribution may pose a layer of regulation by itself. In other words, the cells may contain a system that "measures" the metabolic fluxes that trigger a metabolic reprogramming in a flux-dependent regulation manner (Huberts et al. 2012).

In summary, these examples highlight the unique insights into metabolism obtained using the different flux analysis approaches and its support in integrating the different levels of cellular regulation.

Genome-Scale Kinetic Models

In this chapter, we have presented two divergent modeling strategies to study metabolic networks: (i) Constraint-based modeling and (ii) Kinetic modeling. Typically, constraint-based modeling is used to study large scale metabolic network models. However, this type of approach is unable to give an insight into cellular substrate concentrations and internal metabolic fluxes. On the other hand, kinetic modeling approaches aim to characterize fully the mechanics of each enzymatic reaction enabling a more precise prediction of the intracellular metabolic fluxes. However, this approach is limited in size because parameterizing mechanistic models is both costly and time-consuming, which makes this approach unsuitable for a large-scale analysis.

In order to cope with these limitations, several approaches have been recently developed to combining both approaches. In this sense Smallbone et al. (2010) proposed one of the first large-scale kinetic constructions using an approach that combined linlog approximation (Visser and Heijnen 2003) that was applied on a consensus model of yeast metabolism and FBA for kinetic model reconstruction based solely on reaction stoichiometries (Smallbone et al. 2010). Here the linlog approximation describes the effect of metabolites level on metabolic flux as a linear sum of logarithmic terms where the elasticities represent the kinetic parameters which are estimated from stoichiometric considerations. Finally, fluxes through the system are estimated by FBA according to linlog kinetics.

Li et al. (2010) developed a workflow where a known metabolic network of yeast is systematically enriched with kinetic rate laws from SABIO-RK database (Rojas et al. 2007) or a database of experimental results and with a generic rate law when suitable rate laws are not available (Li et al. 2010). This approach was evaluated for its ability to construct a parameterized model of yeast glycolysis.

Adiamah and Schwartz (2012) presented a workflow to build a genome-scale kinetic model of Mycobacterium tuberculosis metabolism, using generic equations, given a genome-scale flux distribution derived from FBA (Adiamah and Schwartz 2012). This approach uses GraPe software (Adiamah et al. 2010) that uses a genetic algorithm to estimate kinetic parameters using flux values. Finally, the model was validated using COPASI (Hoops et al. 2006). In Stanford et al. (2013) the authors developed a strategy for Genome-scale kinetic models construction. This work-flow is based on a logical layering of multiple and heterogeneous data such as reaction fluxes, metabolite concentrations and kinetic constants into a large-scale metabolic network model (Stanford et al. 2013). This methodology was applied to transform a yeast consensus metabolic network into a kinetic model which contains realistic standard rate laws and plausible parameters, adheres to the laws of thermodynamics and reproduces a predefined steady state.

Finally, k-OptForce is a procedure developed for large-scale kinetic model reconstruction (Chowdhury et al. 2014). This approach uses the available kinetic descriptions of metabolic reactions (kinetic rate equations) to (re) apportion fluxes in a large-scale stoichiometric network. The method describes two subsets of reactions: (i) reactions with kinetic information and (ii) reactions with only stoichiometric information. The first subset is mathematically described by a system of nonlinear ordinary differential equations (ODEs) incorporating enzyme activity, metabolite concentrations and kinetic parameter values that constraints the space of feasible solutions. The other subset with only stoichiometric information is constrained only by stoichiometric balances and reaction directionality restrictions and the fluxes are inferred so as to be consistent with the predicted fluxes of the reactions in the first subset (reactions with kinetic information). This approach has been successfully applied on the strain design studies of *E. coli*.

These studies pave the way for the integrated analysis of kinetic and stoichiometric models and enables elucidating system-wide metabolic interventions while capturing regulatory and kinetic effects. However, further efforts are needed to develop novel strategies allowing the integration of both constraint-based and kinetic modeling approaches in the study of more complex metabolic systems such as mammalian or plant metabolism with important implications in clinics and in industry.

References

Adiamah, D. A., and J. -M. Schwartz. 2012. "Construction of a Genome-Scale Kinetic Model of Mycobacterium Tuberculosis Using Generic Rate Equations." *Metabolites* 2 (3): 382–397.

Adiamah, D. A., J. Handl, and J. -M. Schwartz. 2010. "Streamlining the Construction of Large-Scale Dynamic Models Using Generic Kinetic Equations." *Bioinformatics* 26 (10): 1324–1331.

Agren, R., S. Bordel, A. Mardinoglu, N. Pornputtapong, I. Nookaew, and J. Nielsen. 2012. "Reconstruction of Genome-Scale Active Metabolic Networks for 69 Human Cell Types and 16 Cancer Types Using INIT." *PLoS Computational Biology* 8 (5): e1002518.

Alper, H., K. Miyaoku, and G. Stephanopoulos. 2005. "Construction of Lycopene-Overproducing *E. coli* Strains by Combining Systematic and Combinatorial Gene Knockout Targets." *Nature Biotechnology* 23 (5): 612–616.

Alves, R., F. Antunes, and A. Salvador. 2006. "Tools for Kinetic Modeling of Biochemical Networks." *Nature Biotechnology* 24 (6): 667–672.

Amaral, A. I., A. P. Teixeira, S. Martens, V. Bernal, M. F. Q. Sousa, and P. M. Alves. 2010. "Metabolic Alterations Induced by Ischemia in Primary Cultures of Astrocytes: Merging ^{13}C NMR Spectroscopy and Metabolic Flux Analysis." *Journal of Neurochemistry* 113 (3): 735–748.

Babaei, P., T. Ghasemi-Kahrizsangi, and S. -A. Marashi. 2014. "Modeling the Differences in Biochemical Capabilities of Pseudomonas Species by Flux Balance Analysis: How Good Are Genome-Scale Metabolic Networks at Predicting the Differences?" *The Scientific World Journal* 2014: 416289.

Barabási, A. -L., N. Gulbahce, and J. Loscalzo. 2011. "Network Medicine: A Network-Based Approach to Human Disease." *Nature Reviews Genetics* 12 (1): 56–68.

Barbash, D. A., and J. G. Lorigan. 2007. "Lethality in Drosophila melanogaster/Drosophila Simulans Species Hybrids Is Not Associated with Substantial Transcriptional Misregulation." *Journal of Experimental Zoology. Part B, Molecular and Developmental Evolution* 308 (1): 74–84.

Baumler, D. J., R. G. Peplinski, J. L. Reed, J. D. Glasner, and N. T. Perna. 2011. "The Evolution of Metabolic Networks of *E. coli*." *BMC Systems Biology* 5: 182.

Bordbar, A., N. E. Lewis, J. Schellenberger, B. Ø. Palsson, and N. Jamshidi. 2010. "Insight into Human Alveolar Macrophage and M. Tuberculosis Interactions via Metabolic Reconstructions." *Molecular Systems Biology* 6: 422.

Bordel, S., R. Agren, and J. Nielsen. 2010. "Sampling the Solution Space in Genome-Scale Metabolic Networks Reveals Transcriptional Regulation in Key Enzymes." *PLoS Computational Biology* 6 (7): e1000859.

Bueschl, C., B. Kluger, F. Berthiller, G. Lirk, S. Winkler, R. Krska, and R. Schuhmacher. 2012. "MetExtract: A New Software Tool for the Automated Comprehensive Extraction of Metabolite-Derived LC/MS Signals in Metabolomics Research." *Bioinformatics* 28 (5): 736–738.

Capellades, J., M. Navarro, S. Samino, M. Garcia-Ramirez, C. Hernandez, R. Simo, M. Vinaixa, and O. Yanes. 2016. "geoRge: A Computational Tool To Detect the Presence of Stable Isotope Labeling in LC/MS-Based Untargeted Metabolomics." *Analytical Chemistry* 88 (1): 621–628.

Caspi, R. et al. 2010. "The MetaCyc Database of Metabolic Pathways and Enzymes and the BioCyc Collection of Pathway/genome Databases." *Nucleic Acids Research* 38 (Database issue): D473–D479.

Chakrabarti, A., L. Miskovic, K. Cher Soh, and V. Hatzimanikatis. 2013. "Towards Kinetic Modeling of Genome-Scale Metabolic Networks without Sacrificing Stoichiometric, Thermodynamic and Physiological Constraints." *Biotechnology Journal* 8 (9): 1043–1057.

Chandrasekaran, S., and N. D. Price. 2010. "Probabilistic Integrative Modeling of Genome-Scale Metabolic and Regulatory Networks in *Escherichia Coli* and Mycobacterium Tuberculosis." *Proceedings of the National Academy of Sciences of the United States of America* 107 (41): 17845–17850.

Chokkathukalam, A., A. Jankevics, D. J. Creek, F. Achcar, M. P. Barrett, and R. Breitling. 2013. "mzMatch–ISO: An R Tool for the Annotation and Relative Quantification of Isotope-Labelled Mass Spectrometry Data." *Bioinformatics* 29 (2): 281–283.

Chowdhury, A., A. R. Zomorrodi, and C. D. Maranas. 2014. "K-OptForce: Integrating Kinetics with Flux Balance Analysis for Strain Design." *PLoS Computational Biology* 10 (2): e1003487.

Colijn, C., A. Brandes, J. Zucker, D. S. Lun, B. Weiner, M. R. Farhat, T.- Y. Cheng, D. Branch Moody, M. Murray, and J. E. Galagan. 2009. "Interpreting Expression Data with Metabolic Flux Models: Predicting Mycobacterium Tuberculosis Mycolic Acid Production." *PLoS Computational Biology* 5 (8): e1000489.

Conde-Pueyo, N., A. Munteanu, R. V. Solé, and C. Rodríguez-Caso. 2009. "Human Synthetic Lethal Inference as Potential Anti-Cancer Target Gene Detection." *BMC Systems Biology* 3: 116.

Conrad, T. M., N. E. Lewis, and B. Ø. Palsson. 2011. "Microbial Laboratory Evolution in the Era of Genome-Scale Science." *Molecular Systems Biology* 7: 509.

Crown, S. B., and M. R. Antoniewicz. 2012. Selection of Tracers for ^{13}C-Metabolic Flux Analysis Using Elementary Metabolite Units (EMU) Basis Vector Methodology. *Metabolic Engineering* 14 (2): 150–161

Dreyfuss, J. M., J. D. Zucker, H. M. Hood, L. R. Ocasio, M. S. Sachs, and J. E. Galagan. 2013. "Reconstruction and Validation of a Genome-Scale Metabolic Model for the Filamentous Fungus Neurospora Crassa Using FARM." *PLoS Computational Biology* 9 (7): e1003126.

Duarte, N. C., S. A. Becker, N. Jamshidi, I. Thiele, M. L. Mo, T. D. Vo, R. Srivas, and B. Ø. Palsson. 2007. "Global Reconstruction of the Human Metabolic Network Based on Genomic and Bibliomic Data." *Proceedings of the National Academy of Sciences of the United States of America* 104 (6): 1777–1782.

Duggal, R., B. Minev, U. Geissinger, H. Wang, N. G. Chen, P. S. Koka, and A. A. Szalay. 2013. "Biotherapeutic Approaches to Target Cancer Stem Cells." *Journal of Stem Cells* 8 (3–4): 135–149.

Fan, T. W. M., L. L. Bandura, R. M. Higashi, and A. N. Lane. 2005. "Metabolomics-Edited Transcriptomics Analysis of Se Anticancer Action in Human Lung Cancer Cells." *Metabolomics: Official Journal of the Metabolomic Society* 1 (4): 325–339.

Fendt, S. -M., and U. Sauer. 2010. "Transcriptional Regulation of Respiration in Yeast Metabolizing Differently Repressive Carbon Substrates." *BMC Systems Biology* 4: 12.

Fisher, J., and T. A. Henzinger. 2007. "Executable Cell Biology." *Nature Biotechnology* 25 (11): 1239–1249.

Fong, S. S., A. P. Burgard, C. D. Herring, E. M. Knight, F. R. Blattner, C. D. Maranas, and B. O. Palsson. 2005. "In Silico Design and Adaptive Evolution of *Escherichia Coli* for Production of Lactic Acid." *Biotechnology and Bioengineering* 91 (5): 643–648.

Fowler, Z. L., W. W. Gikandi, and M. A. G. Koffas. 2009. "Increased Malonyl Coenzyme A Biosynthesis by Tuning the *Escherichia Coli* Metabolic Network and Its Application to Flavanone Production." *Applied and Environmental Microbiology* 75 (18): 5831–5839.

Galhardo, M., L. Sinkkonen, P. Berninger, J. Lin, T. Sauter, and M. Heinäniemi. 2014. "Integrated Analysis of Transcript-Level Regulation of Metabolism Reveals Disease-Relevant Nodes of the Human Metabolic Network." *Nucleic Acids Research* 42 (3): 1474–1496.

Gille, C. et al. 2010. "HepatoNet1: A Comprehensive Metabolic Reconstruction of the Human Hepatocyte for the Analysis of Liver Physiology." *Molecular Systems Biology* 6: 411.

Gonçalves, E., R. Pereira, I. Rocha, and M. Rocha. 2012. "Optimization Approaches for the in Silico Discovery of Optimal Targets for Gene Over/underexpression." *Journal of Computational Biology: A Journal of Computational Molecular Cell Biology* 19 (2): 102–114.

Guzmán, G. I., J. Utrilla, S. Nurk, E. Brunk, J. M. Monk, A. Ebrahim, B. O. Palsson, and A. M. Feist. 2015. "Model-Driven Discovery of Underground Metabolic Functions in *E. coli*." *Proceedings of the National Academy of Sciences of the United States of America* 112 (3): 929–934.

Heinemann, M., A. Kümmel, R. Ruinatscha, and S. Panke. 2005. In Silico Genome-Scale Reconstruction and Validation of the Staphylococcus Aureus Metabolic Network. *Biotechnology and Bioengineering* 92 (7): 850–864.

Hoops, S., S. Sahle, R. Gauges, C. Lee, J. Pahle, N. Simus, M. Singhal, L. Xu, P. Mendes, and U. Kummer. 2006. "COPASI—A Complex Pathway Simulator." *Bioinformatics* 22 (24): 3067–3074.

Huang, X., Y. -Jr Chen, K. Cho, I. Nikolskiy, P. A. Crawford, and G. J. Patti. 2014. "X^{13}CMS: Global Tracking of Isotopic Labels in Untargeted Metabolomics." *Analytical Chemistry* 86 (3): 1632–1639.

Huberts, D. H. E. W., B. Niebel, and M. Heinemann. 2012. "A Flux-Sensing Mechanism Could Regulate the Switch between Respiration and Fermentation." *FEMS Yeast Research* 12 (2): 118–128.

Hulit, J., A. Howell, R. Gandara, M. Sartini, H. Arafat, and G. Bevilacqua. 2013. "Creating a Tumor-Resistant Microenvironment." *Cell Cycle* 12 (3): 480–490.

Imam, S., S. Schäuble, A. N. Brooks, N. S. Baliga, and N. D. Price. 2015. "Data-Driven Integration of Genome-Scale Regulatory and Metabolic Network Models." *Frontiers in Microbiology* 6: 409.

Jensen, P. A., and J. A. Papin. 2011. "Functional Integration of a Metabolic Network Model and Expression Data without Arbitrary Thresholding." *Bioinformatics* 27 (4): 541–547.

Jerby, L., L. Wolf, C. Denkert, G. Y. Stein, M. Hilvo, M. Oresic, T. Geiger, and E. Ruppin. 2012. "Metabolic Associations of Reduced Proliferation and Oxidative Stress in Advanced Breast Cancer." *Cancer Research* 72 (22): 5712–5720.

Jerby, L., T. Shlomi, and E. Ruppin. 2010. "Computational Reconstruction of Tissue-specific Metabolic Models: Application to Human Liver Metabolism." *Molecular Systems Biology* 6 (1): 401.

Jouhten, P., E. Pitkänen, T. Pakula, M. Saloheimo, M. Penttilä, and H. Maaheimo. 2009. "13 C-Metabolic Flux Ratio and Novel Carbon Path Analyses Confirmed That Trichoderma Reesei Uses Primarily the Respirative Pathway Also on the Preferred Carbon Source Glucose." *BMC Systems Biology* 3 (1): 104.

Keseler, I. M., J. Collado-Vides, S. Gama-Castro, J. Ingraham, S. Paley, I. T. Paulsen, M. Peralta-Gil, and P. D. Karp. 2005. "EcoCyc: A Comprehensive Database Resource for *Escherichia Coli*." *Nucleic Acids Research* 33 (Database issue): D334–D337.

Kim, H. U. K., S. Young Kim, H. Jeong, T. Yong Kim, J. Jong Kim, H. E. Choy, K. Yang Yi, J. Haeng Rhee, and S. Yup Lee. 2011. "Integrative Genome-scale Metabolic Analysis of Vibrio Vulnificus for Drug Targeting and Discovery." *Molecular Systems Biology* 7 (1): 460.

Kim, J., and J. L. Reed. 2010. "OptORF: Optimal Metabolic and Regulatory Perturbations for Metabolic Engineering of Microbial Strains." *BMC Systems Biology* 4: 53.

Kumar, V. S., and C. D. Maranas. 2009. "GrowMatch: An Automated Method for Reconciling In Silico/In Vivo Growth Predictions." *PLoS Computational Biology* 5 (3): e1000308.

Kümmel, A., S. Panke, and M. Heinemann. 2006. "Putative Regulatory Sites Unraveled by Network-Embedded Thermodynamic Analysis of Metabolome Data." *Molecular Systems Biology* 2: 2006.0034.

Lazar, M. A., and M. J. Birnbaum. 2012. "De-Meaning of Metabolism." *Science* 336 (6089): 1651–1652.

Lee, J. W., D. Na, J. Myoung Park, J. Lee, S. Choi, and S. Yup Lee. 2012. "Systems Metabolic Engineering of Microorganisms for Natural and Non-Natural Chemicals." *Nature Chemical Biology* 8 (6): 536–546.

Lee, K. H., J. Hwan Park, T. Yong Kim, H. Uk Kim, and S. Yup Lee. 2007. "Systems Metabolic Engineering of *Escherichia Coli* for L-Threonine Production." *Molecular Systems Biology* 3: 149.

Lewis, N. E., H. Nagarajan, and B. O. Palsson. 2012. "Constraining the Metabolic Genotype–phenotype Relationship Using a Phylogeny of in Silico Methods." *Nature Reviews. Microbiology* 10 (4): 291–305.

Li, P., J. O. Dada, D. Jameson, I. Spasic, N. Swainston, K. Carroll, W. Dunn et al. 2010. "Systematic Integration of Experimental Data and Models in Systems Biology." *BMC Bioinformatics* 11: 582.

Marin de Mas I., V. A. Selivanov, S. Marin, J. Roca, M. Orešič, L. Agius, and M. Cascante. 2011. "Compartmentation of Glycogen Metabolism Revealed from 13 C Isotopologue Distributions." *BMC Systems Biology* 5 (1): 175.

Mardinoglu, A., and J. Nielsen. 2012. "Systems medicine and metabolic modelling." *Journal of Internal Medicine* 271 (2): 142–154.

Mardinoglu, A., F. Gatto, and J. Nielsen. 2013. "Genome-Scale Modeling of Human Metabolism—a Systems Biology Approach." *Biotechnology Journal* 8 (9): 985–996.

Matsuda, F., C. Furusawa, T. Kondo, J. Ishii, H. Shimizu, and A. Kondo. 2011. "Engineering Strategy of Yeast Metabolism for Higher Alcohol Production." *Microbial Cell Factories* 10: 70.

Monteiro, P. T. et al. 2008. "YEASTRACT-DISCOVERER: New Tools to Improve the Analysis of Transcriptional Regulatory Associations in *Saccharomyces Cerevisiae*." *Nucleic Acids Research* 36 (Database issue): D132–D136.

Murabito, E., E. Simeonidis, K. Smallbone, and J. Swinton. 2009. "Capturing the Essence of a Metabolic Network: A Flux Balance Analysis Approach." *Journal of Theoretical Biology* 260 (3): 445–452.

Nargund, S., and G. Sriram. 2014. "Mathematical Modeling of Isotope Labeling Experiments for Metabolic Flux Analysis." *Methods in Molecular Biology* 1083: 109–131.

Oberhardt, M. A., B. Ø. Palsson, and J. A. Papin. 2009. Applications of Genome-Scale Metabolic Reconstructions. *Molecular System Biology* 5: 320.

Oberhardt, M. A., J. B. Goldberg, M. Hogardt, and J. A. Papin. 2010. "Metabolic Network Analysis of Pseudomonas Aeruginosa during Chronic Cystic Fibrosis Lung Infection." *Journal of Bacteriology* 192 (20): 5534–5548.

Orth, J. D., I. Thiele, and B. Ø. Palsson. 2010. "What Is Flux Balance Analysis?" *Nature Biotechnology* 28 (3): 245–248.

Pál, C., B. Papp, M. J. Lercher, P. Csermely, S. G. Oliver, and L. D. Hurst. 2006. "Chance and Necessity in the Evolution of Minimal Metabolic Networks." *Nature* 440 (7084): 667–670.

Palsson, B., and K. Zengler. 2010. "The Challenges of Integrating Multi-Omic Data Sets." *Nature Chemical Biology* 6 (11): 787–789.

Palsson, B. 2002. "In Silico Biology Through 'omics.'" *Nature Biotechnology* 20 (7): 649–650.

Papp, B., R. A. Notebaart, and C. Pál. 2011. "Systems-Biology Approaches for Predicting Genomic Evolution." *Nature Reviews. Genetics* 12 (9): 591–602.

Park, J. M., T. Yong Kim, and S. Yup Lee. 2009. "Constraints-Based Genome-Scale Metabolic Simulation for Systems Metabolic Engineering." *Biotechnology Advances* 27 (6): 979–988.

Peinado, H., D. Olmeda, and A. Cano. 2007. "Snail, Zeb and bHLH factors in Tumour Progression: An Alliance against the Epithelial Phenotype?" *Nature Reviews Cancer* 7 (6): 415–428.

Pey, J., C. Theodoropoulos, A. Rezola, A. Rubio, M. Cascante, and F. J. Planes. 2011. "Do Elementary Flux Modes Combine Linearly at the 'atomic' Level? Integrating Tracer-Based Metabolomics Data and Elementary Flux Modes." *Bio Systems* 105 (2): 140–146.

Ranganathan, S., P. F. Suthers, and C. D. Maranas. 2010. "OptForce: An Optimization Procedure for Identifying All Genetic Manipulations Leading to Targeted Overproductions." *PLoS Computational Biology* 6 (4): e1000744.

Ren, S., B. Zeng, and X. Qian. 2013. "Adaptive Bi-Level Programming for Optimal Gene Knockouts for Targeted Overproduction under Phenotypic Constraints." *BMC Bioinformatics* 14 (Suppl 2): S17.

Resat, H., L. Petzold, and M. F. Pettigrew. 2009. "Kinetic Modeling of Biological Systems." *Methods in Molecular Biology* 541: 311–335.

Rojas, I., M. Golebiewski, R. Kania, O. Krebs, S. Mir, A. Weidemann, and U. Wittig. 2007. "SABIO-RK: A Database for Biochemical Reactions and Their Kinetics." *BMC Systems Biology* 1 (S–1): S6.

Schellenberger, J., and B. Ø. Palsson. 2009. "Use of Randomized Sampling for Analysis of Metabolic Networks." *The Journal of Biological Chemistry* 284 (9): 5457–5461.

Schmidt, B. J., A. Ebrahim, T. O. Metz, J. N. Adkins, B. Ø. Palsson, and D. R. Hyduke. 2013. "GIM3E: Condition-Specific Models of Cellular Metabolism Developed from Metabolomics and Expression Data." *Bioinformatics* 29 (22): 2900–2908.

Selivanov, V. A., P. Vizán, F. Mollinedo, T. W. M. Fan, P. W. N. Lee, and M. Cascante. 2010. "Edelfosine-Induced Metabolic Changes in Cancer Cells That Precede the Overproduction of Reactive Oxygen Species and Apoptosis." *BMC Systems Biology* 4: 135.

Shlomi, T., M. N. Cabili, M. J. Herrgård, B. Ø. Palsson, and E. Ruppin. 2008. "Network-Based Prediction of Human Tissue-Specific Metabolism." *Nature Biotechnology* 26 (9): 1003–1010.

Smallbone, K., E. Simeonidis, N. Swainston, and P. Mendes. 2010. "Towards a Genome-Scale Kinetic Model of Cellular Metabolism." *BMC Systems Biology* 4: 6.

Stanford, N. J., T. Lubitz, K. Smallbone, E. Klipp, P. Mendes, and W. Liebermeister. 2013. "Systematic Construction of Kinetic Models from Genome-Scale Metabolic Networks." *PloS One* 8 (11): e79195.

Stelling, J., S. Klamt, K. Bettenbrock, and S. Schuster. 2002. "Metabolic Network Structure Determines Key Aspects of Functionality and Regulation." *Nature.* 420 (6912): 190.

Thiele, I. et al. 2013. "A Community-Driven Global Reconstruction of Human Metabolism." *Nature Biotechnology* 31 (5): 419–425.

Van Rijsewijk, B. R. H., A. Nanchen, S. Nallet, R. J. Kleijn, and U. Sauer. 2011. "Large-Scale ^{13}C-Flux Analysis Reveals Distinct Transcriptional Control of Respiratory and Fermentative Metabolism in *Escherichia Coli*." *Molecular Systems Biology* 7 (1): 477.

Visser, D., and J. J. Heijnen. 2003. "Dynamic Simulation and Metabolic Re-Design of a Branched Pathway Using Linlog Kinetics." *Metabolic Engineering* 5 (3): 164–176.

Vizán, P., G. Alcarraz-Vizán, S. Díaz-Moralli, O. N. Solovjeva, W. M. Frederiks, and M. Cascante. 2009. "Modulation of Pentose Phosphate Pathway during Cell Cycle Progression in Human Colon Adenocarcinoma Cell Line HT29." *International Journal of Cancer.* 124 (12): 2789–2796.

Wang, Y., J. A. Eddy, and N. D. Price. 2012. "Reconstruction of Genome-Scale Metabolic Models for 126 Human Tissues Using mCADRE." *BMC Systems Biology* 6: 153.

Wiechert, W. 2001. "^{13}C Metabolic Flux Analysis." *Metabolic Engineering* 3 (3): 195–206.

Winter, G., and J. O. Krömer. 2013. "Fluxomics—Connecting 'omics Analysis and Phenotypes." *Environmental Microbiology* 15 (7): 1901–1916.

Ye, C., W. Zou, N. Xu, and L. Liu. 2014. "Metabolic Model Reconstruction and Analysis of an Artificial Microbial Ecosystem for Vitamin C Production." *Journal of Biotechnology* 182–183: 61–67.

Yim, H. et al. 2011. "Metabolic Engineering of *Escherichia Coli* for Direct Production of 1,4-Butanediol." *Nature Chemical Biology* 7 (7): 445–452.

Yizhak, K., T. Tuller, B. Papp, and E. Ruppin. 2011. "Metabolic Modeling of Endosymbiont Genome Reduction on a Temporal Scale." *Molecular Systems Biology* 7: 479.

10

Data Sharing and Standards

Reza Salek

CONTENTS

Introduction 235
Minimum Information Reporting 236
Metabolomics Data Sharing 239
Open Format Data Standards 240
  Data-Standardization Initiatives 240
  Data-Exchange Formats 240
  Open Access MS Formats 241
  Open Access NMR Formats 245
  Ontology Developments 246
MSI and Metabolite Identification 246
References 248

Introduction

This chapter provides an overview of current and past efforts in data sharing and data standardization in metabolomics. Reproducibility for metabolomics results is challenging and the topic is vast. Here we focus on best practices and existing community efforts in data handling. Several elements are needed to increase reproducibility. Firstly, a framework for reporting experimental details containing minimum information guidelines is needed. Secondly, the community needs to adhere to such guidelines (and encouraged) for sharing experimental data publically. The reporting framework should also be supported by the journals, reinforced by the reviewers and implemented by the data repositories. Data sharing should follow the standards framework, i.e., use an open access data format (when possible), enriched with metadata and containing the complete set of study files—not just the raw files, but also QC and QA samples, blanks, the run order, etcetera. It is equally important to include data processing parameters for the individual data analysis steps.

Minimum Information Reporting

Standardization efforts started in 2004 with the Architecture for Metabolomics consortium (ArMet) (Jenkins et al., 2004) followed closely by the Standard Metabolic Reporting Structure initiative (SMRS) (Lindon et al., 2005). Some of the standardization efforts were mainly technology-focused, e.g., NMR metabonomics, while others focused on plant-based standardization such as the ArMet effort (Lindon et al., 2006). During a Metabolomics Standards Workshop, held by the US National Institutes of Health, existing standards in metabolomics technologies, methods and data treatments were discussed, addressing the needs and challenges at the time, ultimately leading to the Metabolomics Standards Initiative (MSI) (Fiehn et al., 2006; Sansone et al., 2007) as part of the NIH Roadmap effort (Castle et al., 2006). The MSI published a set of influential manuscripts with recommendations on minimal reporting standards for various aspects of metabolomics. The availability of metabolomics MI or metadata about an experiment is crucial for reporting an experiment, allowing others to understand and to follow the study. MI facilitates understanding, assessing and ultimately reproducing experiments. It includes recommendations from study design to data analysis, a set of "best practices" and complete "reporting standards" which is community-agreed. Many of the MI guidelines were inspired by Minimum Information About a Microarray Experiment (MIAME) (Brazma et al., 2001).

Five working groups (WGs), consisting of experts, researches and industry in the field, got together to address different aspects and challenges in metabolomics: a biological context metadata WG, a chemical analysis WG, a data processing WG, an exchange format WG and an ontology WG. Their primary role was to provide engagement with the community and across communities (e.g., other initiatives such as PSI) on the usage, development and training on the promotion of standards as proposed by MSI guidelines. These working groups have led to significant results: Core Information for Metabolomics Reporting (CIMR): A set of documents was created by the joint WGs, specifying minimal guidelines for reporting metabolomics experiments. The minimum information (MI) reporting covered all application areas related to analysis technologies. Metabolomics Ontology: The ontology WG developed the controlled vocabulary or the ontology based on the CIMR guidelines (Sansone et al., 2007). The exchange format WG worked on data model and data exchange formats, making use of the MSI ontology (Rubtsov et al., 2007). The Biological context metadata WG, consisting of four subgroups, produced the metadata guidelines for four different biological experiment areas: *Mammalian/in vivo* (Griffin et al., 2007), *Microbial and in vitro* (van der Werf et al., 2007), *Plant* (Fiehn et al., 2007) and *Environmental* (Morrison et al., 2007). Additionally, the *Mammalian/in vivo* report subgroup formed two sets of reporting standards, one for the *Mammalian Clinical Trials* and the second one for *Human Studies* and *Pre-clinical* (Figure 10.1).

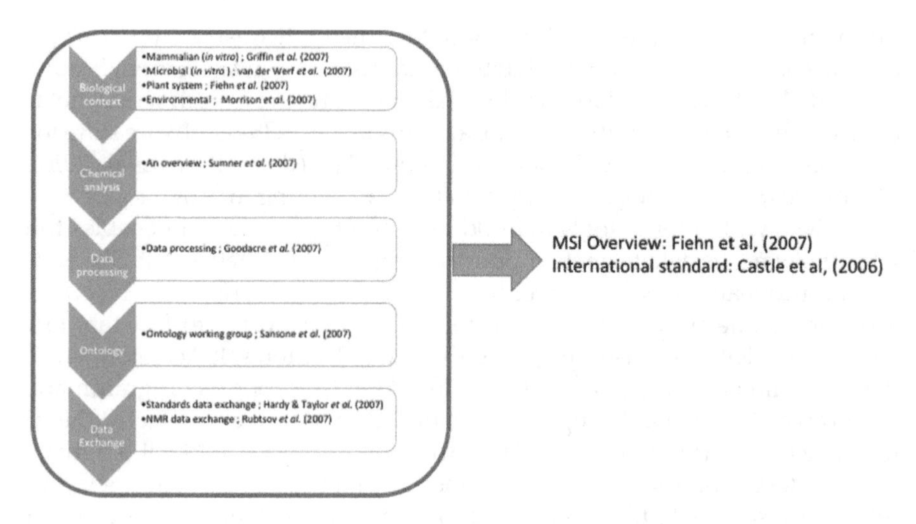

FIGURE 10.1
Relationship and related publication outcome form the five different Metabolomics Society Working Groups and their subgroups. The order is based on study design and not necessarily in the order of the work, as there was significant overlap between and across working groups.

Only a limited application for MI took place between 2007 and 2012 (see below), and journals did not enforce the MSI MI guidelines, nor was it a requirement from the reviewers to check. Some applications were: The Birmingham Metabolite Library, a publicly accessible 1D and 2D NMR reference database of reference metabolite standards (Ludwig et al., 2012), conforming to many of the MSI NMR-based guidelines. The PlantMetabolomics.org, a web portal and database for exploring, visualizing, and downloading mainly GC-MS based plant metabolomics data (Bais et al., 2010) used MS Excel templates for submission and capturing metadata based on MSI guidelines. The MeRy-B, a database for plant NMR-based metabolomics profiling used partial compliance with the MSI guidelines to represent their data sets (Ferry-Dumazet et al., 2011). Finally, Setup-X, a public study design database for metabolomics projects enabled investigators to detail and set up biological experiments and workflows for metabolomics studies (Scholz and Fiehn, 2006).

There were few publications on metadata descriptions for experimental data (see, e.g., [Griffin et al., 2011]). However, most early implementations were focused on one particular technology or limited to a particular species or a particular type of analytical technique. One primary adopter of the MSI reporting (partial implementation) was the EMBL-EBI MetaboLights (http://www.ebi.ac.uk/metabolights). MetaboLights is the first general-purpose database for metabolomics experiments and derived data (Haug et al., 2013). MetaboLights is a curated resource, capturing experimental details, protocols and identified metabolites, as well as a "reference" layer that includes information about individual metabolites, their chemistry and biological roles (Salek et al., 2013a). MetaboLights uses the Investigation/Study/Assay (ISA) tab-delimited format (ISA-Tab) for capturing experimental metadata, tagging metadata with ontological terms using an ISA software suite for curation of standards-compliant experimental data (Rocca-Serra et al., 2010). The ISA-team in 2017 released a Python ISA API, aiming to provide software developers with a set of tools to build ISA objects. The ISA API enables the import of various data formats supporting import of ISA-Tab, ISA JSON, SRA XML (European Nucleotide Archive), Metabolomics Workbench, Biocrates XML and some support for mzML formats, and export it to ISA-Tab, ISA JSON and SRA XML (see https://github.com/ISA-tools/ISAcreator/wiki/API). Beyond enabling I/O of data, the ISA API also supports programmatic creation of ISA content through the Python ISA model objects directly, therefore able to export ISA content which currently is in place at MetaboLights.

The Metabolomics Workbench, another metabolomics data sharing resource, also adopted MSI reporting for capturing metabolomics metadata using Microsoft Excel-based templates or "mwTab" text files and covering a variety of different metabolomics study types. Users capture experimental metadata using the templates and submit the data as well as the metadata including, ideally, both raw and processed experimental data (Sud et al., 2016). MetaboLights, compared to Metabolomics Workbench, has a greater focus on curation efforts linking ontology terms with metadata and has somewhat stricter guidelines (Salek et al., 2013a).

Two other repositories also partially fulfill the MSI guidelines for minimum metadata reporting, MeRy-B and MetaPhen. The Metabolomics Repository Bordeaux (MeRy-B) is a repository mainly for plant metabolomics, storing 1H-NMR and GC-MS data sets (Deborde and Jacob, 2014). The Metabolic Phenotype Database (MetaPhen) (Carroll et al., 2015) is part of the MetabolomeExpress (Carroll et al., 2010), a mainly plant-oriented GC-MS metabolomics data analysis platform with metadata that complies with MSI. MetabolomeExpress is a web server providing processing, statistical analysis, visualization and data storage, while MetaPhen uses statistical similarity matching of metabolic phenotypes for high-confidence detection of functional links within data sets (Carroll et al., 2010, 2015).

An analysis of the publicly available metabolomics metadata showed that despite the progress, full compliance with MSI guidelines is not abided in the repositories mentioned above (Spicer et al., 2017b). Fully MSI-compliant metadata reporting requirement for a repository might make the job of capturing and submitting metadata tedious, therefore limiting its usability (the job of submitting data become too tedious without any automation). Using a subset of "selected" metadata could be a more pragmatic consideration, trying to strike a balance between the effort to add the required meta-information and the time needed to "fulfil it." Bear in mind that publication can cover many descriptive elements hence no need to be repeated. There are few ongoing efforts to address the lack of automation in data submission, for example, a set of Python tools to generate ISA-Tab files, using mzML and nmrML raw data (Larralde et al., 2017) and a set of Python library for access and deposition of data to the Metabolomics Workbench repository (Smelter and Moseley, 2018). Metabolomics as a field has matured since the definition of the MSI guidelines and has accumulated a greater understanding of the techniques, applications and biological backgrounds. A revise of the MI reporting guidelines to adapt and comply with current community needs, as well as those of the near future has been suggested and is ongoing (Spicer et al., 2017a).

Metabolomics Data Sharing

There are now several resources that collect on host metabolomics data sets. Some examples are the EMBL-EBI MetaboLights, the first open access data repository (Haug et al., 2013). MetaboLights generates a stable unique identifier (e.g., MTBLS1) that can be used as a unique identifier of a study for journal publication, referencing the findings and sharing the data sets associated with a publication (Kale et al., 2016). Making metabolomics experimental data publicly available allows peer-reviewed publications to better justify their findings, more opportunities for collaboration and, ultimately, higher impact, visibility and increased citation ("Data producers deserve citation credit," 2009). In the United States, the National Institute of Health (NIH) funded the Metabolomics Workbench as part of the data coordinating the effort, this being the NIH Common Fund Metabolomics Program. The Metabolomics Workbench provides data from the Common Fund Metabolomics Resource Cores, metabolite standards and analysis tools to the wider metabolomics community (Sud et al., 2016) in a way similar to Metabolights. Other resources for metabolomics data are Metabolic Phenotype Database (MetaPhen; https://www.metabolome-express.org/phenometer.php) part of the MetabolomeExpress for storing plant GC-MS metabolomics data (Carroll et al., 2015), MeRy-B, a web knowledge base for the storage, visualization, analysis and annotation of plant NMR metabolomic profiles (Ferry-Dumazet et al., 2011) and Global Natural Products Social Molecular Networking (GNPS; http://gnps.ucsd.edu) (Wang et al., 2016). To help better find metabolomics data, extra resources were set up primarily acting as data aggregators such as MetabolomeXchange (http://www.metabolomexchange.org/) or to discover Metabolomics and its link to other omics data such as Omics Discovery Index (OmicsDI) (Perez-Riverol et al., 2016). The OmicsDI provides a knowledge discovery framework across heterogeneous omics data (genomics, proteomics, transcriptomics and metabolomics) and linking omics public data. OmicsDI uses and links public metabolomics data from GNPS, MetaboLights, MetabolomeExpress and MetabolomicsWorkbench.

Open Format Data Standards

Data-Standardization Initiatives

After several years of slow and scattered progress, the COSMOS project (COordination Of Standards In MetabOlomicS, http://cosmos-fp7.eu) (Salek et al., 2015b) was set up to coordinate data standards among database providers, ontologists, software engineers, journals and instrument vendors working toward open access data standardization. The COSMOS initiative brought together metabolomics and bioinformatics experts worldwide to work standards on open data-exchange formats and data semantics to maximize -omics data interoperability ("It's not about the data," 2012). Following the COSMOS initiative, the Metabolomics Society Data Standards Task Group was established to coordinate efforts to the broader scientific community and to disseminate ongoing efforts on efficient storage data formats, compression methods, terminological annotation and ontology developments within the metabolomics community (Salek et al., 2015a). This task group also linked its efforts with the Human Proteome Organization (HUPO)—Proteomics Standards Initiative (PSI) (http://www.psidev.info) (Mayer et al., 2013) to harmonize its effort across related disciplines using the same MS based technology. Additionally, the Task Group aim was to coordinate the effort with leading vendors, researchers and bioinformaticians, journals and the MSI international communities to develop, support and adopt such open source data-/metadata-exchange formats (Salek et al., 2015a). Recently, the FAIR data movement has gained considerable momentum in Europe, FAIR standing for data being Findable, Accessible, Interoperable and Reusable (Wilkinson et al., 2016). FAIR principles, set forward by stakeholders from academia, industry, funding agencies, and scholarly publishers, aim to improve the infrastructure supporting the reuse of scholarly data (Wilkinson et al., 2016). They put specific emphasis on enhancing the ability of machines to automatically find and use the data, in addition to supporting its reuse by individuals (van Rijswijk et al., 2017) (Figure 10.2).

Data-Exchange Formats

A wide range of different mass spectrometers is used in metabolomics each with a unique design, specifications, performance and data acquisition system. The binary data file formats produced by each mass spectrometer vendor also have a different data type, usually, proprietary, divers and nontransparent making it difficult for bioinformatician to access. To ensure long-term sustainability and to cope with advances in metabolomics analytical technologies, both for mass spectrometry and NMR-based techniques, the need for a standard output format that is vendor-neutral and facilitates data sharing and dissemination, interpretation and data analysis algorithms

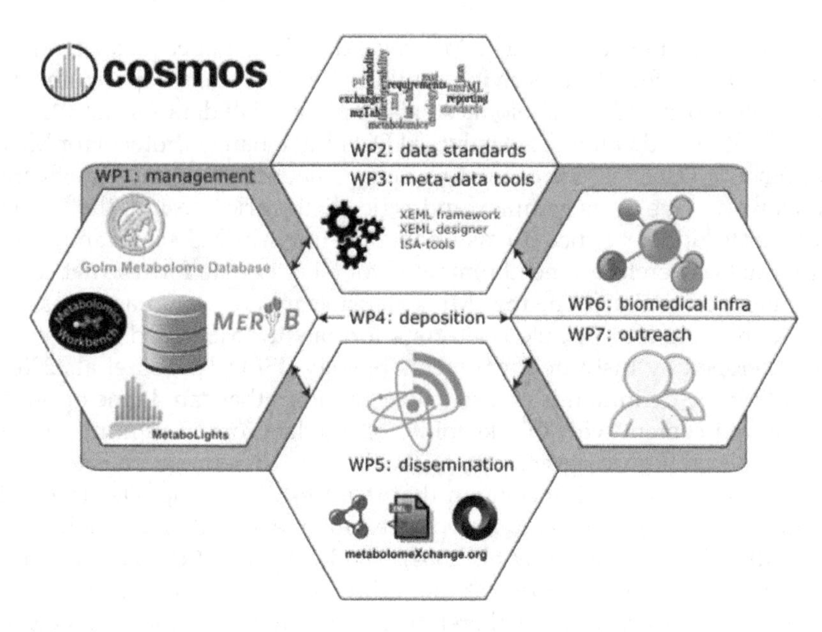

FIGURE 10.2
Overview of the COSMOS initiative activities, within various Work Packages on data and metadata standardization, with links to relevant infrastructures and databases, tools development and outreach activities summarized.

development is paramount. Data standards promote open source software development, open access to data and knowledge, potentially improving scientific reproducibility and quality (Rocca-Serra et al., 2016). Open data has the added advantage of allowing for meta-analysis and re-analysis of data using open access tools and across different vendors and data types (Boulton, 2016). Many such data formats already exist, with different underlying philosophies for mass spectrometry mainly developed by the HUPO-PSI community (Orchard et al., 2003), which also has been used in metabolomics due to similarly of the underlying technology and its application. It is also essential to have the controlled vocabulary and terminology artifacts needed for each domain (i.e., metabolomics) with validation tools to ensure semantic validity, future compatibility, improved support and adoption for the data formats by the community. Below are some of the most common data formats.

Open Access MS Formats

As MS technology is continually evolving with new techniques, new instrument types and new data types, there is a great need for a standard output format that can be shared and analyzed across different instrument vendors (data exchange) (Rocca-Serra et al., 2016). The different proprietary binary data formats produced by individual MS vendors require different sets of software libraries to be able to read and access the data. A standard open format can remedy this. For MS, several different open data standard formats with different underlying philosophies for mass spectrometry-based proteomics and metabolomics exist. The netCDF data format (ASTM E2078-00 "Standard Guide for Analytical Data Interchange Protocol for Mass Spectrometric Data") is one of the first vendor-neutral MS data formats created. However due to its complex and rigid dictionaries design, the format is difficult to update; hence it was unable to store MS/MS scans and other complex data. Therefore, a need remained for a better data format that could be maintained regularly. The mzXML (extensible markup language) was one of the pioneering and flexible open data formats for MS-based proteomics data, developed by Institute for Systems Biology (ISB) (Pedrioli et al., 2004). The mzXML by definition is an extensible language that can define optional and required content with the flexibility of the data format in mind, under an open source license agreement, to facilitate its dissemination, modification and upgrades. The information that requires regular updates (e.g., MS instrument models) have been incorporated in an external ontology instead of mzXML schema (Pedrioli et al., 2004). The XML cannot directly incorporate binary data and converting it into a human readable text would have increased the size significantly; therefore, the m/z intensity binary pairs were encoded in base64, somewhat larger (1.3-fold) than the original binary representation (Pedrioli et al., 2004). Also, since XML adds metadata to the existing data, it also increases the size of the data files. Note that the mzXML does not capture the "original raw binary data" nor the experimental design, therefore might not be sufficient for regular data sharing submission or to fulfill specific computational needs (Pedrioli et al., 2004). Later, the HUPO-PSI community developed a standard incorporating many of the desirable attributes from the previous data formats, adding improvements, controlled vocabulary and validation tools (Lin et al., 2005). The HUPO-PSI mzData format could capture and interchange peak list information, support control vocabulary, collect experimental metadata and store it on the description level, while peak measurements were stored in the acquisition list (Orchard et al., 2005). A number of converters to export data in mzData format from main instrument manufacturers, such as Shimatzu and Bruker, as well as tools such as MASCOT by Matrix Science were developed (Koenig et al., 2008) supported this data format (Orchard et al., 2005). The mzData format philosophy was to be quite flexible, extensively relying on controlled vocabulary for future support (different tools representing data had different dialects), while mzXML in comparison had a more strict schema making any updates incompatible with older versions (Deutsch, 2010b). Still, neither of the two data formats completely fulfilled the need of the community, creating unnecessary confusion in the community and extra complexity for the software development to support both formats (Orchard et al., 2006). As a result, the proteomics community under HUPO-PSI introduced the dataXML in 2006 which a year later was renamed to mzML format, expected to replace mzXML and mzData, but not presumed to replace binary vendor formats completely.

The mzML format was submitted to the PSI document process in November 2007 (version 0.91) wherein it passed through formal internal and then community review and eventually published as version 1.0 in 2008 and 1.1 in 2009 (Deutsch, 2010a). The mzXML is an XML-based format that has an optional indexing scheme allowing random access to spectra inside the XML document, which has been stable at version 1.1 since June 2009 (Martens et al., 2011; Turewicz and Deutsch, 2011). The mzML development had several aims: to keep the format simple, to eliminate alternate ways of encoding the same information, to provide flexibility for encoding new information, and to support the major features of mzData and mzXML. The new format was opened for validation by the community by providing software to read and write the format before its release (Deutsch, 2010b). Version 1.2 of mzML is scheduled to be completed and released in 2018/2019. To ensure consistent usage of the format, improved support for selected reaction monitoring data was implemented to facilitate rapid adoption by the community (Qi et al., 2015). As a result, mzML is a well-tested open-source format for mass-spectrometry output files that can be readily utilized by the community and easily adapted to advances in mass spectrometry technology (Figure 10.3).

FIGURE 10.3
Overview and inside of an mzML file example with the middle segment removed (...) for display purposes. The main part of the mzML document is contained within the <mzML> </mzML> tags, wrapped within an <indexedmzML></indexedmzML> construct, which contains the random access index at the bottom.

Another format with a robust inherent application for MS-based metabolomics (developed for both proteomics, metabolomics and lipidomics) is the mzTab data-exchange format (Griss et al., 2014). The mzTab format facilitates reporting qualitative and semi-quantitative results for proteomics and metabolomics experiments in a flexible and straightforward tab-delimited format, easily viewed by any text editor or spreadsheet program while being lightweight compared to XML-based standard file formats (Griss et al., 2014). The mzTab file can be used for reporting protein, peptide, and metabolite identification as well as experimental metadata. It provides a mechanism to capture results at different levels, ranging from simple summaries to complete evidence-based representations of the results regarding evidence available for metabolite/protein identifications (such as grouping, ms/ms, reference spectra or other evidence used for identification). The mzTab is not intended to store an experiment's complete data/evidence but only its final reported results. The mzTab files can be validated with jmzTab (https://github.com/PRIDE-Utilities/jmzTab), an open-source Java application designed to be embedded in other software packages and providing a flexible framework for maintaining the logical integrity between the metadata and the table-based sections inside the mzTab files (Griss et al., 2014). Since 2017, new efforts are undertaken jointly by the MSI and PSI communities to develop mzTab further and to update its schema. Due to the complex nature of the metabolomics and proteomics experiments and their rapidly evolving technologies, it is no longer possible to have a single flavor of mzTab. In order to better support both sciences, mzTab will evolve into two different versions: one for metabolomics (mzTab-M, and lipidomics) and another for proteomics (mzTab-P), with a common core but separate validators for each platform (see https://github.com/HUPO-PSI/mzTab). The mzTab files can generate results in two ways: "Summary" mode, and "Complete" mode. In "Summary" full results only the final data for the experimental conditions analyzed must be present while in the "Complete" mode, all the results per assay/replicate need to be detailed. There are four data tables that in a single complete mzTab file, which build on each other in the order of the metadata table (MTD), the Small Molecule table (SML), the Small Molecule Feature table (SMF) and the Small Molecule Evidence table (SME). The MTD and SML tables are mandatory, and for a file to contain any evidence about how molecules were quantified or identified by software, then all four tables will be present (see formal mzTab-M specification document https://github.com/HUPO-PSI/mzTab).

The qcML data format in another HUPO-PSI developed standards for metabolomics and proteomics MS-based quality control metrics (Walzer et al., 2014). In the 2016 HUPO-PSI meeting, a new working group was established aiming to provide a community-driven standardized file format for quality control based on the proposed qcML format (Walzer et al., 2014), supporting a variety of use cases for both proteomics and metabolomics applications (Bittremieux et al., 2017). This format will be developed in JSON (JavaScript Object Notation) taking advantage of its API usage and convenient of existing parsers in many programming languages away from historical XML files. In parallel, the CV requirement is undergoing as MIAPE-QC (https://github.com/HUPO-PSI/qcML-development).

For MS imaging data sets, an open access data standards format known as imzML (www.imzml.org) exists. The imzML is a flexible data exchange and data processing format for imaging data produced from different instrument vendor file formats. The imzML is not just limited to MS imaging, but is also useful for other MS applications such as LC-FTMS (Schramm et al., 2012). The imzML consists of two different distinctive files, linked by a universally unique identifier (UUID), where the experimental details are stored in an XML file, described using control vocabulary along with raw data in a binary format (separate spectral data is "only" for performance reasons) (Schramm et al., 2012). The imzML format has similar structure to the HUPO-PSI format mzML (Martens et al., 2011) with metadata based on the mzML format, hence back and forth conversion compatibility. Like other open access formats by using imzML result from different instrument types for MS imaging data, can be displayed and compared within a single software package without the need for a vendor software.

Open Access NMR Formats

The most widely used open source export format for NMR is the Joint Committee on Atomic and Molecular Physical Data (JCAMP-DX, currently at version 6.0). JCAMP originally developed for exchange of infrared spectra and related chemical and physical information between spectrometer across different manufacture (McDonald and Wilks, 1988). The JCAMP-DX could also accommodate Raman, UV, NMR and other types of spectra (X-ray powder patterns, chromatograms, etc.), which required representation of both contours as well as peak position and intensity. However JCAMP did not possess a flexible hierarchical structure and is not easily extendable to capture supplementary information. Such extensibility, as shown in MS-based XML formats, is a highly desired feature of a standard in the fast-paced and dynamic world of NMR metabolomics data-capture. Earlier in 2007, the MSI workgroups proposed detailed suggestions about the metadata to be captured for NMR-based experiments (Rubtsov et al., 2007). In particular, the MSI put forth recommendations to report instrument descriptions and configurations, instrument-specific sample preparation and data acquisition parameters related to NMR (Rubtsov et al., 2007). These efforts resulted in the first round of NMR XML data standard development, focusing on raw and processed one- and two-dimensional NMR experiments and associated metadata. The BML-NMR repository of metabolite spectra has already implemented these suggestions with relatively minimal (and fully documented) modifications. A parallel MSI-approved effort (http://www.metabolomicscentre.ca/exchangeformats) also provided an open XML encoding but focused on instrument raw data (FID, processed spectrum XY graph) in addition to instrument metadata (acquisition frequency, acquisition delay, etc.).

This and other earlier work on open access XML based NMR data were later joined together as part of the COSMOS initiative, which eventually led to the development of the nmrML data format (Schober et al., 2018). The nmrML is an open XML-based exchange for storage of NMR spectral raw data, data-acquisition parameters and metadata such as chemical structures associated with spectral assignments. Several converters and validators for the nmrML for Bruker, Agilent/Varian and JOEL vendor formats were developed by the COSMOS consortium and endorsed by the MSI with wide and growing support from NMR data analysis tools developers (Schober et al., 2018).

Ontology Developments

The field of semantics controlled vocabulary (ontologies) has progressed rapidly over the past decade and since early 2007 proposals, with many interoperable tools, infrastructure for support, maintenance and release of controlled vocabularies now available providing resolvable URI for annotation terms. For metabolomics, the Metabolomics Standards Initiative Ontology (MSIO) (https://github.com/MSI-Metabolomics-Standards-Initiative/MSIO), an application ontology aggregating key semantic descriptors from several resources, covering instrument descriptions and settings, statistical methods, study design, etc. could be an effective strategy. Of particular relevance is its coverage of terminology for consistent description QA/QC standards. Other relevant ontologies for metabolites are: the Chemical Entities of Biological Interest ontology (ChEBI) (Hastings et al., 2013), Ontology for Biomedical Investigations (OBI) (Bandrowski et al., 2016), LipidMAPS (Fahy et al., 2007), HMDB (Wishart et al., 2018) with server tools developed to take advantage of such resources for compounds classification such as ClassyFire (Djoumbou Feunang et al., 2016) and MS-FINDER for structure elucidation (Tsugawa et al., 2016). The chemical ontology is not only useful for the known compounds but also are used with for unknown structures when the structure is partially known or if the class of the compound is known.

MSI and Metabolite Identification

Identification of metabolites remains one of the bottlenecks in data analysis, as only limited structural information can be obtained from the features particularly in untargeted metabolomics. Metabolite annotation is a process by which a tentative metabolite candidate is assigned to a signal based on matching a mass with a database or library entries. In a target metabolomics approach or in the case of the GC-MS analysis, metabolite identification is more straightforward primarily due to available of large commercial libraries (e.g., the NIST 2017 reference library). Most compound identifications are based on properties data such as *m/z* values and retention time using similarity to public or commercial spectral libraries. To address ambiguity and to facilitate confidence in metabolite identification, the MSI metabolite identification WG proposed four different metabolite identification levels (Sumner et al., 2007). For a metabolite to be identified as Level 1, it must be compared to an authentic chemical standard analyzed in the same laboratory, using the same analytical techniques as the experimental data; the comparison is based on two orthogonal molecular properties. Often, pure standards are not available, and annotations can only be provided as putative annotations (level 2), or features can be assigned to putatively characterized compound classes (level 3). If the metabolite remains unknown it would be assigned as a level 4 annotation (Salek et al., 2013b). The classification of MSI levels lacks the flexibility to separate many cases with similar but not unique levels of identification (e.g., isomers), and some researchers have suggested sub-levels as a way to provide more distinctions in metabolite identification (Creek et al., 2014). One such approach was suggested by Schymanski et al. (2014). Their level 5 indicates that no unequivocal information about the structure or molecular formula is available; level 4 means that an unambiguous molecular formula has been obtained using the spectral information (but there is insufficient information to propose possible structures); level 3 indicates that information is sufficient to suggest tentative candidates with evidence for possible structures, but it is insufficient to decide which one exactly. The additional levels indicating a structural identification are level 2a—library spectrum match (the spectrum-structure matching is unambiguous) and level 2b—no other structure fits the experimental information, but no standard or literature information is available for confirmation either. Finally, the ideal situation, where the proposed structure has been confirmed using a reference standard with MS, MS/MS and retention time matching correspond to level 1. It is important that publications do describe the (un-)certainty at which metabolites underlying the findings are described, and it is no shortcoming if not all are identified at level 1.

For NMR spectroscopy, metabolite identification based on the original proposed 4 levels of the MSI guidelines is limited due to the fact that NMR peaks can often be identified (where not too many peaks overlap) using database matching to authentic compounds acquired at the same NMR spectrometer field frequency and peak deconvolution methods (Everett, 2015). Additionally for identification of metabolites based on NMR, seven recommendations for confidence in identification of known metabolites have been suggested, based on comparison with a reference database (Everett, 2015). The recommendations are:

1. The metabolite identification of carbon efficiency (MICE) obtained in an experiment ideally should be ≥1, or the metabolite identification efficiency (MIE) >0.5.
2. The experimental data should fit to reference data spectra within ±0.03 ppm for 1H, and ±0.5 ppm for 13C NMR shifts and ±0.2 Hz for proton (also depending on the field strength).
3. NMR spectral data or resonance should cover of all parts of the molecule.
4. With sufficient (see Chapter 5) signal-to-noise ratio and resolution.
5. Assigning signals in crowded spectral regions needs extra care.
6. 2D NMR data (HSQC) are ideal for orthogonal confirmation.
7. Identifications should be corroborated by HMBC data whenever possible (Dona et al., 2016; Everett, 2015).

It is clear that both for MS and NMR metabolite identification, high-quality curated resources for reference compounds are crucial. The MetaboLights reference layer holds such information about individual metabolites, their chemistry and their spectral data, both for MS and NMR detection (Kale et al., 2016). The Metabolomics Society Data Standards (Salek et al., 2015a), Metabolite Identification (Creek et al., 2014) and Data Quality Assurance (Dunn et al., 2017) Task Groups are working to ensure standards meeting future challenges in metabolomics are being developed.

References

Bais, P. et al. 2010. PlantMetabolomics.org: A Web Portal for Plant Metabolomics Experiments. *Plant Physiology* 152 (4): 1807–1816.

Bandrowski, A. et al. 2016. The Ontology for Biomedical Investigations. *PLoS One* 11 (4): e0154556.

Bittremieux, W., M. Walzer, S. Tenzer, W. Zhu, R. M. Salek, M. Eisenacher, and D. L. Tabb. 2017. The Human Proteome Organization-Proteomics Standards Initiative Quality Control Working Group: Making Quality Control More Accessible for Biological Mass Spectrometry. *Analytical Chemistry* 89 (8): 4474–4479.

Boulton, G. 2016. Reproducibility: International Accord on Open Data. *Nature* 530 (7590): 281.

Brazma, A. et al. 2001. Minimum Information About a Microarray Experiment (MIAME)—Toward Standards for Microarray Data. *Nature Genetics* 29: 365.

Carroll, A. J., M. R. Badger, and M. A. Harvey. 2010. The MetabolomeExpress Project: Enabling Web-Based Processing, Analysis and Transparent Dissemination of GC/MS Metabolomics Datasets. *BMC Bioinformatics* 11: 376.

Carroll, A. J., P. Zhang, L. Whitehead, S. Kaines, G. Tcherkez, and M. R. Badger. 2015. PhenoMeter: A Metabolome Database Search Tool Using Statistical Similarity Matching of Metabolic Phenotypes for High-Confidence Detection of Functional Links. *Frontiers in Bioengineering and Biotechnology* 3: 106.

Castle, A. L., O. Fiehn, R. Kaddurah-Daouk, and J. C. Lindon. 2006. Metabolomics Standards Workshop and the Development of International Standards for Reporting Metabolomics Experimental Results. *Briefings in Bioinformatics* 7 (2): 159–165.

Creek, D. J. et al. 2014. Metabolite Identification: Are You Sure? And How do Your Peers Gauge Your Confidence? *Metabolomics: Official Journal of the Metabolomic Society* 10 (3): 350–353.

Data producers deserve citation credit. 2009. *Nature Genetics* 41 (10): 1045.

Deborde, C., and D. Jacob. 2014. MeRy-B, a Metabolomic Database and Knowledge Base for Exploring Plant Primary Metabolism. In *Plant Metabolism: Methods and Protocols*, edited by G. Sriram. Totowa, NJ: Humana Press, pp. 3–16.

Deutsch, E. W. 2010a. Mass Spectrometer Output File Format mzML. *Methods in Molecular Biology* 604: 319–331.

Deutsch, E. W. 2010b. The Peptideatlas Project. *Methods in Molecular Biology* 604: 285–296.

Djoumbou Feunang, Y. et al. 2016. ClassyFire: Automated Chemical Classification with a Comprehensive, Computable Taxonomy. *Journal of Cheminformatics* 8 (1): 61.

Dona, A. C., M. Kyriakides, F. Scott, E. A. Shephard, D. Varshavi, K. Veselkov, and J. R. Everett. 2016. A Guide to the Identification of Metabolites in NMR-based Metabonomics/Metabolomics Experiments. *Computational and Structural Biotechnology Journal* 14: 135–153.

Dunn, W. B., D. I. Broadhurst, A. Edison, C. Guillou, M. R. Viant, D. W. Bearden, and R. D. Beger. 2017. Quality Assurance and Quality Control Processes: Summary of a Metabolomics Community Questionnaire. *Metabolomics: Official Journal of the Metabolomic Society* 13 (5): 50.

Everett, J. R. 2015. A New Paradigm for Known Metabolite Identification in Metabonomics/Metabolomics: Metabolite Identification Efficiency. *Computational and Structural Biotechnology Journal* 13: 131–144.

Fahy, E., M. Sud, D. Cotter, and S. Subramaniam. 2007. LIPID MAPS Online Tools for Lipid Research. *Nucleic Acids Research*, 35 (Web Server issue): W606–W612.

Ferry-Dumazet, H. et al. 2011. MeRy-B: A Web Knowledgebase for the Storage, Visualization, Analysis and Annotation of Plant NMR Metabolomic Profiles. *BMC Plant Biology* 11: 104.

Fiehn, O. et al. 2006. Establishing Reporting Standards for Metabolomic and Metabonomic Studies: A Call for Participation. *OMICS: A Journal of Integrative Biology* 10 (2): 158–163.

Fiehn, O. et al. 2007. Minimum Reporting Standards for Plant Biology Context Information in Metabolomic Studies. *Metabolomics: Official Journal of the Metabolomic Society* 3 (3): 195–201.

Griffin, J. L., H. J. Atherton, C. Steinbeck, and R. M. Salek. 2011. A Metadata Description of the Data in "A Metabolomic Comparison of Urinary Changes in Type 2 Diabetes in Mouse, Rat, and Human." *BMC Research Notes* 4: 272.

Griffin, J. L. et al. 2007. Standard Reporting Requirements for Biological Samples in Metabolomics Experiments: Mammalian/*In Vivo* Experiments. *Metabolomics: Official Journal of the Metabolomic Society* 3 (3): 179–188.

Griss, J. et al. 2014. The mzTab Data Exchange Format: Communicating Mass-Spectrometry-Based Proteomics and Metabolomics Experimental Results to a Wider Audience. *Molecular & Cellular Proteomics: MCP* 13 (10): 2765–2775.

Hastings, J. et al. 2013. The ChEBI Reference Database and Ontology for Biologically Relevant Chemistry: Enhancements for 2013. *Nucleic Acids Research* 41 (Database issue): D456–D463.

Haug, K. et al. 2013. MetaboLights—An Open-Access General-Purpose Repository for Metabolomics Studies and Associated Meta-Data. *Nucleic Acids Research* 41(Database issue): D781–D786.

It's not about the data. 2012. *Nature Genetics* 44 (2): 111.

Jenkins, H. et al. 2004. A Proposed Framework for the Description of Plant Metabolomics Experiments and Their Results. *Nature Biotechnology* 22 (12): 1601–1606.

Kale, N. S. et al. 2016. MetaboLights: An Open-Access Database Repository for Metabolomics Data. *Current Protocols in Bioinformatics* 53: 14–13.

Koenig, T. et al. 2008. Robust Prediction of the MASCOT Score for an Improved Quality Assessment in Mass Spectrometric Proteomics. *Journal of Proteome Research* 7 (9): 3708–3717.

Larralde, M. et al. 2017. mzML2ISA & nmrML2ISA: Generating Enriched ISA-Tab Metadata Files from Metabolomics XML Data. *Bioinformatics* 33 (16): 2598–2600.

Lindon, J. C., E. Holmes, and J. K. Nicholson. 2006. Metabonomics Techniques and Applications to Pharmaceutical Research & Development. *Pharmaceutical Research*, 23 (6): 1075–1088.

Lindon, J. C. et al. Standard Metabolic Reporting Structures working group. 2005. Summary Recommendations for Standardization and Reporting of Metabolic Analyses. *Nature Biotechnology*, 23 (7): 833–838.

Lin, S. M., L. Zhu, A. Q. Winter, M. Sasinowski, and W. A. Kibbe. 2005. What is mzXML Good for? *Expert Review of Proteomics* 2 (6): 839–845.

Ludwig, C. et al. 2012. Birmingham Metabolite Library: A Publicly Accessible Database of 1-D 1H and 2-D 1H J-resolved NMR Spectra of Authentic Metabolite Standards (BML-NMR). *Metabolomics: Official Journal of the Metabolomic Society* 8 (1): 8–18.

Martens, L. et al. 2011. mzML—A Community Standard for Mass Spectrometry Data. *Molecular & Cellular Proteomics: MCP* 10 (1): R110.000133.

Mayer, G. et al. HUPO-PSI Group. 2013. The HUPO Proteomics Standards Initiative-Mass Spectrometry Controlled Vocabulary. *Database: The Journal of Biological Databases and Curation* 2013: bat009.

McDonald, R. S., and P. A. Wilks. 1988. JCAMP-DX: A Standard form for Exchange of Infrared Spectra in Computer Readable form. *Applied Spectroscopy* 42 (1): 151–162.

Morrison, N. et al. 2007. Standard Reporting Requirements for Biological Samples in Metabolomics Experiments: Environmental Context. *Metabolomics: Official Journal of the Metabolomic Society* 3 (3): 203–210.

Orchard, S. et al. 2006. Proteomics and Beyond: A Report on the 3rd Annual Spring Workshop of the HUPO-PSI 21–23 April 2006, San Francisco, CA, USA. *Proteomics* 6 (16): 4439–4443.

Orchard, S., H. Hermjakob, and R. Apweiler. 2003. The Proteomics Standards Initiative. *Proteomics*, 3 (7): 1374–1376.

Orchard, S. et al. 2005. Further Steps Towards Data Standardisation: The Proteomic Standards Initiative HUPO 3(rd) Annual Congress, Beijing 25–27th October, 2004. *Proteomics* 5 (2): 337–339.

Pedrioli, P. G. A. et al. 2004. A Common Open Representation of Mass Spectrometry Data and its Application to Proteomics Research. *Nature Biotechnology* 22 (11): 1459–1466.

Perez-Riverol, Y. et al. 2016. Omics Discovery Index-Discovering and Linking Public Omics Datasets. *bioRxiv*. doi:10.1101/049205.

Qi, D., C. Lawless, J. Teleman, F. Levander, S. W. Holman, S. Hubbard, and A. R. Jones. 2015. Representation of Selected-Reaction Monitoring Data in the mzQuantML Data Standard. *Proteomics* 15 (15): 2592–2596.

Rocca-Serra, P. et al. 2010. ISA Software Suite: Supporting Standards-Compliant Experimental Annotation and Enabling Curation at the Community Level. *Bioinformatics* 26 (18): 2354–2356.

Rocca-Serra, P. et al. 2016. Data Standards can Boost Metabolomics Research, and if There is a will, There is a Way. *Metabolomics: Official Journal of the Metabolomic Society* 12: 14.

Rubtsov, D. V. et al. 2007. Proposed Reporting Requirements for the Description of NMR-Based Metabolomics Experiments. *Metabolomics: Official Journal of the Metabolomic Society* 3 (3): 223–229.

Salek, R. M. et al. 2015a. Embedding Standards in Metabolomics: The Metabolomics Society Data Standards Task Group. *Metabolomics: Official Journal of the Metabolomic Society* 11 (4): 782–783.

Salek, R. M. et al. 2013a. The MetaboLights Repository: Curation Challenges in Metabolomics. *Database: The Journal of Biological Databases and Curation* 2013: bat029.

Salek, R. M. et al. 2015b. COordination of Standards in MetabOlomicS (COSMOS): Facilitating Integrated Metabolomics Data Access. *Metabolomics: Official Journal of the Metabolomic Society* 11 (6): 1587–1597.

Salek, R. M., C. Steinbeck, M. R. Viant, R. Goodacre, and W. B. Dunn. 2013b. The Role of Reporting Standards for Metabolite Annotation and Identification in Metabolomic Studies. *GigaScience* 2 (1): 13.

Sansone, S.-A. et al. Ontology Working Group Members. 2007. Metabolomics Standards Initiative: Ontology Working Group Work in Progress. *Metabolomics: Official Journal of the Metabolomic Society* 3 (3): 249–256.

Schober, D. et al. 2018. nmrML: A Community Supported Open Data Standard for the Description, Storage, and Exchange of NMR Data. *Analytical Chemistry* 90 (1): 649–656.

Scholz, M., and Fiehn, O. 2006. SetupX–A Public Study Design Database for Metabolomic Projects. In *Biocomputing 2007*. Maui, HI: World Scientific, pp. 169–180.

Schramm, T., A. Hester, I. Klinkert, and J. P. Both. 2012. imzML—A Common Data Format for the Flexible Exchange and Processing of Mass Spectrometry Imaging Data. *Journal of Proteomics* 75 (16): 5106–5110.

Schymanski, E. L., J. Jeon, R. Gulde, K. Fenner, M. Ruff, H. P. Singer, and J. Hollender. 2014. Identifying Small Molecules Via High Resolution Mass Spectrometry: Communicating Confidence. *Environmental Science & Technology* 48 (4): 2097–2098.

Smelter, A., and H. N. B. Moseley. 2018. A Python Library for FAIRer Access and Deposition to the Metabolomics Workbench Data Repository. *Metabolomics: Official Journal of the Metabolomic Society* 14 (5): 64.

Spicer, R. A., R. Salek, and C. Steinbeck. 2017a. A Decade After the Metabolomics Standards Initiative it's Time for a Revision. *Scientific Data* 4: 170138.

Spicer, R. A., R. Salek, and C. Steinbeck. 2017b. Compliance with Minimum Information Guidelines in Public Metabolomics Repositories. *Scientific Data* 4: 170137.

Sud, M. et al. 2016. Metabolomics Workbench: An International Repository for Metabolomics Data and Metadata, Metabolite Standards, Protocols, Tutorials and Training, and Analysis Tools. *Nucleic Acids Research* 44 (D1): D463–D470.

Sumner, L. W. et al. 2007. Proposed Minimum Reporting Standards for Chemical Analysis. *Metabolomics: Official Journal of the Metabolomic Society* 3 (3): 211–221.

Tsugawa, H. et al. 2016. Hydrogen Rearrangement Rules: Computational MS/MS Fragmentation and Structure Elucidation Using MS-FINDER Software. *Analytical Chemistry* 88 (16): 7946–7958.

Turewicz, M., and E. W. Deutsch. 2011. Spectra, Chromatograms, Metadata: mzML-The Standard Data Format for Mass Spectrometer Output. In*Data Mining in Proteomics: From Standards to Applications*, edited by M. Hamacher, M. Eisenacher, and C. Stephan. Totowa, NJ: Humana Press, pp. 179–203.

van der Werf, M. J. et al. 2007. Standard Reporting Requirements for Biological Samples in Metabolomics Experiments: Microbial and *In Vitro* Biology Experiments. *Metabolomics: Official Journal of the Metabolomic Society* 3: 189–194.

van Rijswijk, M. et al. 2017. The Future of Metabolomics in ELIXIR. *F1000 Research* 6. doi:10.12688/f1000research.12342.2.

Walzer, M. et al. 2014. qcML: An Exchange Format for Quality Control Metrics from Mass Spectrometry Experiments. *Molecular & Cellular Proteomics: MCP* 13 (8): 1905–1913.

Wang, M. et al. 2016. Sharing and Community Curation of Mass Spectrometry Data with Global Natural Products Social Molecular Networking. *Nature Biotechnology* 34 (8): 828–837.

Wilkinson, M. D. et al. 2016. The FAIR Guiding Principles for Scientific Data Management and Stewardship. *Scientific Data* 3: 160018.

Wishart, D. S. et al. 2018. HMDB 4.0: The Human Metabolome Database for 2018. *Nucleic Acids Research* 46 (D1): D608–D617.

11

Conclusion

All Authors

CONTENTS

Summary 253
Future Paths 255

Summary

Metabolomics is a vibrant field, developing fast and attracting many scientists from adjacent disciplines. Its multi-disciplinarity bases are one of the defining features of metabolomics. In order to quantify and understand the diversity of the metabolite composition of biological samples one needs chemistry, biology, mathematics and statistics, and computer science (to name just the most central disciplines). Such multi-disciplinarity has been reflected in the programs of many metabolomics conferences. Metabolomics conferences often contain parallel or breakout sessions on topics that may be entirely different from one another (e.g., mass-spectrometry-based or NMR-based metabolomics, computational metabolomics, medical applications, environmental applications, etcetera). Such diversity makes it difficult to obtain a bird's eye view, especially for scientists new to the field. Whatever one's background, one needs to know the basics of many of these fields, and this is where we had set the primary function of this work. We aim to provide a concise yet global overview of the field of metabolomics. If this book can contribute to a better understanding of each other's viewpoints and vocabulary used, the most important goal has been achieved.

The speed at which new developments are arriving has forced us to make choices—there are many exciting developments, especially on the instrumentation level, that have not been included in the book as several texts already exist on these topics. To get an idea of the speed of progress, let us briefly look back for a period of, say, ten years, and assess what changes have occurred since 2008. In LCMS, for example, HPLC has been replaced by UPLC and gradients have become much faster, often saving 50% in time. In 2008 most MS detectors were triple or single-quadrupole instruments, whereas now many more high-resolution instruments are used. As a result, data processing techniques have changed as well: unique algorithms for peak detection for high-resolution MS data started about ten years ago. The role of NMR, which was quite prominent at the beginning of metabolomics, has gradually decreased as compared to mass spectrometry. NMR cannot compete with MS in terms of the number of metabolites detected. However, NMR remains widely used as a fully quantitative and reproducible approach thus compensating for its lack of sensitivity relative to MS. With the advent of large-scale phenotyping and the use of metabolomics in biobanking, we predict that the use of NMR will increase significantly. The main reason for that is in its ability to operate reproducibly over long periods in automation, with no history effects and no metabolite detection bias that remains invaluable.

Our selection is based on an assessment of what a beginning researcher should learn/know, necessarily limited to only the most important items. Apart from the description of the technical aspect of a metabolomics experiment, we have also tried to convey knowledge that is not so easy to extract from scientific literature. Setting up an experiment forces one to think about the questions that require an answer—too often samples were inserted into the instrument just because they were available. Also (or maybe better: even) in untargeted metabolomics experiments one should try to define what information one would like to obtain in as much detail as possible to avoid wasting valuable human and instrument time, as well as money. This involves thinking of what kind of conclusions can be drawn in terms of generality and validity: interestingly, these questions often can be answered before doing the experiment itself, the topic of Chapter 2. Often the most important question is about the sampling unit: what kind of variation are we really interested in? Thinking of what the most important sources of variation are will also help determine which, and how many, samples to take.

For outsiders, it sometimes comes as a surprise to learn that there is also a thing called "data processing noise," i.e., variation introduced by the data processing and analysis. Ask two people independently of each other to analyze the same data set and you will (*you will*) get different results. They will make different assumptions, use different algorithms, use different software and usually a combination of these parameters. Annoyingly, there is usually no clear right or wrong answer. Statistics rarely, if ever, will provide a definite clear-cut answer. The proof of the pudding, as always, is in eating it: if the analysis leads to an understanding of the biological system and is validated in independent follow-up experiments, the original experiment and data processing have been successful.

In the steps from data to information to knowledge, the last step is usually made by combining different parts of information: the relation between metabolites, e.g., coming from a particular pathway, is essential for the generalization of individual pieces to the whole jigsaw, or at least a part of it. This step needs to be taken with utmost care. There is a lot of information we do not know, and in many cases we are simply feeling our way through a very dark room. Nevertheless, it is the way forward: networking approaches as shown in Chapter 8 not only serve to visualize the data in a completely natural way, but also allow interesting modeling approaches and enhanced knowledge discovery. In addition, current metabolic network modeling approaches enable the integration of a variety of "omics" scale data, such as proteomics or transcriptomics, from different high-throughput technologies in order to decipher molecular mechanisms underlying the systems behavior. This integration will be crucial. It will also benefit the annotation process, where information from network structures will provide useful prior knowledge. In addition, annotation will also be more tightly coupled to quantification, so that it is possible to not only obtain statistics on individual features or patterns, but also to look for common biochemical themes within these features.

Since so many of the final elements in a metabolomics experiment depend on external databases (e.g., containing metabolite spectra and reactions) it is of utmost importance that these databases are as complete as possible, and at the very least do not contain incorrect information. Data stewardship has become one of the most important topics in science nowadays—a large part is also involved in making data findable, accessible, interoperable and reusable, as mentioned in Chapter 10. A particularly relevant example is the naming of metabolites: even though several unique identifier systems exist (e.g., InChi keys), the community is slow to adopt them. Another example is the trend that nowadays many journals are requiring paper submissions to be accompanied by the raw data, along with the data processing and data analysis scripts. By doing so everyone is strengthening the field as a whole. Although it is becoming rapidly impossible for scientists to read all papers appearing in a certain area, automatic information retrieval methods will be coming to the rescue identifying relevant publications and data sets. Again, it is the responsibility of the metabolomics community as a whole to make use of these possibilities and to stimulate their development.

Future Paths

Defining metabolomics (metabonomics) as the science occupied with the "the quantitative measurement of the multiparametric metabolic response of living systems to pathophysiological stimuli or genetic modification," there are several routes in which important developments can be expected. Here is our top list for the next five to ten years:

1. Metabolite annotation, arguably the biggest obstacle in many practical applications, will receive a lot of attention. Currently—especially with mass spectrometric detection—it is not uncommon to see less than 10% of all features in untargeted metabolomics experiments annotated. The classical approach, considering all features as unknowns,

finding a subset of features that are correlated with the trait of interest (whether continuous or discrete) and focusing on annotating only this subset, is in many cases too time consuming to be practical. Improvements in instrumentation will lead to better compound separation and higher spectral resolution, making unambiguous annotation easier. Databases containing compound spectra, currently rather incomplete, will continue to grow and will become essential tools in annotating experimental data. Networking approaches will serve to include prior knowledge in the annotation process.

2. With increased annotation rates, multivariate analysis of metabolomics data becomes much easier: the number of metabolites is usually at least an order of magnitude smaller than the number of features. We should still be wary of "believing" our multivariate models too much: if you have 20 independent samples, that is what you have got: 20 independent samples. There is no way that you can reliably estimate all entries of a 10,000-by-10,000 correlation or covariance matrix. In such a situation, methods like PLS are able to give you answers at the expense of uniqueness: there is an infinite number of models, different from the PLS solution, that will give you exactly the same answers. With increased awareness of this fact and improved experimental throughput capacity, the era of only ten or twenty samples in one experiment will be left behind, and scientists will routinely measure larger numbers including true biological replicates.
3. The sensitivity of modern instruments, in particular mass spectrometers, implies that we are measuring many signals that are not coming from the biological sample but from the lab environment, e.g., the atmosphere, solvent, reagents, equipment and many other influences. Separating this molecular background from the molecular signal is an important task, and strategies for dealing with this, incorporating blank and QC samples in the measurement sequence, are becoming increasingly widely adopted.
4. Different measurement platforms will be combined in a more routine fashion. Every detection system only presents us with a partial view of the metabolome, exactly the reason why we use multiple detectors in the first place. Apart from the struggle to bring data from very different machines and data formats together, the data have quite different error structures as well. The field of data fusion (Chapter 7) is trying to come to grips with this but is hampered by the ever more unfavorable variable-to-sample ratio in these cases. More and more, scientists will start assessing the behavior of metabolites captured by different platforms and validate their findings on several different platforms.

5. The importance of more holistic approaches such as networks and pathway-level analyses will continue to grow. All databases are incomplete, more so if you are working on less common organisms (with this we mean anything that is not yeast, *E. coli* or human)—and finding downright incorrect information is a realistic possibility, too. However, the current stress on reproducibility and data sharing means that the availability of large and high-quality databases will grow quickly. One should keep in mind that in most cases the integration of metabolomic data (or any other type of data) into a metabolic model-driven analysis provides a range of feasible flux solutions, which means that different metabolic flux profiles can fit the experimental data equally well. Choosing the "correct" one (provided such a thing exists) may not be easy. Yet, it is the only way forward from the current state of having quite a lot of data from only few samples.
6. Better networks will be obtained by the increase in the number of experiments in which several time points are taken into account, driven by faster scan times and higher sensitivities. These time-resolved experiments are necessary to go beyond a static representation and to find connections between the metabolites and other measurement objects across the different omic levels. The analysis of these types of experiments will lead to new computational modeling methods which have to become "multilevel" to profit from more complex and structured experimental designs.

The number of publications on metabolomics describing really interesting biology, the number of success stories, is steadily increasing. Also the number of data sets made available through generally accessible repositories is increasing. More and more young scientists are attracted to this multidisciplinary field. We are only beginning to learn how to work together and to speak each other's language (one of the main reasons for writing this book!). More and more software is distributed as open source, allowing new generations of scientists to directly build on what has been done before. Instrument manufacturers are more and more opening up their software products as well, in some cases allowing user provided modules to interact with proprietary software. Some exciting innovations in instrumentation are on their way (ion mobility mass spectrometry being one of them), possibly overtaking some of the data handling strategies described in this book. Localization of compounds has already become an active topic, and imaging approaches (MRI, MS-imaging) are basically constituting research fields in their own right. Finally, the awareness of the importance of knowing more about the metabolome in many different fields (environmental science, health sciences, food science, plant science) is increasing as well. For cases where environmental factors do not play a major factor, the link of the metabolome with the other omics sciences (proteomics, transcriptomics) is going to be essential in turning data into information. Where environmental factors do play a role, this is not always the case.

What we have sidestepped (deliberately!) is the essential part of the scientific process, the iterative putting together of the jigsaw pieces, combining more and more information, validating side steps, and revising conclusions until a satisfying answer has been reached. To be able to do this, however, a thorough understanding of the complexity of the data analysis part in metabolomics is indispensable, and we hope that this book has contributed to that.

Appendix MTBLS1: NMR Data Set—A Metabolomic Study of Urinary Changes in Type 2 Diabetes in Human Compared to the Control Group

Reza Salek and Jules Griffin

Background and Summary

Type 2 diabetes mellitus is the result of a combination of impaired insulin secretion and reduced insulin sensitivity of target tissues. There are an estimated 150 million affected individuals worldwide, of whom a large proportion remains undiagnosed because of a lack of specific symptoms early in this disorder and inadequate diagnostics. In this study, NMR-based metabolomic analysis, in conjunction with multivariate statistics, was applied to examine the urinary metabolic changes in two rodent models of type 2 diabetes mellitus as well as unmedicated human sufferers. The db/db mouse and obese Zucker (fa/fa) rat have autosomal recessive defects in the leptin receptor gene, causing type 2 diabetes. 1H-NMR spectra of urine were used in conjunction with uni- and multivariate statistics to identify disease-related metabolic changes in these two animal models and human sufferers. This study demonstrates metabolic similarities between the three species examined, including metabolic responses associated with general systemic stress, changes in the tricarboxylic acid (TCA) cycle, and perturbations in nucleotide metabolism and in methylamine metabolism. All three species demonstrated profound changes in nucleotide metabolism, including that of N-methylnicotinamide and N-methyl-2-pyridone-5-carboxamide, which may provide unique biomarkers for following type 2 diabetes mellitus progression (Salek et al., 2007).

Methods

Extraction. For the human studies, midstream urine samples (~15 mL) were collected from each volunteer and frozen. In total, 84 samples were collected from 12 healthy volunteers (7 time points, 8 males and 4 females) and 50 samples from 30 T2DM patients (1–3 time points, 17 males and 13 females) with well-controlled blood glucose maintained at normal concentrations by diet, following the guidelines issued by the American Diabetes Association, rather than medication. The healthy subjects were aged 18–55 year; had a body mass index (BMI) ≥19 and ≤30 kg/m^2; a body mass ≥50 kg and ≤113 kg; were free from any major diseases; and were notr pregnant. The T2DM patients were aged 30–65 year (mean 56 ± 9 year); had a BMI >25 and <40 kg/m^2; weighed between 65 and 140 kg (mean 95 ± 19 kg); and were taking at most one oral anti-diabetic drug. T2DM patients agreed to stop treatment with oral anti-diabetic agents during the study. Subjects went through a washout period of 4 week before sample collection and abstained from alcohol during the study. Diet was controlled throughout the study.

NMR sample. Aliquots of 400 μL urine samples were made up to 600 μL with phosphate buffer (0.2 M, pH 7.4) and any precipitate removed by centrifugation. In total, 500 μL of supernatant were transferred to 5mm NMR tubes with 100 μL of sodium 3-trimethylsilyl-(2,2,3,3-2H4)-1-propionate (TSP)/D_2O/sodium azide solution (0.05% weight/volume TSP in D_2O and 1% wt/vol sodium azide).

NMR spectroscopy. The spectra of human urine samples were acquired with a Bruker DRX700 NMR spectrometer using a 5 mm TXI ATMA probe at a proton frequency of 700.1 MHz and an ambient temperature of 27°C.

NMR assay. A 1D NOESY pre-saturation pulse sequence was used to analyze the urine samples. For each sample, 128 transients were collected into 64k data points using a spectral width of 14.005 kHz (20 ppm) and an acquisition time of 2.34 s per FID.

Data transformation. Spectra were processed using ACD/1D NMR Manager 8.0 with Intelligent Bucketing Integration (Advanced Chemistry Development, Toronto, ON, Canada). Spectra were integrated 0.20–9.30 ppm excluding water (4.24–5.04 ppm), glucose (3.19–3.99 ppm, 5.21–5.27 ppm), and urea (5.04–6.00 ppm). Intelligent bucketing ensures that bucket edges do not coincide with peak maxima, preventing resonances from being split across separate integral regions; a 0.04 ppm bucket width and a 50% looseness factor were used. All spectra were normalized to total area excluding the water, urea, and glucose regions.

Metabolite identification. Assignments were confirmed by two-dimensional spectroscopy including homonuclear 1H-1H Correlation Spectroscopy (COSY), 1H-13C Heteronuclear Signal Quantum Coherence (HSQC), and 1H-13C Heteronuclear Multiple Bond Correlation (HMBC) Spectroscopy.

Sample collection. For the human studies, midstream urine (~15 mL) samples were collected from each volunteer and frozen. In total, 84 samples were collected from 12 healthy volunteers (7 time points, 8 males and 4 females) and 50 samples from 30 T2DM patients (1–3 time points, 17 males and 13 females) with well-controlled blood glucose maintained at normal concentrations by diet, following the guidelines issued by the American Diabetes Association, rather than medication. T2DM patients agreed to stop treatment with oral anti-diabetic agents during the study. Subjects went through a washout period of 4 week before sample collection and abstained from alcohol during the study. Diet was controlled throughout the study.

Data Records

The experimental metadata was primarily described and structured using the ISA-Tab (Rocca-Serra et al., 2010) format using the ISAcreator tool. The data set is publicly available via the EMBL-EBI MetaboLights repository (Haug et al., 2013) under accession number MTBLS1: https://www.ebi.ac.uk/metabolights/mtbls1.

Additionally, the link to the raw files are available at OMICS-DI (Perez-Riverol et al., 2016), https://www.omicsdi.org/dataset/metabolights_dataset/MTBLS1.

All the raw files for each sample are compressed individually; for example, ADG10003u_007.zip is for sample ADG10003u_007. The zip folder includes both raw and processed data. Additionally, all the raw data available as nmrML format (Schober et al., 2018) (e.g., ADG10003u_007.nmrML, also see Figure A.1).

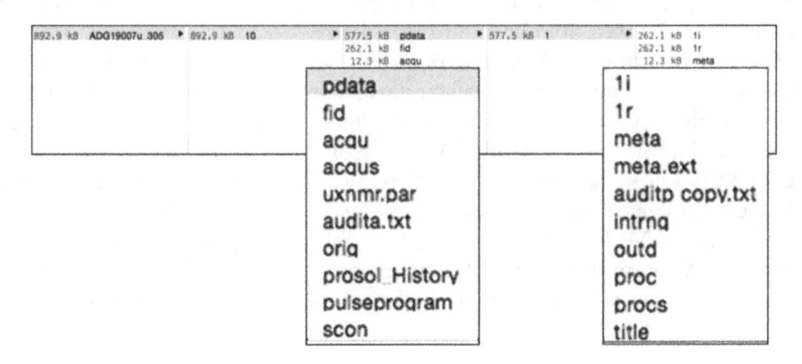

FIGURE A.1
An example for the content within an NMR zipped file, here ADG19007u_306.zip.

Several tools used this data set (MTBLS1) for testing and validation of their methods; for example, rDolphin: a GUI R package for proficient automatic profiling of 1D 1H-NMR spectra of study data sets (Cañueto et al., 2018a); ASICS: an R package for a whole analysis workflow of 1D 1H NMR spectra (Lefort et al., 2019) and a new analysis method using this data set to improve sample classification by harnessing the potential of 1 H-NMR signal chemical shifts (Cañueto et al., 2018b). The same data set was also used in the PhenoMeNal as an example analysis for processing and analysis of metabolomics data in the cloud (Peters et al., 2018).

References

Cañueto, D., Gómez, J., Salek, R. M., Correig, X., and Cañellas, N. (2018a). rDolphin: A GUI R package for proficient automatic profiling of 1D 1H-NMR spectra of study datasets. *Metabolomics: Official Journal of the Metabolomic Society, 14*(3), 24.

Cañueto, D., Salek, R. M., Correig, X., and Cañellas, N. (2018b). Improving sample classification by harnessing the potential of 1H-NMR signal chemical shifts. *Scientific Reports, 8*(1), 11886.

Haug, K., Salek, R. M., Conesa, P., Hastings, J., de Matos, P., Rijnbeek, M. et al. (2013). MetaboLights: An open-access general-purpose repository for metabolomics studies and associated meta-data. *Nucleic Acids Research, 41*(Database issue), D781–D786.

Lefort, G., Liaubet, L., Canlet, C., Tardivel, P., Père, M.-C., Quesnel, H. et al. (2019). ASICS: An R package for a whole analysis workflow of 1D 1H NMR spectra. *Bioinformatics.* doi:10.1093/bioinformatics/btz248.

Perez-Riverol, Y., Bai, M., Leprevost, F., Squizzato, S., Park, Y. M., Haug, O. K. et al. (2016). Omics Discovery Index - Discovering and Linking Public Omics Datasets (p. 049205). doi:10.1101/049205.

Peters, K., Bradbury, J., Bergmann, S., Capuccini, M., Cascante, M., de Atauri, P., … Steinbeck, C. (2018). PhenoMeNal: Processing and analysis of metabolomics data in the cloud (p. 409151). doi:10.1101/409151.

Rocca-Serra, P., Brandizi, M., Maguire, E., Sklyar, N., Taylor, C., Begley, K. et al. (2010). ISA software suite: Supporting standards-compliant experimental annotation and enabling curation at the community level. *Bioinformatics, 26*(18), 2354–2356.

Salek, R. M., Maguire, M. L., Bentley, E., Rubtsov, D. V., Hough, T., Cheeseman, M. et al. (2007). A metabolomic comparison of urinary changes in type 2 diabetes in mouse, rat, and human. *Physiological Genomics, 29*(2), 99–108.

Schober, D., Jacob, D., Wilson, M., Cruz, J. A., Marcu, A., Grant, J. R. et al. (2018). nmrML: A community supported open data standard for the description, storage, and exchange of NMR data. *Analytical Chemistry, 90*(1), 649–656.

Appendix MTBLS18: Metabolic and Transcriptional Response of Arabidopsis Thaliana Wildtype and Mutants to Phytophthora Infestans

Tilo Lübken, Michaela Kopischke, Kathrin Geissler, Lore Westphal, Dierk Scheel, Steffen Neumann and Sabine Rosahl

Background and Summary

The oomycete *Phytophthora infestans* is the causal agent of late blight, the potato disease that triggered the devastating famine in Ireland in the middle of the nineteenth century and which, even today, results in estimated worldwide annual losses of six billion US$. Hence, the understanding of resistance mechanisms can be of important commercial and humanitarian interest. The nonhost plant *Arabidopsis thaliana* is not colonized by *P. infestans* and serves as a model for successful defense against the oomycete.

The first layer of defense is impaired in several *A. thaliana* mutants. Thus, *P. infestans* can penetrate the plant if the *PEN2* gene is mutated (Lipka et al. 2005). The *PEN2* gene encodes an enzyme involved in indole glucosinolate metabolism (Bednarek et al. 2009). Although *P. infestans* can penetrate the epidermis of *pen2* mutant plants, it is unable to colonize the plant. In an ethyl methanesulfonate (EMS) mutagenesis screen, additional mutations in the *pen2* background were identified, which showed an enhanced defense response after inoculation with *P. infestans* (Kopischke et al. 2013). These were named *pen2erp1* and *pen2erp2*. Backcrossing of the mutants to the wild type *gl1* resulted in the single mutants *erp1* and *erp2*.

This appendix describes the data obtained in a combined transcriptomics and metabolomics experiment on the response of several *A. thaliana* mutants at different time points after inoculation with *P. infestans*. The experimental design comprises several experimental factors:

1. The **genotype** was one of the following six: *gl1* as the wildtype, the mutant *pen2*, two mutants with enhanced defense response *pen2erp1*, *pen2erp2* and the two backcrossed lines *erp2* and *erp1*.
2. The plants were subjected to a **treatment** with either a *P. infestans* zoospore suspension or water as mock treatment.
3. Plants were harvested at **time point** 6 and 12 h after the respective treatment.
4. Three independent **replicates** of the experiments were performed, each with a newly prepared and grown *P. infestans* culture.

The combination of these four factors resulted in overall $6 \times 2 \times 2 \times 3 = 72$ samples. The analysis of the experimental data can thus provide insight into processes related to the defense genes upstream and downstream of the mutations and the dynamics of the response. The replicate design allows to detect or compensate variation between unintendedly slightly different growth conditions and different *P. infestans* cultures.

Several secondary metabolites are known to be involved in the plant defense response (Piasecka et al. 2015) and we used LC/MS based metabolite profiling which can routinely detect semi-polar plant metabolites from major biosynthetic classes including glucosinolates, indolic compounds, phenylpropanoids, benzenoids, flavonoids, terpenes and fatty acid derivatives (Böttcher et al. 2008). We also performed gene expression analysis to identify the gene(s) activated in the stress response.

Methods

Plant-Growth Conditions and Infection Experiments

Arabidopsis plants (Kopischke et al. 2013) were grown on steam sterilized soil in a mix with vermiculite (3:1) in a phytochamber with 8 h light at 22°C and 16 h dark at 20°C for five weeks. Plants were drop-inoculated by placing 10 µL of a *P. infestans* zoospore suspension (5×10^5 mL^{-1}) onto the adaxial side of the leaves. Subsequently, plants were placed in a phytochamber with 16 h light at 20°C and 8 h dark at 18°C.

Leaves were harvested 6 and 12 h after inoculation. Six leaves were pooled, frozen in liquid nitrogen and stored at −80°C.

The combination of the four experimental factors resulted in overall $6 \times 2 \times 2 \times 3 = 72$ samples. The samples for the gene expression analysis and metabolite profiling were prepared from the same plant material.

Gene Expression Data

From each sample, the total RNA was isolated according to Kopischke et al. (2013), Ahn (2009) and purified using the RNeasy Plant Mini Kit (Qiagen). The samples were hybridized to 72 Affymetrix ATH1 GeneChip arrays. The actual hybridization, image acquisition and creation of CEL files was performed by AROS APPLIED BIOTECHNOLOGY (Aarhus, Denmark). The processing of the CEL files was performed with the R-package simpleaffy (Wilson and Miller 2005). The result was a 72×22810 data matrix of gene expression intensities.

LC-MS Metabolite Profiling and Data Processing

Metabolite profiling was performed on an UPLC-ESI-QqTOF/MS system. The chromatographic separations were performed on an Acquity UPLC system (Waters) equipped with a HSS T3 column (100 × 1.0 mm, particle size 1.8 μm; Waters) with a flow rate of 200 μL/min at 40°C column temperature. The following gradient program was used: 0–60 s, isocratic 95% A (water/formic acid, 99.9/0.1 (v/v)), 5% B (acetonitrile/formic acid, 99.9/0.1 (v/v)); 60–360 s, linear from 5% to 30% B; 360–600 s, linear from 30% to 95% B; 360–720 s isocratic 95% B. The injection volume was 2.0 μL (full loop injection). Eluted compounds were detected at an acquisition rate of 3 Hz from *m/z* 100–1000 using a micrOTOF-Q-II (Bruker Daltonics, Bremen) equipped with an Apollo II electrospray ion source (also Bruker Daltonics). In positive ion mode the following instrument settings were used: nebulizer gas, nitrogen, 1.2 bar; dry gas, nitrogen, 8 L/min, 190°C; capillary, −4500 V; end plate offset, −500 V; funnel 1 RF, 200 Vpp; funnel 2 RF, 200 Vpp. Mass calibration of individual raw data files was performed on lithium formate cluster ions obtained by automatic infusion of 20 μL 10 mM lithium hydroxide in isopropanol/water/formic acid, 49.9/49.9/0.2 (v/v/v) at a gradient time of 720 s using a diverter valve.

Samples were measured in technical triplicates, after removing measurements with technical issues resulting in a total of 202 chromatograms in positive and 202 in negative mode. The native Bruker data files were converted to the mzML format with the Bruker CompassXport software (version 3.0.9).

The converted mzData files were processed with xcms (Smith et al. 2006). Mass spectral features were detected with the centWave (Tautenhahn et al. 2008) algorithm; redundancy from features corresponding to the same metabolite were eliminated with the CAMERA (Kuhl et al. 2012) package.

XCMS settings for processing LC/MS data were prefilter = 3,500; snthr = 3; ppm = 25, peakwidth = 5,12. For alignment group.density function with parameters minfrac = 0.75 and bw = 5 was used. For retention time grouping using CAMERA, we increased the retention time window to twice the full-width-at-half-maximum of a peak.

Data Records

All data from this experiment was submitted to the respective repositories at the European Bioinformatics Institute (EBI, Hinxton, UK). Since the assay data was obtained from the same sample material, we also submitted the sample information separately to the EBI sample DB. The experiment was primarily described and structured using the ISA-Tab format using the ISAcreator tool, and then modified or converted and submitted to the respective databases.

Data Record 0: Sample Information

The sample DB data record describes the plants, treatments and all of the experimental design factors that were part of the experiment, and not specific to the actual assay performed.

The Samples have been registered in SampleDB at the EBI under the accession SAMEG179892 at http://www.ebi.ac.uk/biosamples/group/SAMEG179892. For each of the 72 samples, a sample identifier (e.g., SAMEA2630372), the experimental factors "Sampling time," "genotype," and "Treatment" are provided. The "Sample Name" refers to the "Source Name" in both ArrayExpress and MetaboLights. Genotype, Terms and factors have been annotated using NCBI taxonomy (NCBITaxon), Plant Ontology (EFO) and Experimental factor ontology (EFO). Chemicals were annotated using the Chemicals of Biological Interest (ChEBI) and quantitative values with Unit Ontology (UO) terms.

Data Record 1: Gene Expression Data Processing

Array Express (Kolesnikov et al. 2015) is a species and technology independent repository for gene expression data sets at the EBI. The submission of the gene expression data and description was exported in the MAGEtab format from within the ISAcreator tool (Rocca-Serra et al. 2010).

The gene expression data has been submitted to ArrayExpress under the accession E-MTAB-3287 and is available at https://www.ebi.ac.uk/arrayexpress/experiments/E-MTAB-3287/.

The Array Express accession E-MTAB-3287 contains the Affymetrix CEL files as obtained from AROS APPLIED BIOTECHNOLOGY (Aarhus, Denmark). From ArrayExpress, the data can be downloaded locally, or transferred to GenomeSpace and re-analyzed with many of the tools available there.

Data Record 2: LC/MS Metabolite Profiling and Data Processing

The Metabolights database (Haug et al. 2013) is the first open, species and technology independent metabolomics data repository. Data sets can be described and submitted using the ISA-Tab format. The LC/MS metabolite profiling data is available from MetaboLights under the accession MTBLS18 at http://www.ebi.ac.uk/metabolights/MTBLS18.

The Metabolights accession MTBLS18 contains the raw data files in the mzData format, as well as the data after processing with XCMS and CAMERA. XCMS settings for processing LC/MS data were prefilter = 3,500; snthr = 3; ppm = 25, peakwidth = 5,12. For alignment group.density function with parameters minfrac = 0.75 and bw = 5 was used. MTBLS18 also contains the LC-UV data, extracted from the raw data via a custom CompassXtract program, read into XCMS and saved in netCDF format.

Acknowledgements

This work was in part supported by DFG SPP122 (Microbial Reprogramming of Plant Cell Development, http://gepris.dfg.de/gepris/projekt/14250334).

References

Ahn, Ji H. 2009. "RNA Extraction from *Arabidopsis* for Northern Blots and Reverse Transcriptase-PCR." *Cold Spring Harbor Protocols* 2009 (9): db.prot5295.

Bednarek, P. et al. 2009. "A Glucosinolate Metabolism Pathway in Living Plant Cells Mediates Broad-Spectrum Antifungal Defense." *Science* 323 (5910): 101–106.

Böttcher, C., E. von Roepenack-Lahaye, J. Schmidt, C. Schmotz, S. Neumann, D. Scheel, and S. Clemens. 2008. "Metabolome Analysis of Biosynthetic Mutants Reveals a Diversity of Metabolic Changes and Allows Identification of a Large Number of New Compounds in Arabidopsis." *Plant Physiology* 147 (4): 2107–2120.

Haug, K. et al. 2013. "MetaboLights—an Open-Access General-Purpose Repository for Metabolomics Studies and Associated Meta-Data." *Nucleic Acids Research* 41 (D1): D781–D786.

Kolesnikov, N. et al. 2015. "ArrayExpress Update—Simplifying Data Submissions." *Nucleic Acids Research* 43 (D1): D1113–D1116.

Kopischke, M. et al. 2013. "Impaired Sterol Ester Synthesis Alters the Response of Arabidopsis Thaliana to Phytophthora Infestans." *The Plant Journal: For Cell and Molecular Biology* 73 (3): 456–468.

Kuhl, C., R. Tautenhahn, C. Böttcher, T. R. Larson, and S. Neumann. 2012. "CAMERA: An Integrated Strategy for Compound Spectra Extraction and Annotation of Liquid Chromatography/mass Spectrometry Data Sets." *Analytical Chemistry* 84 (1): 283–289.

Lipka, V. et al. 2005. "Pre- and Postinvasion Defenses Both Contribute to Nonhost Resistance in Arabidopsis." *Science* 310 (5751): 1180–1183.

Piasecka, A., N. Jedrzejczak-Rey, and P. Bednarek. 2015. "Secondary Metabolites in Plant Innate Immunity: Conserved Function of Divergent Chemicals." *The New Phytologist* 206 (3): 948–964.

Rocca-Serra, P. et al. 2010. "ISA Software Suite: Supporting Standards-Compliant Experimental Annotation and Enabling Curation at the Community Level." *Bioinformatics* 26 (18): 2354–2356.

Smith, C. A., E. J. Want, G. O'Maille, R. Abagyan, and G. Siuzdak. 2006. "XCMS: Processing Mass Spectrometry Data for Metabolite Profiling Using Nonlinear Peak Alignment, Matching, and Identification." *Analytical Chemistry* 78 (3): 779–787.

Tautenhahn, R., C. Böttcher, and S. Neumann. 2008. "Highly Sensitive Feature Detection for High Resolution LC/MS." *BMC Bioinformatics* 9: 504.

Wilson, C. L., and C. J. Miller. 2005. "Simpleaffy: A BioConductor Package for Affymetrix Quality Control and Data Analysis." *Bioinformatics* 21 (18): 3683–3685.

Index

Note: Page numbers in italic and bold refer to figures and tables, respectively.

A

ab initio approach, 185–187
abundance sensitivity, 42
alkaptonuria, 9
AMDIS (Automated Mass spectrometry Deconvolution and Identification System), 86
Analysis of Covariance (ANCOVA), 141
Analysis of Variance (ANOVA), 141
annotated data matrices, 84–88
apodization, 106
Arabidopsis thaliana, 263
Architecture for Metabolomics consortium (ArMet), 236
Array Express, 266–267
Automated Mass spectrometry Deconvolution and Identification System (AMDIS), 86
automatic reconstruction, 179–180
autoscaling, 148

B

batch effects, 22, 27, 30
biochemistry, 3
bioengineering, 221–222
biological variation, 18, *19*
bipartite graph, 190, *191*
biplot, 147
blind source separation (BSP), 122
blocking, 22–23
block randomization, 23
block scaling, 173
Bonferroni correction, 137
bucketing/binning, 110–111

C

capillary electrophoresis (CE-MS), 39
^{13}C fluxomics, 206–208
carrier gas, 38
CBM, *see* constraint-based modeling (CBM)
central-limit theorem, 131
cerebrospinal fluid (CSF), 159
chemical fingerprint, 187
chemical noise, 80
chemical shift (δ), 51
  calibration, 108–109
  misalignments correction, 109
chromatographic resolution, 38–39
common/distinct information methods, 164
  DISCO-SCA, 165–166
  JIVE, 166
  OnPLS, 167
  orthogonality aspects, 168
common *vs.* distinctive variation, 159–160
compound graph, 190, *191*
comprehensive biomarkers, 159
confidence intervals, 133–134, *134*
constraint-based modeling (CBM), 203
  application, 216–218, *217*
  mathematical basis, 211–213
Core Information for Metabolomics Reporting (CIMR), 236
COSMOS initiative, 240
covariate adaptive randomization, 23
creatinine, 85
crossover design, 24
cross-validation, 150
cycle, 194

D

data
  -driven methods, 201
  -exchange formats, 240–241
  integration methods, 157–158
  matrices, 80–84
  pre-processing, 174–175
  processing, 265–266

data (*Continued*)
processing noise, 254
sharing, 235, 239
stewardship, 255
system, 42–44
transformation, 260
data analysis
preparations, 28–30
reproducibility, 30–32
databases, 181–183
data fusion
definition of, 157–159
goals, 159–160
issues in SCA type, 172–174
data fusion methods, 160
correlation-based, 160–162, *162*
SCA type, 162–163, *164*
data records, 266
data processing, 267
gene expression data processing, 266–267
LC/MS metabolite profiling, 267
sample information, 266
degree of node, 194
digital resolution, 104
DIMS (Direct Infusion Mass Spectrometry), 78
diode array (DA) detection, 66–68
Direct Infusion Mass Spectrometry (DIMS), 78
Discriminant Analysis, 141
drug discovery, 218–221, 225
dynamic flux-map computation, 204–208
dynamic metabolic modeling, 206–208
dynamic range, 5, 41–42, 56

E

edge, 194
electronic noise, 80
electrospray ionization (ESI), 84–85
elementary metabolite units (EMU), 224
enzyme, 3
Enzyme Commission (EC) numbers, 182
epigenome, 2
ethyl methanesulfonate (EMS) mutagenesis screen, 263
exchange fluxes, 10
experimental designs, 21–25
exploratory analysis, 159
extensible markup language (mzXML), 242–243
extracted ion chromatogram (EIC), 81, 86
extraction and pretreatment, 26

F

FBA (flux balance analysis), 204, 207–208, 211, 218
features, 77–80
FID (free induction decay), 51, 56, 58, 102
fingerprinting approach, 110–111
Fisher's exact test, 189
Flow Injection Mass Spectrometry (FIMS), 78
flux balance analysis (FBA), 204, 207–208, 211, 218
fluxes, 9–10
fluxomics, 9
Fourier transformation (FT), 105
free induction decay (FID), 51, 56, 58, 102
full factorial design, 24

G

gas chromatography (GC-MS), 38–39, *74*
gas-phase ion sources, 40
GC-MS metabolomics, 89–91
gene, 2
gene expression data, 265
gene-protein-reactions (GPRs), 179, 182
genome, 2
genome-scale kinetic models, 227–229
genome-scale metabolic models (GSMMs), 211, *212*, 216
Genome-Scale Metabolic Networks, 182
graph modeling, 189–194
network visualization, 194–195
pathway mapping, 187–189
reconstruction, *see* genome-scale reconstruction
genome-scale reconstruction
databases, 181–183
HRMS, 185–186
limitation, 184–185
principle, 178–181
SBML, 183–184

genotype, 264
graph, 194
graph modeling, 189–191
from global metabolic network, 192–193
GSMMs (genome-scale metabolic models), 211, *212*, 216

H

haphazard sampling, 20–21
hexokinase activity, 209
hierarchical drawing, 195
high-resolution magic angle spinning (HR-MAS), 121–122
high resolution mass spectrometry (HRMS), 185–186
horizontal gene transfer, 223
HR-MAS (high-resolution magic angle spinning), 121–122
HRMS (high resolution mass spectrometry), 185–186
HumanCyc, 182, *183*
hypergraphs, 190–191, *191*

I

imzML, 245
incomplete block design, 24
inference, 133
confidence intervals, 133–134, *134*
statistical testing, 134–140, *138*, *139*
infrared (IR) spectroscopy, 64–65
in silico tools, 91, *92*
inter-omics applications, 159
inter-platform applications, 158
inter-sample applications, 159
intracellular fluxes, 10
ionization, 39–40
IR (infrared) spectroscopy, 64–65
isodyn, 207
Isotopologue Parameter Optimisation (IPO), 84

J

Joint and Individual Variation Explained (JIVE) method, 166

K

KEGG, 182–183
kinetic modeling, 203
application, 208–210, *209*
genome-scale, 227–229
mathematical basis, 204–206, *205*
Kovats index (KI), 89

L

Lambda, 108
large-scale ^{13}C-constrained FBA, 226–227
large-scale flux-map computation/ constraint-based modeling, 211–218, 225–226
Larmor frequency (ω_0), 52
latent variables, 150
LC-MS, *see* liquid chromatography (LC-MS)
least-squares regression, 140
linear equation, 208
linear regression, 140–142, **142**
liquid chromatography (LC-MS), 39
data processing in, *74*
LC-MS metabolite profiling, 265–266
liquid-phase ion sources, 40

M

manual curation, 178–181
mass accuracy, 41
mass range, 41
mass resolution, 41
mass spectral databases, 89–91
mass spectrometry (MS), **6**
advantages and disadvantages, 47
data system, 42–44
detection, 41–42
ionization, 39–40
MS/MS, 44–46
separation techniques, 36–39
MCR (multivariate curve resolution), 67, *68*
measurement technologies
DA detection, 66–68
MS, 36–47
NMR spectroscopy, 47–62
vibrational spectroscopy, 62–66

metabolic fluxes, 203, 207
metabolic flux modeling
  bioengineering, 221–222
  drug discovery, 218–221
  dynamic flux-map computation/kinetic modeling, 204–210
  evolution, 222–223
  future challenges, 223–229
  large-scale flux-map computation/constraint-based modeling, 211–218
  network analysis, model-driven strategies, 201–203
metabolic network, 159
metabolic pathway, 2, 182
metabolic phenotypic analysis (MPA), 225
metabolic reaction, 182
MetaboLights, 237–239
metabolism, 2
metabolite
  dynamic range, 4–5
  identification, 246–248, 260
  incorporating network information, 94
  pathways, 9–11
metabolite assignment/identification, 114–115
  NMR and MS, 117–118
  online resource, 116–117
  STOCSY, 117
metabolite identification of carbon efficiency (MICE), 248
metabolome, 2
metabolomics, 1–2, 253
  multidisciplinary nature, *7*
  targeted *vs.* untargeted, 7–9
Metabolomics Standards Initiative (MSI), 119, 236, 246–248
MetaNetter, *186*
missing values, 29, 82–83
mobile phase, 37
model-driven strategies, 201–203
MS, *see* mass spectrometry (MS)
MS data processing, 73–74
  annotated data matrices, 84–88
  features to data matrices, 80–84
  identified metabolites, 88–94
  raw data to features, 77–80
  vendor formats to mzML, 74–77
MSI (Metabolomics Standards Initiative), 119, 236, 246–248
multidimensional NMR spectroscopy, 60–62
multivariate curve resolution (MCR), 67, *68*
multivariate statistics, 142
  PCA regression, 145–148, *147*
  PLS regression, 148–152, 256
mzData format, 242
mzML files, 74–77
mzTab data-exchange format, 244

N

natural linewidth, NMR, 104
network visualization, 194–195
NMR, *see* nuclear magnetic resonance (NMR) spectroscopy
NMR data analysis, 102–104
  applications, 119–122
  file formats and standards, 118–119
  metabolite assignment/identification, 114–118
  pipeline schema, *103*
  raw data to spectra, 105–109
  spectra to data matrices, 109–114
NMR/MS Translator, 118
node, 194
non-informative variables, 29
nuclear magnetic resonance (NMR) spectroscopy, **6**, 259–260
  advantages/disadvantages, 62
  assay, 260
  hardware, 55–58
  LC-NMR, 59–60
  multidimensional, 60–62
  1D ^{1}H-NMR spectrum, *49–50*, 58–59
  overview, 47–51
  parameters, 59
  proton NMR spectrum, 53–54
  relaxation times, 55
  sample, 260
  sample preparation, 58
  signal intensities, 54–55
  spin, 52–53
  structural elucidation, 120–121
  zipped file, *261*
null hypothesis (H0), 135–140

O

objective function, 213, 216
ODE (ordinary differential equation), 208, 210–211
omics data integration, 213–216
Omics Discovery Index (OmicsDI), 239
1D ^{1}H-NMR spectrum, *49–50*, 58–59
OnPLS, 167–172
open format data standards
  data-exchange formats, 240–241
  initiatives, 240
  ontology developments, 246
  open access MS formats, 241–245
  open access NMR formats, 245–246
ordinary differential equation (ODE), 208, 210–211
overfitting, 150

P

parametric tests, 133
Pareto scaling, 148
partial correlations, 144
Partial Least Squares Discriminant Analysis (PLSDA), 151
Partial Least Squares (PLS) regression, 148–152
path, 194
pathway enrichment analysis, 188–189
pathway mapping, 187–189
PCA (Principal Component Analysis), 145–148, *147*
PCR (Principal Component Regression), 149
peak detection, 78
peak extraction, 86
phasing and baseline correction, 106–108
phosphofructokinase, 209
physiology, 3
Phytophthora infestans, 263
PLS (Partial Least Squares) regression, 148–152
post-translational modification (PTM), 3
Principal Component Analysis (PCA), 145–148, *147*
Principal Component Regression (PCR), 149
Probabilistic Quotient Normalisation (PQN), 112
profile *vs.* centroid data, *43*
proteins, 2
proton NMR spectroscopy, 53–54
pseudoreplicates, 21

Q

qcML data format, 244
QSRRs (quantitative structure-retention relationships), 93–94
quality control (QC) strategy, 27
quantile-quantile (QQ) plots, 131, *132*
quantitative structure–retention relationships (QSRRs), 93–94

R

Raman spectroscopy, 63–64
randomization, 23
raw data to spectra
  frequency domain, 106–109
  pre-processing, 105–106
Rayleigh scattering, 63
reaction graph, 190, *191*
receiver overload, 107
regions of interest (ROIs) approach, 111–112
regression coefficients, 149
resolution, 56
retention index (RI), 89, 93–94
retention time (RT)
  creatinine, 85
  definition, 37
  feature, 79
  GC-MS metabolomics, 89
  prediction of, 93–94
RI (retention index), 89, 93–94
ROIs (regions of interest) approach, 111–112
RT, *see* retention time (RT)
run-order effects, 83

S

sample collection/handling, 25–26
sampling units, 20–21

SBML (Systems Biology Markup Language), 183–184
SCA (simultaneous component analysis), 160, 162–163, 165
scanning mode, 41
scan speed, 42
score matrix, 146
scree plots, 148
selected-ion monitoring (SIM), 41
selected-reaction monitoring (SRM), 41
signal-to-noise ratio (S/N), 106
simple randomization, 23
simultaneous component analysis (SCA), 160, 162–163, 165
singular value decomposition (SVD), 146, 167
SIRM (Stable Isotope-Resolved Metabolomics), 119–120
solid-state ion sources, 40
SOP (standard operating procedure), 25
sparse methods, 155
spectral libraries, 88
spectra to data matrices
- data normalization and scaling, 112–113
- fingerprinting approach, 110–111
- region exclusion, 112–113
- ROIs approach, 111–112
- targeted profiling approach, 113–114

spin, 52–53
spin-spin (J) coupling, 51, 54–55
split-plot design, 24
Stable Isotope-Resolved Metabolomics (SIRM), 119–120
stable isotope tracing, *11*, 120, 208
stable isotopic labeling, 206
standard deviation, 135, 148
standard operating procedure (SOP), 25
stationary phase, 37–39
Statistical HeterospectroscopY (SHY), 161
statistical hypothesis, 130, 135
statistical testing, 134–140, *138*, *139*
Statistical Total Correlation Spectroscopy (STOCSY), 117, 161
steady-state mass-balance constraints, 212
stoichiometric matrix, 211, 213, 216
stoichometry, 203
stratified randomization, 23
study design and preparation, 15–20
- collection/handling, 25–26
- data analysis, 28–32
- experimental design, 21–25
- extraction/pretreatment, 26
- measurements, 26–28
- sampling units, 20–21

SVD (singular value decomposition), 146, 167
Systems Biology Markup Language (SBML), 183–184

T

tandem mass spectrometry (MS/MS), 42, 44–46
targeted metabolomics, 7–8
targeted profiling approach, 113–114
target loading, 165
temperature program, 38
thermodynamics, 203–205
transcription, 3
transcript/mRNA, 2
transcriptomics block, 169
translation, 3
tree, 194
tricarboxylic acid (TCA) cycle, 259
type 2 diabetes mellitus, 259

U

untargeted metabolomics, 15–16, *28*
- data analysis pipeline, *31*
- targeted *vs.*, 7–9, *8*

V

validation, 181
vapor pressure, 38
variation, 18, *19*
vibrational spectroscopy, 62

advantages and disadvantages, 65–66
applications of, 66
IR, 64–65
Raman spectroscopy, 63–64

W

Wilkinson–Rogers notation, 141

X

XCMS, 81, 83, 266–267

Z

zero-filling, 106